BIFURCATIONS IN PIECEWISE-SMOOTH CONTINUOUS SYSTEMS

WORLD SCIENTIFIC SERIES ON NONLINEAR SCIENCE

Editor: Leon O. Chua
University of California, Berkeley

Series A. MONOGRAPHS AND TREATISES*

Volume 54: Dynamics of Crowd-Minds
A. Adamatzky

Volume 55: Control of Homoclinic Chaos by Weak Periodic Perturbations
R. Chacón

Volume 56: Strange Nonchaotic Attractors
U. Feudel, S. Kuznetsov & A. Pikovsky

Volume 57: A Nonlinear Dynamics Perspective of Wolfram's New Kind of Science
L. O. Chua

Volume 58: New Methods for Chaotic Dynamics
N. A. Magnitskii & S. V. Sidorov

Volume 59: Equations of Phase-Locked Loops
J. Kudrewicz & S. Wasowicz

Volume 60: Smooth and Nonsmooth High Dimensional Chaos and the Melnikov-Type Methods
J. Awrejcewicz & M. M. Holicke

Volume 61: A Gallery of Chua Attractors (with CD-ROM)
E. Bilotta & P. Pantano

Volume 62: Numerical Simulation of Waves and Fronts in Inhomogeneous Solids
A. Berezovski, J. Engelbrecht & G. A. Maugin

Volume 63: Advanced Topics on Cellular Self-Organizing Nets and Chaotic Nonlinear Dynamics to Model and Control Complex Systems
R. Caponetto, L. Fortuna & M. Frasca

Volume 64: Control of Chaos in Nonlinear Circuits and Systems
B. W.-K. Ling, H. H.-C. Lu & H. K. Lam

Volume 65: Chua's Circuit Implementations: Yesterday, Today and Tomorrow
L. Fortuna, M. Frasca & M. G. Xibilia

Volume 66: Differential Geometry Applied to Dynamical Systems
J.-M. Ginoux

Volume 67: Determining Thresholds of Complete Synchronization, and Application
A. Stefanski

Volume 68: A Nonlinear Dynamics Perspective of Wolfram' New Kind of Science (Volume III)
L. O. Chua

Volume 69: Modeling by Nonlinear Differential Equations
P. E. Phillipson & P. Schuster

Volume 70: Bifurcations in Piecewise-Smooth Continuous Systems
D. J. Warwick Simpson

*To view the complete list of the published volumes in the series, please visit:
http://www.worldscibooks.com/series/wssnsa_series.shtml

WORLD SCIENTIFIC SERIES ON NONLINEAR SCIENCE Series A Vol. 70

Series Editor: Leon O. Chua

BIFURCATIONS IN PIECEWISE-SMOOTH CONTINUOUS SYSTEMS

David John Warwick Simpson

University of British Columbia, Canada

World Scientific

NEW JERSEY • LONDON • SINGAPORE • BEIJING • SHANGHAI • HONG KONG • TAIPEI • CHENNAI

Published by

World Scientific Publishing Co. Pte. Ltd.
5 Toh Tuck Link, Singapore 596224
USA office: 27 Warren Street, Suite 401-402, Hackensack, NJ 07601
UK office: 57 Shelton Street, Covent Garden, London WC2H 9HE

British Library Cataloguing-in-Publication Data
A catalogue record for this book is available from the British Library.

BIFURCATIONS IN PIECEWISE-SMOOTH CONTINUOUS SYSTEMS
World Scientific Series on Nonlinear Science, Series A — Vol. 70

ISBN-13 978-981-4293-84-6
ISBN-10 981-4293-84-9

Printed in Singapore by Mainland Press Pte Ltd.

Preface

Many real-world systems involve a discontinuity or sudden change, such as impacting in mechanical systems and switching in electrical circuits. Smooth dynamical systems generally do not provide ideal mathematical models for such situations. It becomes necessary to incorporate a nonsmooth component into the model. Often this yields a piecewise-smooth system.

Studies of piecewise-smooth systems prior to about twenty years ago are quite rare. Perhaps this is because the 1970's and 80's saw significant advances in the theory of smooth dynamical systems, such as an understanding of chaos. Nowadays smooth dynamical systems theory has a firm footing (though many open questions remain, e.g. Hilbert's 16th problem) and piecewise-smooth systems are a popular topic of research. The advancement of piecewise-smooth theory has been led most notably by the Bristol school which has recently produced a unique and important textbook in this area [di Bernardo *et al.* (2008a)].

Although numerically computed solutions of a mathematical model at particular parameter values can give some understanding of the dynamics associated with the model, a more complete understanding relies on determining parameter values at which the qualitative features of these solutions change in some fundamental way, i.e. *bifurcations*. Of particular interest in this book is the nature of those bifurcations that are unique to piecewise-smooth systems. For instance in a piecewise-smooth system an attracting solution may lose stability and undergo an instantaneous transition to chaos. This had been observed in DC/DC converters for some time, see for instance [Fossas and Olivar (1996)], but explained only recently by nonsmooth bifurcation theory [di Bernardo *et al.* (1998b); Yuan *et al.* (1998)]. Unlike period-doubling cascades, period-adding sequences (which may occur in smooth systems) are not completely understood. However recent studies have explained such sequences in one-dimensional, piecewise-smooth maps [Avrutin *et al.* (2007); Halse *et al.* (2003); Kawczyński and Strizhak (2000)]. Also, resonance tongues (or mode-locking regions) that display a curious lens-chain structure have been observed in models of DC/DC converters [Zhusubaliyev and Mosekilde (2003)] and in a trade cycle model [Sushko and Gardini (2006)]. A novel description of the dynamics near points of zero width of such tongues is described here in Chapter 6.

Bifurcations unique to piecewise-smooth systems differ in many fundamental aspects to those in smooth systems. Often the dynamical behavior local to a bifurcation of a piecewise-smooth system is determined by only linear terms of an appropriate series expansion. This yields two immediate consequences. First, if nonlinear terms are put aside, the resulting approximation is piecewise-linear and relatively easy to study. This is because solutions of linear systems may be determined explicitly. In fact some researchers choose to study piecewise-linear approximations of smooth, nonlinear systems, for example [Bergami *et al.* (2006)]. Second, invariant sets that are created at such a bifurcation generically grow in size or separate linearly with respect to parameter values. This is different to the behavior of familiar smooth bifurcations, such as the Andronov-Hopf bifurcation [Kuznetsov (2004); Marsden and McCracken (1976)] at which a periodic orbit is created that grows in size as the square root of the bifurcation parameter.

An important characteristic of any bifurcation is its *codimension* - the number of parameters that need to be varied in order for the bifurcation to occur. In an arbitrary parameter space of a dimension equal to the codimension of the bifurcation, an *unfolding* of a bifurcation is a description of all dynamical behavior that may generically occur near the bifurcation. A large component of this book is the unfolding of codimension-two bifurcations. The *normal form* of a local bifurcation is a simple system that exhibits all aspects of the unfolding. For smooth systems the concept of dimension reduction via center manifold analysis allows bifurcations in systems of any number of dimensions to be transformed to their normal forms. This technique is crucial to understanding complex dynamical systems of high dimension. However in piecewise-smooth systems a lack of differentiability often disallows dimension reduction. Consequently little is known about bifurcations in piecewise-smooth systems of arbitrary dimension.

This book is essentially an updated and revised version of my PhD thesis [Simpson (2008)]. The book concerns piecewise-smooth systems that are continuous and autonomous. Both continuous-time systems (ODEs) and discrete-time systems (maps) are investigated. The smooth components of the systems will be assumed to contain no singularities that affect the dynamical behavior of interest (in particular, maps with a square-root or 3/2-type singularity [di Bernardo *et al.* (2001b)] are not considered). A brief overview of each chapter is now given.

Chapter 1 provides a background. Section 1.1 discusses applications; the remaining sections derive the piecewise equations used in later chapters and summarize key results of previous researchers. A variety of terminology used below and throughout this book is introduced in this chapter.

The next three chapters investigate piecewise-smooth, continuous ODE systems. Chapter 2 looks at discontinuous bifurcations in planar systems. Section 2.1 discusses periodic orbits. Section 2.2 studies a discontinuous analogue of a Hopf bifurcation. The main result of this section, Theorem 2.1, details the creation of a periodic orbit at this bifurcation in a general setting. A practical summary is given in Sec. 2.3 where it is shown that there are exactly four distinct discontinuous bifurcations in planar systems.

Chapter 3 presents unfoldings of three different codimension-two, discontinuous bifurcations. In the order given they are the simultaneous occurrence of a (smooth) saddle-node bifurcation and a discontinuous bifurcation, the simultaneous occurrence of a Hopf bifurcation and a discontinuous bifurcation, and a discontinuous Hopf bifurcation of indeterminable criticality. The first unfolding is accomplished for a piecewise-smooth, continuous ODE system of arbitrary dimension. The second scenario is completely determined for a two-dimensional system, then a partial result is obtained in N dimensions. The third unfolding is carried out in two dimensions.

Chapter 4 applies the theoretical results of the previous two chapters to a real physical system. The system studied is an eight-dimensional, cybernetic model of the continuous cultivation of *Saccharomyces cerevisiae* (a common yeast) formulated by Jones & Kompala [Jones and Kompala (1999)]. As detailed in Sec. 4.1, the model is piecewise-smooth, continuous, due to the key model assumption that the yeast switches between metabolic pathways in a manner that maximizes its growth rate. Nonsmoothness and other basic properties are discussed in Sec. 4.2. A bifurcation analysis is presented in Sec. 4.3. Here a variety of codimension-two, discontinuous bifurcations are found. These are shown to be in full agreement with the results of Chapter 3. Stable oscillations which arise via Hopf and discontinuous Hopf-like bifurcations are described in Sec. 4.4. These correspond to spontaneous oscillations observed in experiments. By a detailed inspection of time series plots, an explanation as to the cause of the oscillations is determined.

The next three chapters study discrete-time systems. Chapter 5 analyzes codimension-two, border-collision bifurcations. It is convenient that the piecewise-smooth ODE systems that describe discontinuous bifurcations, have the same form as the piecewise-smooth maps describing border-collision bifurcations. Much of the basic analysis of the continuous-time case may be carried over to the discrete-time case. This is most evident for the codimension-two, simultaneous occurrence of a saddle-node bifurcation and a border-collision bifurcation detailed in Sec. 5.1. Sec. 5.2 studies the simultaneous occurrence of a period-doubling bifurcation and a border-collision bifurcation. A complete description is obtained in one dimension; a partial result is constructed for higher dimensions.

Chapter 6 looks at periodic solutions of piecewise-smooth, continuous maps near border-collision bifurcations at a smooth switching manifold. Symbolic dynamics enable periodic solutions to be identified by symbol sequences that give their itinerary relative to the switching manifold. As a result, any periodic solution may be expressed as the solution to a linear system which is constructed in manner determined by the associated symbol sequence. Two important matrices appear in the given formulation of the linear system: the *stability matrix* and the *border-collision matrix*. Bifurcations of periodic solutions and thus resonance tongue boundaries are described in terms of the multipliers of the stability matrix and the singularity of the border-collision matrix. A new class of symbol sequences, called "rotational symbol sequences" is introduced in Sec. 6.4. Such symbol sequences correspond to periodic solutions that lie on an invariant circle. A reason for the success of the symbolic approach is that no assumptions need to be made about the existence of invariant circles. Algebraic properties of rotational symbol sequences are used to explain the lens-chain geometry of resonance tongues observed in two-parameter bifurcation diagrams (bifurcation sets). Points where adjacent lenses connect, i.e. where the resonance tongues have zero width, are termed "shrinking points". A rigorous unfolding of shrinking points is performed in the final section.

Chapter 7 studies border-collision bifurcations in a planar, piecewise-smooth, continuous map for which the multipliers associated with the fixed point are complex and "jump" from inside to outside the unit circle at the bifurcation. These bifurcations are shown to exhibit a wide variety of Neimark-Sacker-like and nonsmooth behavior. An invariant circle may be created at the bifurcation. Mode-locking on the invariant circle corresponds to a point in a resonance tongue. The theoretical results of the previous chapter are compared with numerically computed lens-chain-shaped resonance tongues. Additional complex phenomena are also discussed. A curve of shrinking points is seen to be a boundary between chaotic and non-chaotic dynamics. Periodic solutions with non-rotational symbol sequences are found to belong to resonance tongues that do not display a lens-chain structure. Unlike at a generic border-collision bifurcation in a one-dimensional, piecewise-smooth, continuous map, multiple attractors may coexist. Also, in some cases at the bifurcation the fixed point is a saddle and no invariant circle is created.

There are four appendices. Appendix A contains proofs of lemmas and theorems that are deemed too lengthy for inclusion in the main text. The hope is that this segregation allows the general ideas of the book to be more easily assimilated. Extra figures intended as additional references are included in Appendix B. Appendix C overviews so-called adjugate matrices that are utilized throughout the book. Lastly Appendix D lists parameter values for the *S. cerevisiae* growth model studied in Chapter 4.

A quick note with regards to notation: $O(k)$ [$o(k)$] will be used to denote terms that are order k or larger [larger than order k] in all variables and parameters of a given expression.

Finally, it should be noted that six research articles contain much of the work presented here. The codimension-one, discontinuous Hopf bifurcation discussed in Sec. 2.2 is the subject of [Simpson and Meiss (2007)]. Similarly, the coincidence of a Hopf bifurcation and a discontinuous bifurcation is described in [Simpson and Meiss (2008b)] and the coincidence of a period-doubling bifurcation and a border-collision bifurcation is described in [Simpson and Meiss (2009b)]. The bifurcation analysis of the *S. cerevisiae* model studied in Chapter 4 is presented in [Simpson *et al.* (2009)]. The N-dimensional unfoldings of Sec. 3.1 and Sec. 3.2 may also be found in this paper. Most of Chapter 6 and Chapter 7 may be found in [Simpson and Meiss (2009a)] and [Simpson and Meiss (2008a)] respectively.

D. J. W. Simpson

Acknowledgments

This book could not have been accomplished without the help of numerous people who deserve more credit than I am able to give them here. The discussions and suggestions of Jim Meiss have been fundamental to all the material presented here. Many thanks go to Dhinakar Kompala concerning the application of the theoretical results to a biochemical model. I am also very grateful for the valuable help and comments of the remaining members of my PhD thesis committee, David Bortz, Jim Curry and Keith Julien.

Special thanks to two professors who recently visited The University of Colorado, Holger Dullin and Robert MacKay, for many fruitful discussions relating to the dynamics of maps and the properties of slow manifolds. I am very appreciative of the ideas and comments of Bob Easton and Jim Howard who made special efforts to help me. Thanks also to graduate students, too many to name here, who I've shared an office with or worked with in the computer lab, for much miscellaneous help.

Finally, thanks to friends, family and my wife Shannon (whose ability to find for me seemingly obscure but invaluable references knows no end) for love and support.

Contents

Chapter 1

Fundamentals of Piecewise-Smooth, Continuous Systems

A system of ordinary differential equations (ODEs), $\dot{x} = F(x)$, or a map, $x' = F(x)$, is said to be *piecewise-smooth* if the phase space can be partitioned into countably many regions where F has a different smooth (i.e. C^k for some $k \in \mathbb{N}$) functional form. Nonsmoothness occurs on codimension-one region boundaries which are called *switching manifolds*. The last two decades have seen an explosion of interest in piecewise-smooth systems. Some recent textbooks in this area include [di Bernardo *et al.* (2008a); Leine and Nijmeijer (2004); Zhusubaliyev and Mosekilde (2003); Tse (2003); Banerjee and Verghese (2001); Brogliato (1999)].

It is useful to classify piecewise-smooth, ODE systems by the severity of the discontinuities at the switching manifolds:

(1) **Hybrid systems.** The system is a combination of differential equations and maps [Johansson (2003); Van der Schaft and Schumacher (2000)]. Such systems naturally arise when modeling hard impact phenomena [Wiercigroch and De Kraker (2000); Brogliato (1999); Popp (2000); Blazejczyk-Okolewska *et al.* (1999)]. This is because impacts result in the sudden velocity change of an object; when velocity is a system variable, a map is required to execute a jump in phase space.
(2) **Filippov systems.** The vector field, F, is discontinuous at switching manifolds. These systems are named after Aleksei Filippov who was perhaps the first to resolve ambiguities arising from *sliding* which occurs when the vector field on both sides of a switching manifold points either towards or away from the switching manifold [Filippov (1964, 1988)].
(3) **Piecewise-smooth, continuous systems.** F is continuous throughout the phase space but the Jacobian of F is discontinuous

at the switching manifolds.

(4) **Systems with higher order discontinuities.** Derivatives of F up to some degree n are continuous but of degree $n+1$ are not. Systems of this type are rarely encountered. As a general rule they exhibit the same bifurcations as smooth systems except non-degeneracy conditions dependent on high-order derivatives of F may not be computable.

Piecewise-smooth maps may be classified in a similar manner but this is not done here since many interesting piecewise-smooth maps contain a square-root or 3/2-type singularity. Other formulations exist for non-smooth systems such as *mixed logic dynamical systems* - a very general class of systems that use logic components [Ferrari-Trecate *et al.* (2003); Bemporad and Morari (1999)], *linear complementarity systems* - systems with inequality constraints [Heemels and Brogliato (2003); Heemels *et al.* (2000)], systems with differential inclusions [Deimling (1992)], and systems with delay [Sieber (2006); Sieber *et al.* (unpublished); Barton *et al.* (2006); Senthilkumar and Lakshmanan (2005)].

In piecewise-smooth systems the interaction of invariant sets with switching manifolds often produces bifurcations that are forbidden in smooth systems. These are collectively known as *discontinuity induced bifurcations.* Whereas period-doubling cascades are common in smooth systems, in piecewise-smooth systems periodic orbits may undergo what are called period-adding sequences or transition directly to chaos. Dynamical behavior local to a discontinuity induced bifurcation is often determined purely by linear terms of an appropriate piecewise series expansion. Consequently invariants commonly grow in size linearly with respect to a bifurcation parameter, such as the Hopf cycle in the discontinuous analogue of an Andronov-Hopf bifurcation, see Sec. 2.2.

Much is known about discontinuity induced bifurcations in systems of low dimension ($N = 1, 2$). However, the technique of dimension reduction via center manifold analysis that is hugely useful in studying smooth systems does not apply at non-differentiable points in piecewise-smooth systems. More dimensions provide more allowable geometries and only scattered results exist for such bifurcations in systems of any dimension. Furthermore, in piecewise-smooth, continuous, ODE systems, invariant manifolds of equilibria may be conical instead of planar [Carmona *et al.* (2005b)].

Perhaps the most basic discontinuity induced bifurcations are those that arise from the perturbation of a piecewise-smooth, ODE system that has an equilibrium located exactly on a switching manifold. Such bifurcations are commonly referred to as *boundary equilibrium bifurcations*. In hybrid or Filippov systems these bifurcations may involve what are known as *pseudo-equilibria* due to a lack of continuity in the vector field (these are not discussed in this book, the reader is referred to [di Bernardo *et al.* (2008a); Kuznetsov *et al.* (2003); Giannakopoulos and Pliete (2001); di Bernardo *et al.* (2008b); Buzzi *et al.* (2006)]). In piecewise-smooth, continuous systems, they are further known as *discontinuous bifurcations* and are either analogues of familiar smooth bifurcations, or novel and unique to piecewise-smooth systems. The discrete analogue, that is a bifurcation resulting from the perturbation of a piecewise-smooth, continuous map with a fixed point on a switching manifold, is known as a *border-collision bifurcation*.

The remainder of this chapter is organized as follows. Section 1.1 presents an overview of physical systems that have been modeled by piecewise-smooth systems. Section 1.2 introduces a general form commonly used to investigate discontinuous and border-collision bifurcations. Here it is shown that nonlinear terms do not affect structurally stable local dynamics. Equilibria and fixed points of the general form are then computed in Sec. 1.3. When particular non-degeneracy conditions are satisfied, the general form may be put into a canonical form involving companion matrices, Sec. 1.4. Section 1.5 summarizes current theory relating to discontinuous bifurcations. The discrete equivalent, border-collision bifurcations, are discussed in Sec. 1.6. Section 1.7 describes Poincaré maps of piecewise-smooth, ODE systems. Generically, the Poincaré maps are smooth unless a tangency occurs, in which case so-called discontinuity maps are used to control any singularities. Four common tangency scenarios are described in detail. The concept of sliding is also discussed in this section. Section 1.8 describes period-adding cascades and finally Sec. 1.9 discusses smooth approximations to piecewise-smooth systems.

1.1 Applications

This section discusses applications of piecewise-smooth systems. To be concordant with the focus of this book an emphasis is placed on physical systems that are well-modeled by piecewise-smooth, continuous, ODE systems and maps. However, also of interest are piecewise-smooth, discontinuous, ODE systems that exhibit oscillatory behavior and are of a form that leads to piecewise-smooth, continuous Poincaré maps. As detailed in

Sec. 1.7, Filippov systems with a corner collision or a sliding bifurcation generically lead to piecewise-smooth, continuous Poincaré maps, as do some non-autonomous systems.

Piecewise-smooth systems are ideal for modeling circuits that contain a switching component such as a diode or a transistor [Banerjee and Verghese (2001); Zhusubaliyev and Mosekilde (2003); Tse (2003)]. Many times models are piecewise-linear, ODE systems. The absence of nonlinear terms allows for analytical expressions to be obtained for trajectories between switching manifolds which is particularly useful in the construction of Poincaré maps [Freire *et al.* (2007); Carmona *et al.* (2005a)]. Simple relay control systems are often well modeled by equations of the form

$$\begin{aligned} \dot{x} &= Ax + bu \;, \\ \xi &= g(x,t) \;, \\ u &= \mathrm{sgn}(\xi) \;, \end{aligned} \tag{1.1}$$

where x is a vector of state variables (such as currents and voltages), ξ is the input signal to the relay element or switch, u is the output signal of the relay and the matrix A and the vector b are constants. The combined effect of feedback and nonsmoothness in (1.1) can result in extremely complicated dynamics [Zhusubaliyev and Mosekilde (2003)].

One of the simplest circuit systems known to exhibit chaos is Chua's circuit [Chua and Lin (1990, 1991)], commonly modeled by a piecewise-linear, continuous, ODE system. Conceived over 20 years ago it has since been studied extensively in a variety of different forms, see for instance [Chua (1994); Yang and Chua (2000)]. DC/DC buck converters are also well-studied and have been shown to exhibit a variety of bifurcations unique to piecewise-smooth systems [Zhusubaliyev and Mosekilde (2008a, 2006b); Zhusubaliyev *et al.* (2006); Zhusubaliyev and Mosekilde (2006a); Zhusubaliyev *et al.* (2007); Benadero *et al.* (2003); di Bernardo *et al.* (2002a); Fang and Abed (2001)], see Fig. 1.1. Other examples include the Colpitts oscillator [Maggio *et al.* (2000)] and the Wien bridge oscillator [López *et al.* (2005); Carmona *et al.* (2002)].

Vibro-impacting systems are commonly modeled by piecewise-smooth systems [Wiercigroch and De Kraker (2000); Brogliato (1999); Leine and Nijmeijer (2004); Popp (2000); Awrejcewicz and Lamarque (2003); Blazejczyk-Okolewska *et al.* (1999)]. Fig. 1.2(a) shows perhaps the simplest vibro-impacting system: a mass on a spring that collides with a solid wall. The associated equations of motion form a piecewise-smooth impacting system. When forced this system may exhibit chaotic dynamics

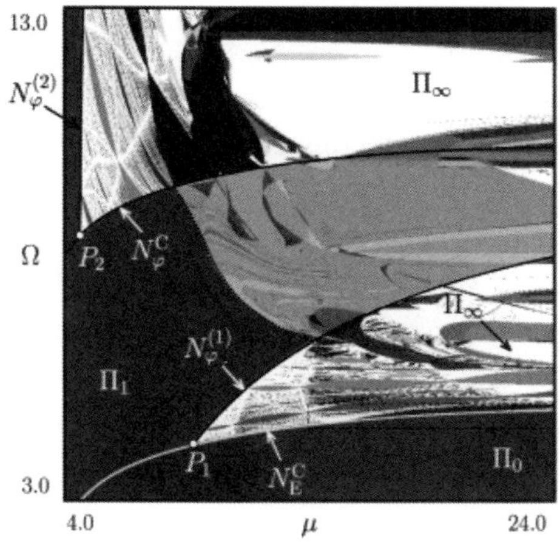

Fig. 1.1: This figure is taken from [Zhusubaliyev and Mosekilde (2008a)]. It shows resonance tongues (or mode-locking regions) for a two-dimensional, piecewise-linear, continuous map modeling a DC/DC power converter. In some areas resonance tongues exhibit a lens-chain geometry. Neimark-Sacker and Neimark-Sacker-like bifurcations are labeled N. A further description of these curves is presented in [Zhusubaliyev and Mosekilde (2008a)].

[Peterka and Vacik (1992)]. Other basic scenarios include a pendulum with impacts [Piiroinen *et al.* (2004); Awrejcewicz *et al.* (2004)] and a bouncing ball [Ivanov (2000); Luo and Han (1996)]. These simple situations have immediate applications to more complex mechanical systems such as rattling in gears [Halse *et al.* (2007); Wiercigroch and De Kraker (2000)], walking robots [Garcia *et al.* (1998); Aoi and Tsuchiya (2006)], printers [Dankowicz and Piiroinen (2002); Tung and Shaw (1988)], and models of rocking blocks [Lenci and Rega (2005); Hogan (1994); Lipscombe and Pellegrino (1993)] which are useful to earthquake-resistant building design.

(a) (b)

Fig. 1.2: Schematic diagrams of simple mechanical systems exhibiting non-smooth behavior. In panel (a), a mass on a spring impacts a wall. In panel (b), a mass on a spring is subject to dry friction.

Soft impacts, characterized by a fast but continuous velocity change, are commonly modeled by Filippov systems. Consider the scenario depicted in Fig. 1.3(a). Here a mass on a spring periodically comes in contact with a non-rigid, massless support. If y represents the vertical displacement of the mass from its rest position, then

$$m\ddot{y} + b\dot{y} + ky = F(t) , \tag{1.2}$$

where m denotes mass, $F(t)$ is a forcing function, and b and k are dependent upon whether or not the mass is in contact with the support:

$$\begin{aligned} \text{no contact}: \quad & b = b_{\rm spr} , & k = k_{\rm spr} , \\ \text{contact}: \quad & b = b_{\rm spr} + b_{\rm sup} , & k = k_{\rm spr} + k_{\rm sup} , \end{aligned}$$

where $b_{\rm spr}$ [$b_{\rm sup}$] and $k_{\rm spr}$ [$k_{\rm sup}$] represent the damping coefficient and spring constant of the spring [support], respectively. Nonsmoothness occurs whenever the mass impacts or detaches from the support. It is assumed the support has a relatively small relaxation time, therefore an impact occurs whenever y decreases to a critical value, $y_{\rm sup}$. A detachment occurs whenever the contact force between the mass and the support, $f = k_{\rm sup}(y - y_{\rm sup}) + b_{\rm sup}\dot{y}$, becomes zero. As shown in the corresponding phase portrait, Fig. 1.3(b), the resulting switching manifold (the solid line) is non-differentiable at $(y, \dot{y}) = (y_{\rm sup}, 0)$. Consequently systems of this type commonly produce

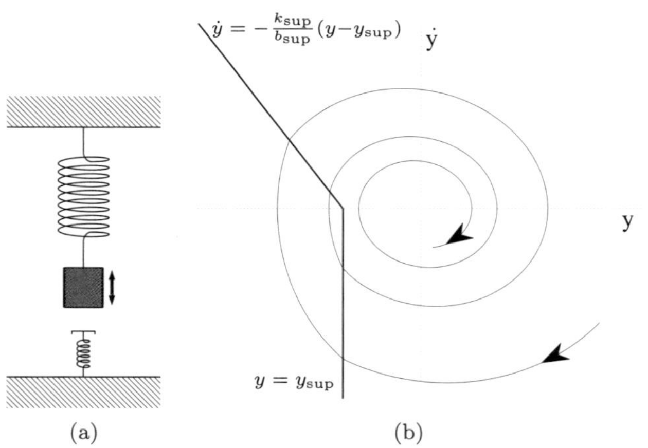

Fig. 1.3: Panel (a) is a schematic diagram of the soft impact system (1.2). A corresponding phase portrait is shown in panel (b). This system may exhibit so-called corner-collisions at the non-differentiable point, $(y_{\rm sup}, 0)$, on the switching manifold.

corner-collisions, see Sec. 1.7. Various versions of this system are studied in [Leine and Van Campen (2002); Leine *et al.* (2000); Ma *et al.* (2006, 2008)].

Nearly all mechanical systems suffer from friction to some degree, in many cases affecting the system qualitatively. There are two types of friction. For surfaces in contact, the force that opposes their movement when they are stationary relative to each other is known as *static friction.* When there is motion, a different force applies, namely *kinetic friction.* Objects that vibrate encounter both forces of friction and are repeatedly at rest for nonzero lengths of time. This is known as *stick-slip* motion, see Fig. 1.2(b). Piecewise-smooth systems are commonly used to model stick-slip systems [Szalai and Osinga (2008); Batako *et al.* (2007); Csernák and Stépán (2006); di Bernardo *et al.* (2003); Dankowicz and Nordmark (2000); Cone and Zadoks (1995)]. Chaotic dynamics abound in such systems, as do sliding bifurcations. Popular applications include brake systems, turbines and oil drilling. For modeling friction see [Wiercigroch and De Kraker (2000); Popp (2000); Awrejcewicz and Lamarque (2003); Abadie (2000)].

Mathematical models of economics systems are often nonsmooth as a result of non-negativity conditions or optimization. For instance individuals may make different investment decisions at different times to maximize wealth [Caballé *et al.* (2006)]. A version of the Hicks' trade cycle model is described by the discrete-time system

$$\begin{aligned} Y_t &= cY_{t-1} + I_t \ , \\ I_t &= \max(a(Y_{t-1} - Y_{t-2}), -d) \ , \end{aligned} \tag{1.3}$$

where Y_t and I_t denote income and investment respectively and a, c and d are positive-valued constants [Puu and Sushko (2006)]. The investment is proportional to the change in income and may be negative, but cannot be so negative that wealth would have to be actively destroyed. This puts a lower bound on investment, which, as in (1.3), may be modeled with a piecewise-linear function.

In production-distribution models, nonsmoothness typically arises when a supplier has insufficient goods to meet demand [Mosekilde and Laugesen (2007); Laugesen and Mosekilde (2006)]. In a model of a socialist economy, there are upper and lower bounds on investment [Hommes *et al.* (1995)]. The examples listed above are discrete-time models (i.e. maps) as is common in economics. Furthermore, each example is piecewise-smooth continuous and has been shown to exhibit a variety of border-collision bifurcations.

Piecewise-smooth models are utilized in many other fields, such as biology and physiology [Casey *et al.* (2006); Zhao *et al.* (2008); Alur *et al.*

(2001); Rosen (1970); Keener and Sneyd (1998)]. The model of yeast growth investigated in Chapter 4 is such an example due to assumptions regarding switching between metabolic pathways. So-called Glass networks are used as a framework to model a variety of biological systems, most notably switching of feedback responses in gene regulation [Killough and Edwards (2005); Edwards (2000); Plahte and Kjøglum. (2005); Gouzé and Sari (2002)]. The characterizing feature of these networks is that variables change continuously but depend only on the sign of the variables. Piecewise-smooth systems have also been utilized in an evolutionary model [Dercole (2005)], a predator-prey system [Kuznetsov *et al.* (2003)], a forest fire model [Dercole and Maggi (2005)] and epidemics [Breban *et al.* (2007)].

1.2 A Framework for Local Behavior

This section describes equations governing dynamical behavior near a generic discontinuous or border-collision bifurcation that occurs at a smooth point on a switching manifold. Bifurcations at nonsmooth points on switching manifolds or at the intersection of multiple switching manifolds require an extra codimension to occur (unless there is a special symmetry, perhaps arising from assumptions concerning the underlying physical system being modeled) and for this reason are not discussed here, see for instance [Leine (2006); Leine and Nijmeijer (2004)].

Local to a single switching manifold a piecewise-smooth, continuous, autonomous, ODE system may be written as

$$\dot{x} = \begin{cases} f^{(L)}(x;\xi), \; H(x;\xi) \leq 0 \\ f^{(R)}(x;\xi), \; H(x;\xi) \geq 0 \end{cases}, \tag{1.4}$$

where $f^{(L)}, f^{(R)} : \mathbb{R}^N \times \mathbb{R}^M \to \mathbb{R}^N$ are C^k and $H : \mathbb{R}^N \times \mathbb{R}^M \to \mathbb{R}$ is C^l. By continuity, $f^{(L)}(x;\xi) = f^{(R)}(x;\xi)$ whenever $H(x;\xi) = 0$. The switching manifold is the parameter dependent set, $\mathcal{M}_\xi = \{x \in \mathbb{R}^N \mid H(x;\xi) = 0\}$. If, when $(x;\xi) = (0;0)$, we have $H = 0$ and $\nabla_x H \neq 0$, then locally $\mathcal{M}_0$ is a curve in phase space that intersects the origin. Using coordinate transformations similar to those given in [di Bernardo *et al.* (2001b)], we may assume to l^{th} order that H is simply equal to, say, x_1. The effects of nonsmoothness of switching manifolds on dynamical behavior is beyond the scope of this book. For this reason it is assumed that l is sufficiently large to not affect local dynamics and for simplicity H is set identically equal to x_1. The switching manifold is then the plane $x_1 = 0$ and in what follows the *left half-system* refers to $f^{(L)}$ and the *right half-system* refers to $f^{(R)}$.

Assume the origin is an equilibrium when $\xi = 0$. Since the origin lies on the switching manifold and (1.4) is continuous, it is an equilibrium of both left and right half-systems. To unfold this scenario with a one-dimensional parameter, we assume that in one half-system (we choose the right half-system without loss of generality) this equilibrium exists for all ξ in a neighborhood of 0. Thus we assume

$$\det(D_x f^{(R)}(0;0)) \neq 0 \,, \tag{1.5}$$

that is, for the right half-system, zero is not an associated eigenvalue of the origin when $\xi = 0$. Then by the implicit function theorem, locally, the right half-system has an equilibrium, $x^{*(R)}(\xi)$, with $x^{*(R)}(0) = 0$ and C^k dependent upon ξ. As is generically the case, we may assume the distance the equilibrium is from the switching manifold varies linearly with some choice of linear change in parameter values. Without loss of generality we may assume ξ_1 (the first component of $\xi \in \mathbb{R}^M$) is a suitable choice. That is

$$\frac{\partial x_1^{*(R)}(0)}{\partial \xi_1} \neq 0 \,. \tag{1.6}$$

Let $\mu = \xi_1$ and $\eta \in \mathbb{R}^{M-1}$ denote the remaining components of ξ. By the implicit function theorem, there is a C^k function $\phi_1 : \mathbb{R}^{M-1} \to \mathbb{R}$ such that $x_1^{*(R)}(\phi_1(\eta), \eta) = 0$. In other words when $\mu = \phi_1(\eta)$, the equilibrium lies on the switching manifold. By performing the nonlinear change of coordinates

$$\mu \mapsto \mu - \phi_1(\eta) \,, \tag{1.7}$$

$$x \mapsto x - x^{*(R)}(\phi_1(\eta), \eta) \,, \tag{1.8}$$

we may factor μ out of the constant term in the system, i.e.

$$f^{(L)}(0; \mu, \eta) = f^{(R)}(0; \mu, \eta) = \mu b(\mu, \eta) + o(k) \,, \tag{1.9}$$

where b is C^{k-1}. Notice the transformation does not alter the switching manifold. The system is now

$$\dot{x} = \begin{cases} f^{(L)}(x; \mu, \eta), & x_1 \leq 0 \\ f^{(R)}(x; \mu, \eta), & x_1 \geq 0 \end{cases} \,, \tag{1.10}$$

with

$$f^{(i)}(x; \mu, \eta) = \mu b(\mu, \eta) + A_i(\mu, \eta)x + O(|x|^2) + o(k) \,, \tag{1.11}$$

where A_L and A_R are $N \times N$ real-valued matrices, C^{k-1} dependent upon μ and η. By continuity, the two half-systems of (1.10) must agree whenever

$x_1 = 0$. Therefore all but possibly the first columns of A_L and A_R are identical, i.e.

$$A_L e_i = A_R e_i, \forall i \neq 1 . \tag{1.12}$$

Thus we have constructed a general system which exhibits a discontinuous bifurcation at $\mu = 0$ (unless no bifurcation occurs, see Sec. 1.5).

The above procedure may also be applied to the map

$$x' = \begin{cases} f^{(L)}(x;\xi), \; H(x;\xi) \le 0 \\ f^{(R)}(x;\xi), \; H(x;\xi) \ge 0 \end{cases} . \tag{1.13}$$

If the Jacobian of the right half-system does not have a multiplier of 1 at $(x;\xi) = (0;0)$ and the distance of the fixed point of the right half-system from the switching manifold varies linearly with some linear change in ξ, then (1.13) may be simplified to

$$x' = \begin{cases} f^{(L)}(x;\mu,\eta), \; x_1 \le 0 \\ f^{(R)}(x;\mu,\eta), \; x_1 \ge 0 \end{cases} , \tag{1.14}$$

with (1.11). The map (1.14) with (1.11) will be used in later chapters to investigate border-collision bifurcations. The parameter, η, is useful only in situations of codimension at least two as studied in later chapters. For the remainder of this chapter the η dependency is omitted for clarity.

An alternate form for (1.11) that replaces the piecewise component of the ODE system or map with an absolute value function, is often convenient. If $A = \frac{1}{2}(A_R + A_L)$ and $\zeta e_1^{\mathsf{T}} = \frac{1}{2}(A_R - A_L)$ (recall (1.12)), then

$$A_i x = Ax + |x_1|\zeta , \tag{1.15}$$

which is sometimes called a Lur'e-like form [Carmona *et al.* (2002)]. The canonical piecewise-linear representation of Chua [Chua and Deng (1988); Kang and Chua (1978)] that is used to describe many circuit systems may be written as a combination of equations of the form (1.15), see [Li *et al.* (2001); Julián *et al.* (1999)].

It is now argued that for small μ and x, structurally stable dynamics of the piecewise-linear approximation to (1.10) are exhibited by (1.10) (the same argument holds for the map (1.14)). Let $\varepsilon > 0$ be given and suppose, $|x|, \mu = O(\varepsilon)$. If $\mu = 0$, (1.10) with (1.11) becomes

$$\dot{x} = \begin{cases} A_L(0)x + O(\varepsilon), \; x_1 \le 0 \\ A_R(0)x + O(\varepsilon), \; x_1 \ge 0 \end{cases} . \tag{1.16}$$

Otherwise when $\mu \neq 0$, let $z = \frac{x}{|\mu|}$. Then

$$\dot{z} = \begin{cases} \operatorname{sgn}(\mu)b(0) + A_L(0)z + O(\varepsilon), \; z_1 \le 0 \\ \operatorname{sgn}(\mu)b(0) + A_R(0)z + O(\varepsilon), \; z_1 \ge 0 \end{cases} , \tag{1.17}$$

where $\mathrm{sgn}(\mu)$ represents the sign of μ. Omitting terms that are at least order ε yields

$$\dot{w} = \begin{cases} \sigma b(0) + A_L(0)w, & w_1 \leq 0 \\ \sigma b(0) + A_R(0)w, & w_1 \geq 0 \end{cases}, \tag{1.18}$$

where $\sigma = -1, 0, 1$. (When $\sigma \neq 0$, some authors choose to call (1.18) piecewise-affine since the linear components do not necessarily intersect the origin.) By definition, structurally stable dynamics of (1.18) are also exhibited by (1.10) for sufficiently small ε. The upshot of this is that to investigate structurally stable dynamics of the nonlinear system (1.10) near $(x; \mu) = (0; 0)$, it suffices to study the piecewise-linear system (1.18) for $\sigma = -1, 0, 1$. Furthermore, since $x = |\mu| z$, any structurally stable, bounded invariant of (1.18) is exhibited by (1.10) and shrinks linearly in size to 0, as $\mu \to 0$. In other words, invariants created at the bifurcation grow in size linearly with respect to μ. This is one of the most characteristic features of discontinuous and border-collision bifurcations and is evident throughout this book.

Not surprisingly many researchers have performed detailed investigations into piecewise-linear systems. Although such systems appear simple (solutions to the linear components may be computed explicitly) they may exhibit very complicated dynamics. For instance a relatively early paper in this area by Colin Sparrow [Sparrow (1981)], describes a three-dimensional system of the form (1.18) exhibiting Sil'nikov homoclinic chaos. Also, determination of the stability of the origin for (1.18) when $\sigma = 0$ is a very difficult problem, see [Casey *et al.* (2006); Iwatani and Shinji (2006); Gonçalves *et al.* (2003)], despite radial symmetry allowing the flow to be projected onto the lower dimensional manifold, $\mathbb{S}^{N-1}$ [Hadeler (1992)]. It has recently been shown in three dimensions, even when the equilibrium in both half-systems is stable (i.e. $A_L(0)$ and $A_R(0)$ are *Hurwitzian*, that is, their eigenvalues all have negative real part) the equilibrium in the full system may not be stable [Carmona *et al.* (2006)]. (For a piecewise-linear, discontinuous system this behavior may occur in two dimensions [Branicky (1998)].) Instead of invariant planes, there may be invariant cones [Küpper (2008); Carmona *et al.* (2005b)]. Similar behavior appears in piecewise-linear, continuous maps. Determining the stability of a fixed point on a switching manifold in only two dimensions is also a difficult problem [Do *et al.* (2008)]. A border-collision bifurcation is said to be *dangerous* if the fixed point on each side of the bifurcation is stable, but at the bifurcation it is not [Ganguli and Banerjee (2005); Do (2007); Do and Baek (2006); Hassouneh *et al.* (2004)].

This section is concluded with an important lemma concerning the piecewise-linear, continuous map

$$w' = \begin{cases} \sigma b + A_L w, \; w_1 \le 0 \\ \sigma b + A_R w, \; w_1 \ge 0 \end{cases} . \tag{1.19}$$

Lemma 1.1. *The map (1.19) is a homeomorphism if and only if* $\det(A_L)\det(A_R) > 0$.

Proof. In order for (1.19) to be a homeomorphism each half-map must be a homeomorphism, thus we require both A_L and A_R to be nonsingular. Assume A_L and A_R are nonsingular and let f denote the map (1.19). By continuity of f, A_L and A_R differ in only possibly the first column, therefore $A_L^{-1}A_R$ and I also differ in only possibly the first column. Hence,

$$e_1^{\mathsf{T}} A_L^{-1} A_R = \frac{\det(A_R)}{\det(A_L)} e_1^{\mathsf{T}} . \tag{1.20}$$

If $\det(A_L)\det(A_R) < 0$, by (1.20) the first component of the vector $A_L^{-1}A_R e_1$ is negative, hence

$$f(A_L^{-1}A_R e_1) = \sigma b + A_L(A_L^{-1}A_R e_1) = \sigma b + A_R e_1 = f(e_1) .$$

Thus f is not a one-to-one function because $A_L^{-1}A_R e_1 \neq e_1$, therefore in this case f is not a homeomorphism.

Conversely, if $\det(A_L)\det(A_R) > 0$, f is a homeomorphism because it has the following well-defined inverse

$$f^{-1}(y) = \begin{cases} A_L^{-1}(y - \sigma b), \; t \le 0 \\ A_R^{-1}(y - \sigma b), \; t \ge 0 \end{cases} ,$$

where $t = e_1^{\mathsf{T}} A_R^{-1}(y - \sigma b)$, because by (1.20), $e_1^{\mathsf{T}} A_L^{-1}(y - \sigma b) = \frac{\det(A_R)}{\det(A_L)} e_1^{\mathsf{T}} A_R^{-1}(y - \sigma b)$ has the same sign as t. □

When (1.19) is non-homeomorphic its range is not the full space, $\mathbb{R}^N$. Various non-homeomorphic, piecewise-linear, continuous maps are discussed in [Kapitaniak and Maistrenko (1999); Sushko and Gardini (2006); Gardini *et al.* (2006a)].

1.3 Existence of Equilibria and Fixed Points

The previous section introduced the systems (1.10) and (1.14), with (1.11), to describe dynamics local to discontinuous and border-collision bifurcations respectively. The current section looks at equilibria of the ODE system (1.10) as in, for instance [Leine and Nijmeijer (2004); Leine (2006);

di Bernardo *et al.* (2008a)]. The given method is easily adapted for the map (1.14). In fact it was for the discrete-time case that the method was originally formulated by Mark Feigin in the 1970's [Feigin (1974, 1978, 1995)].

When $\mu = 0$, the origin is an equilibrium of (1.10) and lies on the switching manifold. By the implicit function theorem, if $\det(A_L(0)), \det(A_R(0)) \neq 0$, each half-system of (1.10) has a unique equilibrium in a neighborhood of $(x;\mu) = (0,0)$ given by

$$x^{*(i)}(\mu) = -A_i(0)^{-1}b(0)\mu + O(\mu^2) \,. \tag{1.21}$$

(If one of $A_L(0)$ or $A_R(0)$ is singular the bifurcation is codimension-two and nonlinear terms become important, see Sec. 3.1.) The equilibrium, $x^{*(L)}$, is an equilibrium of the full system, (1.10), and said to be *admissible* whenever $x_1^{*(L)} \leq 0$, otherwise it is said to be *virtual*. Similarly, $x^{*(R)}$, is admissible exactly when $x_1^{*(R)} \geq 0$, An expression for $x_1^{*(i)}$ to determine values of μ for which each $x^{*(i)}$ is admissible is now derived.

As detailed in Appendix C, the adjugate matrices of A_L and A_R share the same first row,

$$\varrho_{\rm ode}^{\mathsf{T}}(\mu) = e_1^{\mathsf{T}}\mathrm{adj}(A_L(\mu)) = e_1^{\mathsf{T}}\mathrm{adj}(A_R(\mu)) \,. \tag{1.22}$$

Using (C.1) the first component of the equilibrium, (1.21), is

$$x_1^{*(i)}(\mu) = -\frac{\varrho_{\rm ode}^{\mathsf{T}}(0)b(0)}{\det(A_i(0))}\mu + O(\mu^2) \,. \tag{1.23}$$

In the previous section the assumption was made that $x_1^{*(R)}$ varies linearly with respect to $\xi_1 = \mu$, (1.6), therefore

$$\varrho_{\rm ode}^{\mathsf{T}}(0)b(0) \neq 0 \,, \tag{1.24}$$

hence admissibility is determined by the sign of $\frac{\varrho_{\rm ode}^{\mathsf{T}}(0)b(0)}{\det(A_i(0))}$.

There are two cases. If $\det(A_L(0))$ and $\det(A_R(0))$ have the same sign then $x^{*(L)}$ and $x^{*(R)}$ are admissible for different signs of μ. Therefore, in a neighborhood of $(x;\mu) = (0;0)$, (1.10) has a unique equilibrium. Alternatively if $\det(A_L(0))$ and $\det(A_R(0))$ have opposite sign, $x^{*(L)}$ and $x^{*(R)}$ are admissible for the same sign of μ and collide and annihilate at the origin as $\mu \to 0$. The first case is known as *persistence*, the second is called a *nonsmooth fold.*

A similar analysis may be applied to the map (1.14). If $\det(I - A_L(0)), \det(I - A_R(0)) \neq 0$, each half-map of (1.14) has a unique fixed point in a neighborhood of $(x;\mu) = (0,0)$ with first component

$$x_1^{*(i)} = \frac{\varrho_{\rm map}^{\mathsf{T}}(0)b(0)}{\det(I - A_i(0))}\mu + O(\mu^2) \,, \tag{1.25}$$

where

$$\varrho_{\rm map}^{\sf T}(\mu) = e_1^{\sf T}{\rm adj}(I - A_L(\mu)) = e_1^{\sf T}{\rm adj}(I - A_R(\mu)) \; . \tag{1.26}$$

1.4 The Observer Canonical Form

To analyze any local bifurcation from a general point of view it is often a good idea to first obtain a normal form. This section describes the observer canonical form for discontinuous and border-collision bifurcations, as in [Carmona *et al.* (2002); di Bernardo *et al.* (2008a)]. (For a different approach see [Bemporad *et al.* (2000)].) The word *canonical* instead of *normal* is used because in smooth bifurcation theory, normal forms have a precise meaning [Kuznetsov (2004); Arnold (1988)]. This section explains why the canonical form may only be obtained when the linearizations of the left and right half-systems have no eigenspace tangent to the switching manifold at the bifurcation.

Call an $N \times N$ matrix whose $i^{\rm th}$ column is e_{i-1} for all $i \neq 1$, a *companion matrix*. For example

$$C = \left[\begin{array}{c|cccc} -\Delta_1 & 1 & & & \\ -\Delta_2 & & 1 & & \\ \vdots & & & \ddots & \\ -\Delta_{N-1} & & & & 1 \\ -\Delta_N & & & & \end{array} \right] , \tag{1.27}$$

is a companion matrix, where each $\Delta_i \in \mathbb{R}$. Alternatively companion matrices may be defined so that the first $N-1$ columns are standard basis vectors [Dummit and Foote (2004); Hartley and Hawkes (1970)]. Such matrices are similar to (1.27) via transformation by a unitary matrix, hence the difference is immaterial (in fact some authors choose the first $N-1$ rows or the last $N-1$ rows).

The characteristic polynomial of C, (1.27), is

$$p(\lambda) \equiv \det(\lambda I - C) = \lambda^N + \Delta_1 \lambda^{N-1} + \Delta_2 \lambda^{N-2} + \cdots + \Delta_N \; . \tag{1.28}$$

Every $N \times N$ matrix, A, has a unique characteristic polynomial, (1.28), and a unique companion matrix, C, with the same characteristic polynomial. With the idea of applying a similarity transformation, one might like to know when a given matrix A is similar to its companion C. The answer is that A and C are similar if and only if there exists no monic polynomial, p^*, of degree less than N with $p^*(A) = 0$, [Dummit and Foote (2004);

Hartley and Hawkes (1970); Turnbull and Aitken (1932)]. When A is not similar to C, it is instead similar to a block diagonal matrix with each block in companion form. This matrix (else the matrix C, when A and C are similar), is known as the *rational canonical form* of A and is a popular standard matrix form like the Jordan normal form [Dummit and Foote (2004); Hartley and Hawkes (1970); Turnbull and Aitken (1932)].

However, when transforming the piecewise-smooth systems (1.10) and (1.14), we require that the transformation does not perturb the switching manifold. Thus we would like to know when A and C are similar and a similarity transform exists that leaves the surface $x_1 = 0$ unchanged. For example the matrix $A = \begin{bmatrix} 1 & 0 \\ 2 & 3 \end{bmatrix}$ is similar to its companion matrix, $C = \begin{bmatrix} 4 & 1 \\ -3 & 0 \end{bmatrix}$, but any appropriate similarity transform must perturb $x_1 = 0$.

To resolve this question, as in [Carmona *et al.* (2002)], the control theory concept of observability [Sontag (1998); Fairman (1998); Vincent and Grantham (1997); Trentelman *et al.* (2001)] is extended to piecewise-smooth, continuous systems.

Definition 1.1. The ODE system (1.10) or the map (1.14) is said to be *observable* if the *observability matrix*

$$\mathcal{O} = \begin{bmatrix} e_1^{\mathsf{T}} A_L(0)^{N-1} \\ \vdots \\ e_1^{\mathsf{T}} A_L(0) \\ e_1^{\mathsf{T}} \end{bmatrix}, \tag{1.29}$$

is nonsingular.

Note, it is easily shown that invertibility of (1.29) is independent of the first column of $A_L(0)$. Consequently, by continuity, (1.12), it is irrelevant which of A_L and A_R is used in the above definition.

The following lemma provides a geometric interpretation of the notion of observability.

Lemma 1.2. *The ODE system (1.10) [the map (1.14)] is observable if and only if $A_L(0)$ has no eigenspace orthogonal to e_1.*

Proof. Let $A = A_L(0)$. Suppose A has an eigenspace orthogonal to e_1. Let $v \neq 0$ be an element of such an eigenspace. Then $A^k v$ is also an element of this eigenspace for all $k \in \mathbb{N}$. Thus $e_1^{\mathsf{T}} A^k v = 0$ for all k. Thus $\mathcal{O} v = 0$ and therefore $\mathcal{O}$ is singular.

Conversely, suppose $\mathcal{O}$ is singular. Then $\exists v \neq 0$ with $\mathcal{O}v = 0$. Thus $e_1^{\mathsf{T}} A^k v = 0$ for $k = 0, \ldots, N-1$. Thus the N vectors $v, \ldots, A^{N-1}v$ are contained in a $(N-1)$-dimensional linear subspace and are therefore linearly dependent. Let m be the maximum value of k for which the vectors $v, \ldots, A^k v$ are linearly independent. Then $A^k v \in \mathrm{span}\{v, \ldots, A^m v\} \equiv \Lambda$, for all k, since by an inductive argument, if $A^{k-1}v \in \Lambda$, then $A^k v = A(A^{k-1}v) = A(\alpha_0 v + \cdots + \alpha_m A^m v) = \alpha_0 Av + \cdots + \alpha_m A^{m+1} v \in \Lambda$. It follows that Λ is either an eigenspace or the span of several eigenspaces orthogonal to e_1. □

Observability is easy to verify for two-dimensional systems. If $A_L(0) = \begin{bmatrix} a & b \\ c & d \end{bmatrix}$, then $\mathcal{O} = \begin{bmatrix} a & b \\ 1 & 0 \end{bmatrix}$, thus $\det(\mathcal{O}) = -b$. Therefore the system is observable if and only if $b \neq 0$. Notice that $b \neq 0$ whenever $A_L(0)$ has complex-valued eigenvalues, thus we have the following result (which also follows immediately from Lemma 1.2).

Corollary 1.1. *If the ODE system (1.10) [the map (1.14)] is two-dimensional and $A_L(0)$ has complex-valued eigenvalues then (1.10) [(1.14)] is observable.*

Corollary 1.1 will be used in later chapters. The next result introduces the transformation matrix required in Theorem 1.1.

Lemma 1.3. *Let C_L be the unique companion matrix with the same characteristic polynomial as $A_L(0)$. Then there exists a nonsingular, real-valued, $N \times N$ matrix, T, with*

$$e_1^{\mathsf{T}} T = e_1^{\mathsf{T}} , \tag{1.30}$$

$$T A_L(0) T^{-1} = C_L , \tag{1.31}$$

if and only if the system (1.10) [the map (1.14)] is observable.

A proof of this result is given in Appendix A and is essentially taken from the control theory textbook of Stephen Barnett [Barnett (1975)]. Also, note that by continuity (1.12), it follows the matrix T provides a similarity transformation for $A_R(0)$ to its companion matrix C_R.

Theorem 1.1. *Suppose the ODE system (1.10) [the map (1.14)] is observable and $\varrho^{\mathsf{T}}(0) b(0) \neq 0$. Then there exists a nonlinear transformation*

$$\hat{x} = \hat{T}(\mu)(x - \psi(\mu)) , \tag{1.32}$$

$$\hat{\mu} = \alpha(\mu) , \tag{1.33}$$

with $e_1^{\mathsf{T}} \hat{T}(\mu) = e_1^{\mathsf{T}}$, $\psi_1(\mu) = 0$ *and* $\alpha'(0) \neq 0$ *such that (1.11) becomes the* C^k *function*

$$\hat{f}(\hat{x}; \hat{\mu}) = \hat{\mu} e_N + C_i(\hat{\mu})\hat{x} + O(|\hat{x}|^2) + o(k) , \tag{1.34}$$

where C_i *is a companion matrix, for sufficiently small* μ.

See Appendix A for a proof. The ODE system (1.10) or the map (1.14) with (1.34) in place of (1.11) is known as the *observer canonical form.* Some authors choose a vector other than e_N in (1.34) as part of their canonical form. For the moment, call this vector q. It becomes apparent in the proof of Theorem 1.1, that in the continuous-time case q may be any vector not orthogonal to e_N, whereas in the discrete-time case q may be any vector not orthogonal to $(1, \ldots, 1)^{\mathsf{T}}$ and in this case often $q = e_1$ is chosen instead. For this book the vector $q = e_N$ is chosen because it may be used in both cases.

Whenever (1.10) or (1.14) is not observable it may be decomposed into two lower-dimensional decoupled systems [Fairman (1998)]. One of these systems will be piecewise-smooth, the other will be smooth but possibly non-autonomous. Consequently, nonsmoothness for a non-observable system occurs inside a space of dimension less than N.

Theorem 1.1 allows investigations into discontinuous and border-collision bifurcations to be greatly simplified. Dynamical behavior of the general system (1.4) local to a discontinuous bifurcation that is non-degenerate in the sense that, a) it may be transformed to (1.10) with (1.11), b) the dynamics are structurally stable, and c) it is observable, are described by a piecewise-linear system of the form,

$$\dot{x} = \begin{cases} \mu e_N + C_L x, & x_1 \leq 0 \\ \mu e_N + C_R x, & x_1 \geq 0 \end{cases} . \tag{1.35}$$

In other words, the system (1.35) may exhibit any such non-degenerate discontinuous bifurcation. The elements of the first columns of the $N \times N$ matrices C_L and C_R are the only values that may be altered, therefore a complete analysis of non-degenerate discontinuous bifurcations may be accomplished by considering all variations of these $2N$ values. Since there is a one-to-one relation between companion matrices and characteristic polynomials, this is equivalent to considering all possible characteristic polynomials for both C_L and C_R. Note, the eigenvalues of C_L and C_R are the associated eigenvalues of the left and right equilibria of (1.35):

$$x^{*(i)} = -\mu C_i^{-1} e_N , \tag{1.36}$$

when they are well-defined. Consequently we have the following statement: dynamical behavior local to a non-degenerate, discontinuous bifurcation [border-collision bifurcation] is uniquely determined by the associated eigenvalues [multipliers] of the left and right equilibria (1.36) [fixed points].

1.5 Discontinuous Bifurcations

As described earlier, a bifurcation arising from the perturbation of a piecewise-smooth, continuous, ODE system with an equilibrium on a switching manifold is known as a discontinuous bifurcation. The previous three sections have derived a generically valid canonical form, (1.10) with (1.34), and shown that local structurally stable dynamics are determined by linear terms. Either a single equilibrium exists on each side of the bifurcation (persistence) or two equilibria exist on exactly one side (non-smooth fold). Other invariants, such as periodic orbits, may be created at the bifurcation. This section discusses the the typical nature of discontinuous bifurcations. In general, a discontinuous bifurcation is either the analogue of a well-known, smooth bifurcation or novel and unique to piecewise-smooth systems.

For smooth systems a local bifurcation occurs when one or more eigenvalues associated with an equilibrium cross the imaginary axis as a system parameter is continuously varied, Fig. 1.4(a). In contrast, as an equilibrium is followed through a discontinuous bifurcation, its associated eigenvalues may change discontinuously (panel (b)). The Jacobian of the equilibrium of each half-system of (1.10), (1.34), is given by

$$J_i(\mu) = C_i(0) + O(1) \,. \tag{1.37}$$

Therefore the critical eigenvalues are those of $C_L(0)$ and $C_R(0)$.

Here the eigenvalue path concept of Remco Leine and coworkers that attempts to classify discontinuous bifurcations is introduced. As in [Leine (2006); Leine and Nijmeijer (2004)], since the Jacobian is undefined at $\mu = 0$, we instead consider the generalized Jacobian

$$\begin{aligned} \hat{J} &= \text{convex hull } (J_L(0), J_R(0)) \\ &= \left\{ (1-s)J_L(0) + sJ_R(0) \;\middle|\; s \in [0,1] \right\} . \end{aligned} \tag{1.38}$$

We can treat the eigenvalues of $\hat{J}$ as defining a unique path of eigenvalues at the discontinuity as s is varied from 0 to 1, see Fig. 1.4(b). This is

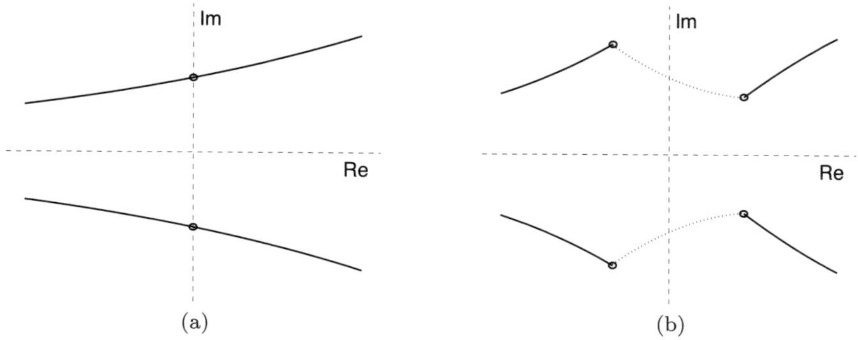

Fig. 1.4: Sketches of eigenvalues under variation of a system parameter near a local bifurcation in a smooth system, panel (a), and a piecewise-smooth, continuous system, panel (b).

called the *eigenvalue path*. Let $\hat{n}$ denote the number of values of s for which $(1 - s)A_L + sA_R$ has one or more eigenvalues on the imaginary axis. The idea is that $\hat{n}$ represents the possible degree of complexity of the discontinuous bifurcation.

It is conjectured that if $\hat{n} = 0$ then no bifurcation occurs at $\mu = 0$. This has been verified for many examples but to the author's knowledge has yet to be proved [Leine (2006); Leine and Nijmeijer (2004)]. If $\hat{n} = 1$, the discontinuous bifurcation is called a *single-crossing bifurcation.* Leine *et al.* observe that single-crossing bifurcations invariably appear to be the analogue of some familiar smooth bifurcation. Fig. 1.5 illustrates discontinuous analogues of saddle-node and Hopf bifurcations (see [Leine and Nijmeijer (2004)] for transcritical and pitchfork-like bifurcations). Finally if $\hat{n} > 1$, the bifurcation is called a *multiple-crossing bifurcation* and may not have a smooth counterpart, see Sec. 2.3 for examples. Such bifurcations can sometimes be regarded as the coincidence of one or more smooth bifurcations but this is not always the case.

However, until conjectures regarding the distinction between single-crossing and multiple-crossing bifurcations can be resolved in systems of arbitrary dimension, it is not obvious that the generalized Jacobian provides the most useful connection between the left and right limiting eigenvalues. Any path connecting the limiting eigenvalues is in some sense artificial. It seems reasonable that the ideal theory of discontinuous bifurcations should be path independent.

In one and two-dimensional systems codimension-one, discontinuous bi-

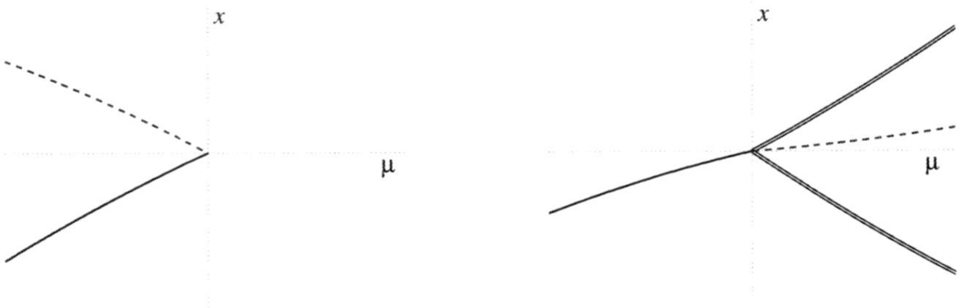

(a) (b)

Fig. 1.5: Schematic diagrams of discontinuous analogues of a saddle-node bifurcation, panel (a), and a supercritical Hopf bifurcation, panel (b). Stable [unstable] equilibria are indicated by a solid [dashed] line. Maximum and minimum values of a periodic orbit are shown with double lines.

furcations are well-understood. The 1998 paper of Freire *et al.* [Freire *et al.* (1998)], derives all generic discontinuous bifurcation scenarios for a general planar piecewise-linear system. These are summarized in Chapter 2 where selected results are extended to piecewise-smooth systems. As mentioned in Sec. 1.2, in higher dimensions many complications can arise.

1.6 Border-Collision Bifurcations

The pioneering 1992 paper of Nusse and Yorke [Nusse and Yorke (1992)] coins the term *border-collision bifurcation* (see also [Hommes and Nusse (1991)]) to refer to a bifurcation resulting from the collision of a fixed point with a switching manifold in a piecewise-smooth, continuous map. (More recently this term has also been used to include similar bifurcations in piecewise-smooth maps with an alternate nonsmoothness classification.) Outside the western world, in the 1970's Mark Feigin studied these bifurcations and gave them the name *C-bifurcations* [Feigin (1970)] for the Russian word *švejnye* for "sewing". This section restates the result in Sec. 1.3 concerning the existence of fixed points of (1.14) in terms of critical multipliers as originally derived by Feigin. Known phenomena relating to border-collision bifurcations is then overviewed.

From (1.25) it is seen that for small values of μ, the fixed point, $x^{*(i)}$, exists exactly when $\det(I - A_i(0))$ is nonsingular. If $x^{*(i)}$ exists and $\varrho_{\rm map}^{\sf T}(0)b(0) \neq 0$, then the sign of μ for which the fixed point is admissible is determined by the sign of $\det(I - A_i(0))$. Feigin noticed that this sign depends on the number of real multipliers of $A_i(0)$ greater than 1, call this number σ_i^+. It follows that if $\sigma_L^+ + \sigma_R^+$ is even, then a single fixed point persists (i.e. $x^{*(L)}$ and $x^{*(R)}$ are admissible for opposite signs of μ). Conversely, if $\sigma_L^+ + \sigma_R^+$ is odd then the bifurcation is a nonsmooth fold.

A similar condition exists for 2-cycles comprised of one point in each half-plane. Chapter 6 shows that such a cycle exists exactly when $\det(I - A_L(0)A_R(0))$ is nonsingular and in the absence of nonlinear terms has no points on the switching manifold if and only if $\det(I + A_L(0))$ and $\det(I + A_R(0))$ are both nonsingular. As shown in [di Bernardo *et al.* (1999)], the 2-cycle is admissible for one sign of μ if and only if $\sigma_L^- + \sigma_R^-$ is odd, where σ_i^- denotes the number of real multipliers of $A_i(0)$ less than -1. As with any structurally stable bounded invariant of (1.14) that exists for small values of μ, the 2-cycle shrinks linearly to a point as $\mu \to 0$. Consequently, the border-collision bifurcation may resemble a period-doubling bifurcation. Effective period-doubling via a border-collision bifurcation has recently been suggested to play an important role in cardiac dynamics [Zhao *et al.* (2008); Zhao and Schaeffer (2007)]. Determining the existence and admissibility of cycles of higher period in terms of the multipliers of A_L and A_R is a far more difficult task, see Chapter 6. For an extension of Feigin's results to piecewise-smooth, discontinuous maps, see [Dutta *et al.* (2008)].

Under smooth parameter variation, periodic cycles may collide with a switching manifold. Remarkably, the unfolding of an n-cycle that has one point on a switching manifold leads the to same general form as for fixed points, i.e. (1.14) with (1.11). This is shown in Chapter 6. Hence Feigin's results may also be used to determine the admissibility of n and $2n$-cycles local to this border-collision bifurcation [di Bernardo *et al.* (1999)].

In one dimension, (1.14) is completely understood [di Bernardo *et al.* (2008a)]. Here $a_L = A_L(0)$ and $a_R = A_R(0)$ are scalars and together with μ form a three-dimensional parameter space within which all codimension-one bifurcations are contained. By scaling and symmetry considerations, we may assume $b = 1$, and $a_L \geq 0$. If also $a_R > 0$, then for small values of μ the map is a non-decreasing function and displays only simple dynamics. Thus we assume $a_R \leq 0$. The map exhibits persistence if $a_L < 1$ and a nonsmooth fold if $a_L > 1$. When $\mu \leq 0$, fixed points are the only possible invariant sets. When $\mu > 0$ there is the possibility for complex dynamics.

Only local non-degenerate dynamics are of interest here, hence, as detailed in Sec. 1.2, it suffices to investigate the piecewise-linear approximation

$$x' = \begin{cases} \mu + a_L x, & x \le 0 \\ \mu + a_R x, & x \ge 0 \end{cases} . \tag{1.39}$$

Fig. 1.6 illustrates stable solutions of (1.39) when $\mu > 0$, also described in [di Bernardo *et al.* (2008a, 1999); Nusse and Yorke (1995); Banerjee *et al.* (2000a)]. At any point in the diagram there exists at most one attracting set [di Bernardo *et al.* (2008a)]. The fixed point, $x^{*(L)}$, is admissible whenever $a_L > 1$ but is never stable. The fixed point, $x^{*(R)}$, is always admissible and is stable whenever $a_R > -1$. The only admissible, stable periodic cycles are those with exactly one positive-valued point. Regions of existence are

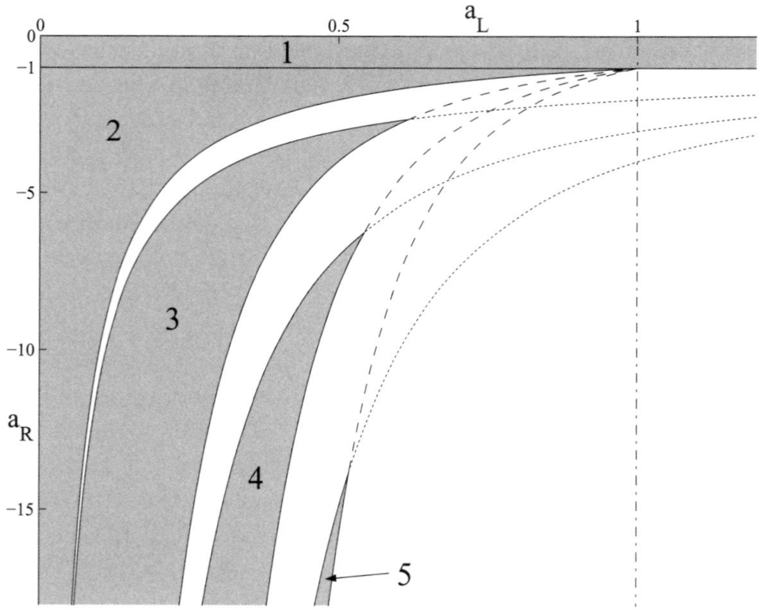

Fig. 1.6: A bifurcation set of the one-dimensional, piecewise-linear, continuous map (1.39) when $\mu > 0$. Gray regions correspond to the existence of an admissible, attracting, periodic solution of the indicated period. Dotted curves given by $a_R = -\frac{1-a_L^{n-1}}{a_L^{n-2}-a_L^{n-1}}$ correspond to border-collision bifurcations of n-cycles. Dashed curves given by $a_R = -\frac{1}{a_L^{n-1}}$ correspond to a loss of stability of the cycles via an associated multiplier attaining the value -1. Inside the white region left of the vertical dash-dot line there exist chaotic attractors.

shown in Fig. 1.6 up to period 5 (periodic cycles for all $n > 5$ exist at larger negative values of a_R than shown in the figure). Stability is lost when an associated multiplier continuously crosses the critical value -1. Due to the piecewise-linear nature of the map, this bifurcation does not correspond to a period-doubling bifurcation. Instead a chaotic attractor is born. The chaotic attractor is robust in the sense that there are no periodic windows [Banerjee *et al.* (1998)] (a property that has only recently been found in smooth maps [Andrecut and Ali (2001)]). In addition to stable periodic solutions, there exist many unstable periodic solutions in accordance with Sharkovskii's Theorem, see [Sharkovskii (1995); Li and Yorke (1975)].

In more than one dimension, (1.14) may exhibit extremely complicated dynamics such as multiple strange or quasiperiodic attractors. Analytical investigations commonly utilize the canonical form, (1.34). There is no restriction on the first columns of C_L and C_R, therefore in N dimensions a complete investigation of (1.14) with (1.34) requires variation of exactly $2N$ independent parameters in addition to μ.

Several recent publications present partial classifications in two dimensions. In [Banerjee and Grebogi (1999)] (and continued [Banerjee *et al.* (2000b)]), the authors describe 11 qualitatively different types of border-collision bifurcations under the physically motivated restriction that (1.14) is dissipative (i.e. $\det(C_L), \det(C_R) < 0$). In the case that the multipliers of C_L and C_R are real-valued a classification of fixed points and 2-cycles is given in [di Bernardo *et al.* (1999)]. In [Zhusubaliyev *et al.* (2006, 2008)] the authors allow for complex-valued multipliers and describe the creation of invariant circles on which the dynamics generically are either quasiperiodic or tend to an attracting periodic solution. In the latter case there also exists a saddle-type solution of the same period. Homoclinic tangles arise when one-dimensional invariant manifolds of the saddle-type solution intersect, implying the existence of homoclinic solutions and Smale horseshoes [Zhusubaliyev *et al.* (2006); Belghith (2000)]. Multiple attractors may coexist; boundaries between basins of attraction often appear fractal [Sushko and Gardini (2006); Dutta *et al.* (1999); Nusse *et al.* (1994)]. A border-collision bifurcation from chaos to hyper-chaos (behavior characterized by two positive Lyapunov exponents) has been described in [Thamilmaran *et al.* (2004)]. Also, a sequence of nested spirals has been observed in a three-dimensional, piecewise-linear, continuous map that models an electrical circuit [Bonatto and Gallas (2008)]. For basic border-collision bifurcations in three-dimensional maps, see [Roy and Roy (2008)].

Particularly when C_L and C_R have complex-valued multipliers, it is of interest to compute regions in parameter space within which periodic solutions of a particular period or rotation number exist and are attracting. As with circle maps, such regions are called resonance tongues (or Arnold tongues). In contrast to smooth systems, resonance tongues in piecewise-smooth, continuous systems typically exhibit a distinctive lens-chain structure. This was first observed for a one-dimensional, piecewise-linear circle map [Yang and Hao (1987)] and has since been described for the map (1.14), in [Zhusubaliyev *et al.* (2006)], for other two-dimensional, piecewise-linear maps [Zhusubaliyev and Mosekilde (2008b); Sushko and Gardini (2006); Gardini *et al.* (2006b); Sushko *et al.* (2004)], and higher dimensional maps [Zhusubaliyev *et al.* (2003, 2002)]. Lens-chains are not observed in Fig. 1.6 because a minimum of two dimensions are required to realize these structures for (1.14). A detailed analysis of lens-chain structures is presented in Chapter 6.

1.7 Poincaré Maps and Discontinuity Maps

Let Π be an $(N-1)$-dimensional manifold, contained in the phase space of an N-dimensional, ODE system. The *Poincaré map* for Π, is a function $g : \Pi \to \Pi$, where for any $p \in \Pi$, $g(p)$ is the next intersection of the trajectory of the ODE system passing through p, if such a point exists. Poincaré maps are often ill-defined globally [Lee *et al.* (2008)]. It is usual to study a Poincaré map locally, say in a neighborhood of a periodic orbit (where $g(p) = p$) and choose the Poincaré section, Π, such that intersections are transversal. Poincaré maps are ubiquitous in dynamical systems, fundamental for studying periodic orbits and more complicated oscillatory behavior such as strange attractors [Glendinning (1999); Robinson (2004); Wiggins (2003)].

This section discusses ODE systems that elicit piecewise-smooth, Poincaré maps. The main focus is on Filippov systems and emphasis is placed on particular scenarios that lead to piecewise-smooth, continuous Poincaré maps of the form (1.14) with (1.11) studied intensively in later chapters. Various non-autonomous systems yield piecewise-smooth, continuous Poincaré maps, see for example [Zhusubaliyev and Mosekilde (2006b, 2008b)] for non-autonomous Filippov models of a DC/DC buck converter and [Sieber *et al.* (unpublished)] for a hybrid system with delay. Poincaré maps for smooth, ODE systems are generally smooth. However they may contain discontinuities when an invariant manifold of a saddle-type invariant set intersects the Poincaré section.

Suppose the ODE system is Filippov and Π is a smooth Poincaré section. Of interest is the degree of smoothness of the induced Poincaré map, g, in a neighborhood of a point $p \in \Pi$. It is assumed that g is well-defined throughout this neighborhood and that the trajectory passing through p, call it Γ, has transversal intersections with Π at the points p and $g(p)$. For a formal treatment of Poincaré maps in piecewise-linear systems, see [Llibre and Teruel (2004)].

The simplest case is when Γ fails to cross any switching manifolds. Here g has a well-defined Taylor series and is therefore smooth. Alternatively, suppose Γ crosses one or more switching manifolds and suppose all such crossings occur transversally (i.e. non-tangentially and without sliding, see below). Then g may be expressed as the composition of a sequence of smooth maps between switching manifolds. Hence g is also smooth in this case. Note if $g(p) = p$, stability of the fixed point, p, may be determined by composition of Jacobian matrices determined by smooth components of the flow with so-called *saltation matrices* which account for sudden angle changes of trajectories at switching manifolds that influence stability [di Bernardo *et al.* (2008a); Aizerman and Gantmakher (1958)].

Now suppose Γ tangentially intersects a switching manifold, Σ, at, say, the origin, see Fig. 1.7. In this case Γ is called a *grazing trajectory*. For a point, $p_0 \in \Pi$, near p, if the trajectory passing through p_0 intersects Σ, as in Fig. 1.7, denote the first and second intersections of this trajectory with Σ by p_2 and p_3, respectively. One may express g as a composition of three maps via the sequence: $p_0 \to p_2 \to p_3 \to g(p)$. However, in this form singularities arising from the local tangency appear in a global component of g (specifically in the map, $p_0 \to p_2$). A typically preferable alternative is to incorporate all singularities into a local map known as a *discontinuity map*, which will now be described.

Locally, Σ divides phase space into two regions. With an eye to Fig. 1.7, to aid the following discussion the region containing the point, p, is referred to as the right half-space and the second region is referred to as the left half-space. Let $\varphi_t^R(x)$ denote the right half-flow (the flow in the right half-space) and $\varphi_t^L(x)$ the left half-flow. Let Ξ be an $(N-1)$-dimensional manifold intersecting the origin and not tangent to Σ at this point. Artificially following the right half-flow from p_0, we arrive at a point p_1 on Ξ. If p_1 were in right half-space, by continuing from p_1 along the right half-flow we would arrive at $g(p_0)$. However if p_1 is in the left half-space, it is necessary to calculate a correction following the left half-flow from a

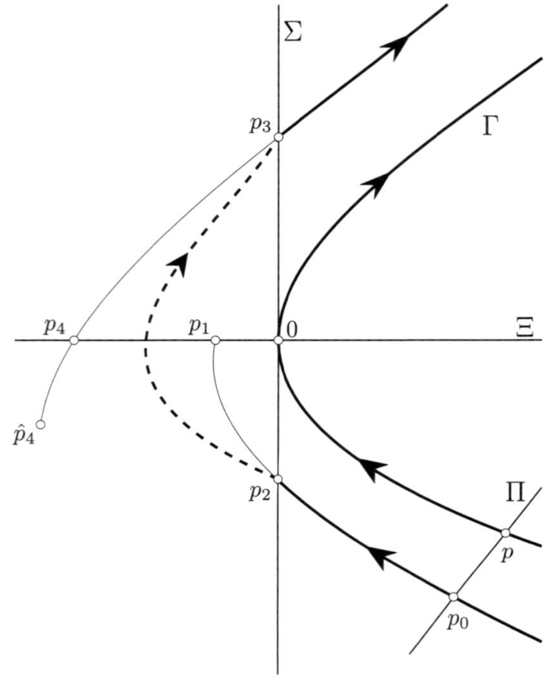

Fig. 1.7: A phase portrait near grazing for a Filippov system. Γ is a trajectory that grazes the switching manifold, Σ, at the origin. A trajectory passing through the Poincaré section, Π, at p_0, transversely crosses Σ at p_2 and p_3. Solution curves following the right [left] half-flow are shown as solid [dashed] curves. Points p_1 and p_4 lie on a cross-section, Ξ. The PDM is the map $p_1 \to p_4$, and the ZDM is the map $p_1 \to \hat{p}_4$, where $\hat{p}_4$ is defined by (1.44).

different point, say p_4 or $\hat{p}_4$ in Fig. 1.7, in order to arrive at $g(p_0)$. The map, $p_1 \to p_4$, where $p_4 \in \Xi$, is known as the *Poincaré-section discontinuity map* (PDM) [Dankowicz and Nordmark (2000); di Bernardo *et al.* (2001b)]. The map, $p_1 \to \hat{p}_4$, where the transition time from $\hat{p}_4$ to p_3 is equal to the transition time from p_2 to p_3 (via the left half-flow) minus the transition time from p_2 to p_1, is known as the *zero-time discontinuity map* (ZDM). Usually the PDM is preferable when the ODE system is autonomous and the ZDM preferable otherwise. The following equations (where each t_i denotes the positive-valued transition time between p_i and p_{i+1} or vice-versa) summarize Fig. 1.7:

$$p_1 = \varphi^R_{t_0}(p_0) \,, \tag{1.40}$$

$$p_2 = \varphi^R_{-t_1}(p_1) \,, \tag{1.41}$$

$$p_3 = \varphi^L_{t_2}(p_2) \,, \tag{1.42}$$

$$p_4 = \varphi^R_{-t_3}(p_3) \,, \tag{1.43}$$

$$\hat{p}_4 = \varphi^R_{t_1 - t_2}(p_3) \,. \tag{1.44}$$

The Poincaré map, g, is obtained via composition of a discontinuity map with one or more global maps. When the global maps are smooth, g will exhibit the same degree of smoothness as the discontinuity map. Smoothness properties of discontinuity maps in different scenarios are now described. Note, generically, the leading order terms of the PDM and the ZDM have the same order, therefore the following discussion applies to both types of discontinuity map.

In a general Filippov system, the vector field at a switching manifold may be simultaneously attracting (or repelling) from both sides. Subsets of switching manifolds where this is the case are known as *sliding regions*. An orbit that intersects a switching manifold at an attracting sliding region will evolve within the switching manifold until it exits the sliding region. This is known as *sliding motion*. Forward evolution of sliding is typically determined by Filippov's method [Filippov (1964); Leine *et al.* (2000)], which takes an average of the two vector fields. In backwards time, the flow is non-unique. In Filippov models of an oscillator subject to dry friction, sliding motion corresponds to the "sticking" phase of the motion [di Bernardo *et al.* (2003); Casini *et al.* (2006); Szalai and Osinga (2008)]. Other studies of sliding include [di Bernardo *et al.* (2001c); Cunha *et al.* (2003); Kowalczyk (2005); Teixeira (1993)].

Consider again the grazing scenario depicted in Fig. 1.7. If in a neighborhood of the grazing point the left half-flow points away from the switching manifold, the discontinuity maps will be discontinuous, assuming there is some mechanism that reinjects the flow back onto the other side of the switching manifold and through the Poincaré section. For recent accounts of discontinuous maps see [di Bernardo *et al.* (2008a); Hogan *et al.* (2007); Avrutin *et al.* (2007); Jain and Banerjee (2003); Rakshit *et al.* (2008)]. Conversely, if the left half-flow points towards the switching manifold near the grazing point, there will exist an attracting sliding region. There are four distinct codimension-one sliding scenarios [di Bernardo *et al.* (2008a)]. The scenario shown in Fig. 1.8(a1) is known as *grazing-sliding*. Here the discontinuity maps will be continuous. Trajectories that enter the attracting

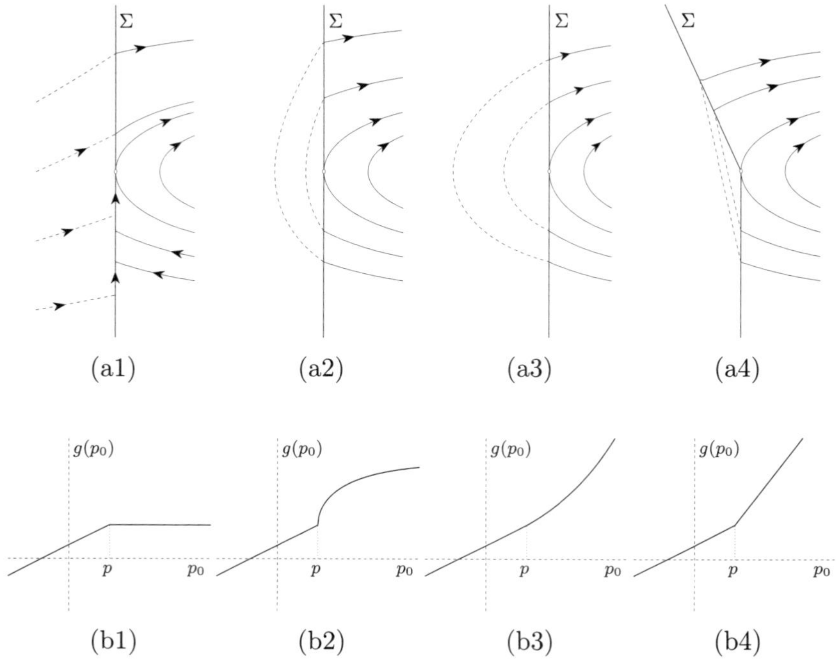

Fig. 1.8: The top panels show phase portraits for Filippov systems with four different grazing scenarios: panel (a1) - grazing-sliding, panel (a2) - regular grazing, panel (a3) - continuous grazing, panel (a4) - grazing at a corner. Solution curves following the right [left] half-flow are shown as solid [dashed] curves. In each case it is assumed the forward and backward orbits of the grazing point transversely intersect a Poincaré section, Π, thus locally there is an induced Poincaré map, g. The generic nonsmoothness property that g possesses in any dimension is illustrated for each case in the corresponding panel below. An exception is regular grazing: in this case a 1/2-type singularity arises only if the flow is at least three-dimensional.

sliding region exit on a generically $(N-2)$-dimensional subset of the switching manifold, Σ. Consequently the corresponding discontinuity map is not invertible. To lowest order, the Poincaré map, g will be of the form (1.14) with (1.11), and one of A_L or A_R will have a multiplier zero [di Bernardo *et al.* (2002b, 2003)]. If one-dimensional, g will resemble Fig. 1.8(b1).

Now suppose there are no sliding regions in a neighborhood of the grazing point. When the grazing trajectory, Γ, is admissible, as in Fig. 1.8(a2), the grazing point is known as a *regular grazing point*. The criterion of no sliding generally corresponds to adding a codimension to the bifurcation, however in practice underlying physical assumptions often eliminate the possibility for sliding motion. Furthermore, many times it is sufficient to assume that only trajectories of interest do not enter sliding regions [di Bernardo *et al.* (2008a)]. If the flow is discontinuous at a regular grazing point, the discontinuity maps and the Poincaré map, g, will have a square-root singularity [di Bernardo *et al.* (2001b)], see Fig. 1.8(b2). (Note however this requires the flow is at least three-dimensional, if two-dimensional the square-root term vanishes since via a time-scaling the scenario may be reduced to Fig. 1.8(a3)). Maps of these form have been shown to exhibit so-called period adding cascades and direct transitions to chaos [Chin *et al.* (1994); Fredriksson and Nordmark (1997); Nordmark (1997, 2001)]. Alternatively if the flow is continuous at the grazing point, there is a 3/2-type singularity [di Bernardo *et al.* (2001b)], see Fig. 1.8(a3). The map is differentiable at the nonsmooth point and generically no bifurcation occurs here, however saddle-node bifurcations have commonly been observed nearby [di Bernardo *et al.* (2008a)]. The same class of Poincaré maps results when the ODE system is piecewise-smooth continuous.

The final situation exemplified by Fig. 1.8 is grazing at a non-differentiable point on Σ, in the absence of sliding. This situation is known as a *corner-collision*. It arises naturally in many physical systems, such as a soft impacting system [Leine and Van Campen (2002); Leine *et al.* (2000)], shown earlier in Fig. 1.3. For further examples see [Budd and Piiroinen (2006); Osorio *et al.* (2008); Sieber *et al.* (unpublished); di Bernardo *et al.* (1997, 1998a)]. As detailed in [di Bernardo *et al.* (2001a)], the Poincaré map associated with the corner-collision in Fig. 1.8(a4) is piecewise-smooth continuous. Note that the collision of a periodic solution with a non-differentiable point on a switching manifold does not require an extra codimension to occur.

1.8 Period Adding

A relation of the well-known period-doubling cascade, a *period-adding cascade* is characterized by successive jumps in the period of a periodic orbit in an ODE system in a manner that forms an approximately arithmetic sequence. The increase in period corresponds to the appearance of an additional excursion of the periodic orbit through phase space. Between windows of periodicity, there are often chaotic regions but this is not always the case. As with period-doubling cascades, the width of periodicity windows may decrease to zero as the period increases over a finite parameter range. In this case the periodic orbit usually limits on a homoclinic, heteroclinic or chaotic orbit.

Period-adding cascades have been observed in a variety of smooth systems, such as a model of internal ocean waves [Tanaka *et al.* (2003)], the Chay and Rose-Hindmarsh models of neuron activity [Yang *et al.* (2006); Holden and Fan (1992)], a model of fluidized bed catalytic reactors [El-nashaie *et al.* (2001)], and the Lorenz equations [Tomita and Tsuda (1980); Tanaka *et al.* (2003)]. In smooth systems, each transition between adjacent periodicity windows involves a collection of familiar bifurcations, such as a period-doubling cascade to chaos over a relatively small parameter range [Tanaka *et al.* (2003); Yang *et al.* (2006)]. However the mechanisms that generate a succession of period-adding phenomena are currently not fully understood.

Poincaré maps provide a useful tool for investigating period-adding behavior. When the Poincaré section is well chosen, periodic orbits in adjacent periodicity windows correspond to periodic solutions of the Poincaré map that differ in period by one. Periodicity windows of the ODE system then correspond to resonance tongues of the Poincaré map. Unlike for smooth maps, resonance tongues in piecewise-smooth maps are often ordered by their associated period. For this reason piecewise-smooth maps readily yield period-adding scenarios.

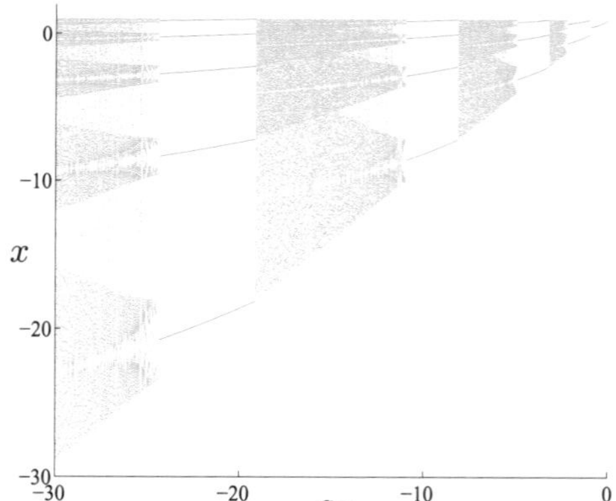

Fig. 1.9: A bifurcation diagram of (1.39) when $a_L = 0.45$ and $\mu = 1$.

For instance resonance tongues of the one-dimensional, piecewise-linear, continuous map (1.39), shown in Fig. 1.6, lie in order of period. Consequently, the corresponding one-parameter bifurcation diagram shown in Fig. 1.9 displays a period-adding sequence. In general period-adding sequences differ from Fig. 1.9 in two major aspects. First, periodicity windows may converge to a point. Different scaling laws apply for different classes of piecewise-smooth maps [Avrutin *et al.* (2007); Halse *et al.* (2003); Kawczyński and Strizhak (2000)]. Second, a lack of smoothness allows for the transition between periodicity windows to occur via a single nonsmooth bifurcation. Thus there may be no regions of chaos between windows. Also windows may overlap resulting in hysteresis. For examples of period-adding in piecewise-smooth, ODE models, see [Piiroinen *et al.* (2004); Piassi *et al.* (2004); Coombes and Osbaldestin (2000)].

1.9 Smooth Approximations

Since piecewise-smooth systems may exhibit exotic, complicated bifurcations, many of which are not fully understood, many authors consider smooth approximations. By mollifying the nonsmooth components of a piecewise-smooth, continuous vector field, V, one obtains a smooth vector field, $V_{\rm smooth}$, and the norm, $||V - V_{\rm smooth}||_\infty$, may be set as small as one likes. This can be useful when smoothness is a desired property. For instance in [Lázaro *et al.* (2001)], the authors study a smooth version of the canonical piecewise-linear form of Chua [Chua and Deng (1988); Kang and Chua (1978)]. The motivation is that global differentiability aids the prediction of intermodulation distortion in some resistor models. A smooth version of the Lozi map (which is a piecewise-linear version of the Hénon map) is studied in [Aziz-Alaoui *et al.* (2001)] to provide a form better suited for some aspects of rigorous analysis. In [Zhao *et al.* (2008)] the authors express a generic period-doubling-like bifurcation of a piecewise-smooth, continuous map as the limit of (smooth) period-doubling bifurcations. In [Leine (2006); Leine and Nijmeijer (2004)], Leine *et al.* describe a variety of discontinuous bifurcations as limits of smooth bifurcations.

However, smooth approximations yield several difficulties. In particular, a jump in derivative will become a rapid change in derivative in a smooth approximation. Consequently smooth approximations are typically stiff and numerically expensive to solve [Leine *et al.* (2000)]. Also, a simple bifurcation in a piecewise-smooth system may correspond to a complex bifurcation sequence in a smooth approximation. For example, Fig. 1.10

shows a bifurcation diagram of the one-dimensional, piecewise-linear map, (1.39). As the value of μ passes through zero, the stable fixed point bifurcates to a stable 3-cycle. By Sharkovskii's Theorem, see [Sharkovskii (1995); Li and Yorke (1975)], orbits of every period coexist with the 3-cycle. For the piecewise-linear map, these orbits are all created at $\mu = 0$. But the equivalent transition for a smooth map requires an infinite sequence of bifurcations. Fig. 1.10 shows a smooth approximation exhibiting an entire period-doubling cascade followed by chaotic solutions and periodicity windows. Thus in this situation, arguably the piecewise-smooth system describes the transition more succinctly than any smooth system is able to.

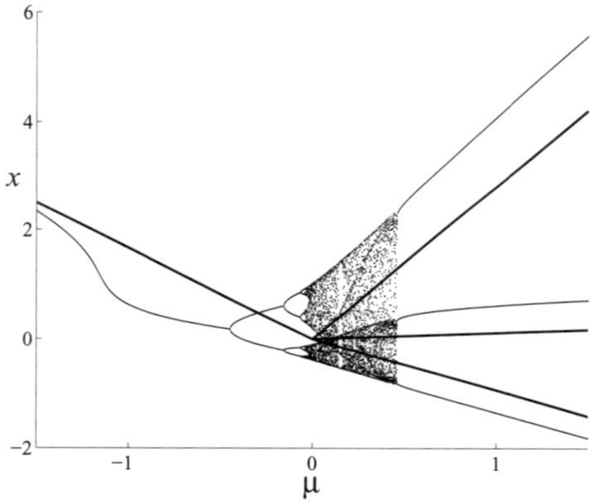

Fig. 1.10: The thick lines form a bifurcation diagram of (1.39) when $a_L = 0.4$ and $a_R = -4$. For $\mu < 0$ there is a stable fixed point, for $\mu > 0$ there is a stable 3-cycle. Superimposed is a bifurcation diagram for a smooth approximation to (1.39) constructed by replacing $|x_1|$ in (1.15) with $x_1 \tanh(K x_1)$, where $K = 1$.

Chapter 2

Discontinuous Bifurcations in Planar Systems

As detailed in Sec. 1.5, a bifurcation resulting from the collision of an equilibrium with a switching manifold in a piecewise-smooth, continuous ODE system is known as a discontinuous bifurcation. Discontinuous bifurcations in systems of dimension three or greater may generate complicated invariant sets, such as homoclinic orbits satisfying Sil'nikov's condition [Llibre *et al.* (2007); Sparrow (1981)]. This chapter presents a complete classification of codimension-one, discontinuous bifurcations in two-dimensional systems by the number and nature of the generated invariant sets. The first serious attempt at such a classification is probably the work by Lum and Chua [Lum and Chua (1991)]. All necessary theoretical results were then completed by Freire *et al.* [Freire *et al.* (1998)]. This chapter provides a novel overview and summary. Also given is a rigorous analysis for an analogue of the Andronov-Hopf bifurcation when nonlinear terms are present.

Dynamical behavior local to a non-degenerate, discontinuous bifurcation is described by an ODE of the form (1.10) with (1.11), see Sec. 1.2. In order to classify discontinuous bifurcations of this system it is necessary to compute all generic invariant sets, i.e. equilibria and periodic orbits, on both sides of the bifurcation. Determination of equilibria is straightforward; this was accomplished in Sec. 1.3. Equilibria of (1.10) are given by (1.21). The computation of periodic orbits, however, is more difficult. It is helpful to assume that only structurally stable periodic orbits are of interest because then it suffices to study the piecewise-linear approximation, (1.18). For a treatment periodic orbits in piecewise-linear systems with multiple switching manifolds, the reader is referred to [Mitrovski and Kocarev (2001); Gonçalves (2005)].

The remainder of this chapter is organized as follows. In Sec. 2.1 periodic orbits are computed for the piecewise-linear system, (1.18), in two dimensions. Section 2.2 studies in full generality the Hopf-like, discontinuous bifurcation characterized by the sudden change of an attracting focus to a repelling focus. Using the information gained in these two sections, Sec. 2.3 argues that there are exactly four distinct possible non-degenerate, discontinuous bifurcations, in addition to no bifurcation occurring. This enables a classification of some codimension-two discontinuous bifurcations.

2.1 Periodic Orbits

A well-known consequence of index theory [Guckenheimer and Holmes (1986); Glendinning (1999)], is that a periodic orbit in a continuous, planar vector field must encircle at least one equilibrium. Moreover, if the equilibria are hyperbolic, the number of equilibria inside the orbit must be odd. Assuming $\det(A_L(0))$ and $\det(A_R(0))$ are nonsingular, the system (1.18) has at most two equilibria. For all codimension-one scenarios that this chapter is interested in, the equilibria will be hyperbolic, hence the number of equilibria inside a periodic orbit of (1.18) is exactly one. Furthermore, such an equilibrium must be a focus (or a center but this special case is ignored in this chapter, see Sec. 3.2) since otherwise the linear nature of the half-systems of (1.18) would imply the presence of invariant, semi-infinite lines contradicting the existence of a periodic orbit.

Consequently, in the search for periodic orbits it suffices to assume that the equilibrium of, say, the left half-system of (1.18) is an admissible, repelling focus. All other cases may be reduced to this situation via flipping the sign of μ or the direction of time or both. Then by Corollary 1.1, (1.18) is observable and may be transformed to the observer canonical form, see Sec. 1.4. In two dimensions and in the absence of nonlinear terms, the canonical form is

$$\begin{bmatrix} \dot{x} \\ \dot{y} \end{bmatrix} = \begin{bmatrix} f(x,y;\mu) \\ g(x,y;\mu) \end{bmatrix} = \begin{cases} \mu \begin{bmatrix} 0 \\ 1 \end{bmatrix} + \begin{bmatrix} \tau_L & 1 \\ -\delta_L & 0 \end{bmatrix} \begin{bmatrix} x \\ y \end{bmatrix}, & x \leq 0 \\ \mu \begin{bmatrix} 0 \\ 1 \end{bmatrix} + \begin{bmatrix} \tau_R & 1 \\ -\delta_R & 0 \end{bmatrix} \begin{bmatrix} x \\ y \end{bmatrix}, & x \geq 0 \end{cases}, \tag{2.1}$$

where x and y are scalar variables and for each $i = L, R$, the real-valued parameters τ_i and δ_i are the trace and determinant of

$$C_i = \begin{bmatrix} \tau_i & 1 \\ -\delta_i & 0 \end{bmatrix}, \tag{2.2}$$

respectively. The equilibrium of each half-system of (2.1) is

$$z^{*(i)} = \begin{bmatrix} x^{*(i)} \\ y^{*(i)} \end{bmatrix} = \frac{\mu}{\delta_i} \begin{bmatrix} 1 \\ -\tau_i \end{bmatrix} , \tag{2.3}$$

when $\delta_i \neq 0$. From the prior assumption that $z^{*(L)}$ is an admissible, repelling focus, it follows

$$\tau_L > 0 \, , \quad \delta_L > \frac{1}{4}\tau_L^2 \, , \quad \mu < 0 \, . \tag{2.4}$$

Two Important Lemmas Relating to Periodic Orbits

The following result (given also in [Llibre and Sotomayor (1996); Freire *et al.* (1998)]) shows that (2.1) cannot exhibit a periodic orbit if $\tau_L \tau_R > 0$. Let Γ be a periodic orbit of (2.1) and let Σ denote the switching manifold. The vector field,

$$H(x, y; \mu) = \begin{bmatrix} -g(x, y; \mu) \\ f(x, y; \mu) \end{bmatrix} , \tag{2.5}$$

is orthogonal to (2.1), thus

$$\oint_\Gamma H \cdot ds = 0 \, .$$

By Green's formula,

$$\iint\limits_{\mathrm{int}(\Gamma)/\Sigma} \nabla \times H \cdot dA = 0$$

where we may subtract Σ from the region being integrated because it has measure zero. We do this because H is not differentiable here. Notice $\nabla \times H = \mathbf{k}(\frac{\partial f}{\partial x} + \frac{\partial g}{\partial y})$ which is equal to τ_L whenever $x < 0$ and τ_R whenever $x > 0$. Let $\mathrm{int}(\Gamma)_L$ and $\mathrm{int}(\Gamma)_R$ denote the subsets of the interior of Γ that lie to the left and right of Σ, respectively. Then

$$\iint\limits_{\mathrm{int}(\Gamma)_L} \tau_L \, dA \; + \iint\limits_{\mathrm{int}(\Gamma)_R} \tau_R \, dA = 0 \, .$$

Let $S_i = \mathrm{area}(\mathrm{int}(\Gamma)_i)$. Then

$$\tau_L S_L + \tau_R S_R = 0 \, . \tag{2.6}$$

Since S_L and S_R are non-negative we have the following result:

Lemma 2.1. *If the piecewise-linear, continuous system (2.1) has a periodic orbit, then* $\tau_L \tau_R \leq 0$.

The next lemma, first conjectured in [Lum and Chua (1991)] then proved in [Freire *et al.* (1998)], provides uniqueness of non-degenerate periodic orbits.

Lemma 2.2. *The piecewise-linear, continuous system (1.18) has at most one limit cycle.*

See [Freire *et al.* (1998)] for a proof. Recall, a limit cycle is a periodic orbit that has a nearby trajectory that limits on the periodic orbit as either $t \to \infty$ or $t \to -\infty$. Freire *et al.* prove Lemma 2.2 by directly computing all the possible periodic orbits of (2.1) and observing that multiple limit cycles never coexist. The analysis presented below treats Lemma 2.2 as fact prior to computing periodic orbits. Consequently the mathematical arguments below are much simpler than those in [Freire *et al.* (1998)], but still instructive and insightful.

Construction and Analysis of Poincaré Maps

Let

$$\Sigma^+ = \{(0, y) \mid y \geq 0\} , \tag{2.7}$$

$$\Sigma^- = \{(0, y) \mid y \leq 0\} , \tag{2.8}$$

denote subsets of the switching manifold. The x-component of the vector field, (2.1), at the switching manifold is simply y. Hence a trajectory crossing the switching manifold at any point in Σ^+ other than the origin travels into the right half-plane whereas a trajectory crossing $\Sigma^- / \{0\}$ travels into the left half-plane. The unique trajectory that passes through the origin grazes the switching manifold at the origin and nearby lies in the left half-plane when $\mu < 0$. Assuming (2.4), a periodic orbit of (2.1) must enter both left and right half-planes. To compute periodic orbits we will construct a Poincaré map, $\mathcal{P} : \Sigma^+ \to \Sigma^+$.

Let $\varphi_t^L(x, y; \mu)$ and $\varphi_t^R(x, y; \mu)$ denote the left and right half-flows of (2.1), respectively. Let

$$\mathcal{P}_L : \Sigma^- \to \Sigma^+ , \tag{2.9}$$

be a map defined by $\mathcal{P}_L(y_1; \mu) = y_2$ where y_2 is the first intersection of $\varphi_t^L(0, y_1; \mu)$ with Σ^+ at some time, $t > 0$. Since we have (2.4), $\mathcal{P}_L$ is well-defined for all $y_1 \leq 0$. Similarly let

$$\mathcal{P}_R : \Sigma^+ \to \Sigma^- , \tag{2.10}$$

be a map defined by $\mathcal{P}_R(y_0;\mu) = y_1$ where y_1 is the first intersection of $\varphi_t^R(0, y_0; \mu)$ with Σ^- at some time, $t \geq 0$. When $\mu < 0$, if $\delta_R \neq 0$, $\mathcal{P}_R$ is well-defined for small $y \geq 0$. Then

$$\mathcal{P} = \mathcal{P}_L \circ \mathcal{P}_R \,. \tag{2.11}$$

A global description of $\mathcal{P}_L$ is easily determined. From the origin, the trajectory $\varphi_t^L(0, 0; \mu)$ spirals clockwise around the equilibrium, $z^{*(L)}$, in the left half-plane. Since $z^{*(L)}$ is repelling, before $\varphi_t^L(0, 0; \mu)$ has completed 360° about the equilibrium it will intersect the switching manifold at some point, $\mathcal{P}_L(0;\mu) = \hat{y}(\mu) > 0$, see Fig. 2.1(a). Since distinct trajectories of (2.1) do not intersect, if $y_1 < y_3 < 0$ then $\mathcal{P}_L(y_1;\mu) > \mathcal{P}_L(y_3;\mu) > \hat{y}(\mu)$. Also, since (2.1) has no nonlinear terms, $\mathcal{P}_L$ has a well-defined inverse for all points on Σ_+ above $\hat{y}(\mu)$. Consequently, $\mathcal{P}_L$ is a monotone decreasing function with no upper bound, Fig. 2.1(b).

A similar determination of the basic properties of $\mathcal{P}_R$ depends upon the nature of the equilibrium of the right half-system, $z^{*(R)}$, (2.3). However, in any case $\mathcal{P}_R(0) = 0$. Also, as with $\mathcal{P}_L$, $\mathcal{P}_R$ is monotone decreasing. Consequently $\mathcal{P}(0) = \hat{y}$ and $\mathcal{P}$ is a monotone increasing function.

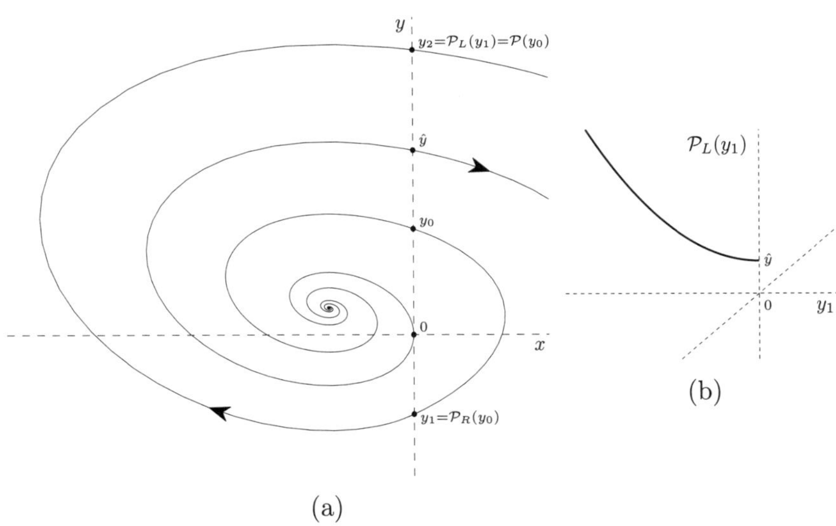

Fig. 2.1: Panel (a) shows a phase portrait of (2.1) in the case that the equilibrium of the left half-system, $z^{*(L)}$, is an admissible, repelling focus. The left and right flows define maps $\mathcal{P}_L$ and $\mathcal{P}_R$ to and from Σ^+, (2.7) and Σ^-, (2.8). The Poincaré map, $\mathcal{P} : \Sigma^+ \to \Sigma^+$, is then given by (2.11). Panel (b) shows a sketch of $\mathcal{P}_L$.

By Lemma 2.1, periodic orbits can only appear when $\tau_R \le 0$. When $\tau_R < 0$ there are three cases to consider (for the case $\tau_R = 0$ see Sec. 3.2). First, $z^{*(R)}$ may be an attracting focus. This case is studied in Sec. 2.2. It is shown that a periodic orbit exists for certain values of the parameters of (2.1).

In the two remaining non-degenerate cases ($z^{*(R)}$ is an attracting node or a saddle) the matrix C_R (2.2) has two real-valued eigenvalues, $\lambda_{1,2}$ ($\lambda_1 \ge \lambda_2$). Since (2.1) has no nonlinear terms, the parts of the associated eigenvectors, $v_{1,2}$, that lie in the right half-plane form invariant, semi-infinite lines, $\gamma_{1,2}$. When $z^{*(R)}$ is admissible, each γ_i intersects $z^{*(R)}$ in the right half-plane, otherwise the straight line extension of each γ_i intersects the virtual equilibrium, $z^{*(R)}$, in the left half-plane. When $\delta_R < \frac{1}{4}\tau_R^2$, the eigenvalues and eigenvectors of C_R are

$$\lambda_{1,2} = \frac{\tau_R}{2} \pm \sqrt{\frac{1}{4}\tau_R^2 - \delta_R}\,, \tag{2.12}$$

$$\text{and} \quad v_{1,2} = \begin{bmatrix} 1 \\ -\frac{\tau_R}{2} \pm \sqrt{\frac{1}{4}\tau_R^2 - \delta_R} \end{bmatrix}. \tag{2.13}$$

The slopes of the eigenvectors are $m_{1,2} = -\frac{\tau_R}{2} \pm \sqrt{\frac{1}{4}\tau_R^2 - \delta_R}$, hence $\gamma_{1,2}$ intersects the switching manifold at the y-value

$$\begin{aligned} c_{1,2} &= y^{*(R)} - m_{1,2}x^{*(R)} \\ &= \frac{\mu}{\delta_R}\left(-\tau_R + \frac{\tau_R}{2} \mp \sqrt{\frac{1}{4}\tau_R^2 - \delta_R}\right) \\ &= -\frac{\mu\lambda_{1,2}}{\delta_R}\,. \end{aligned} \tag{2.14}$$

Let us now examine the case that $z^{*(R)}$ is an attracting node. Since $\mu < 0$ and $\delta_R > 0$, $z^{*(R)}$ lies in the left half plane, so is virtual. The eigenvalues λ_1 and λ_2 are both negative, thus by (2.14), $c_2 < c_1 < 0$, see Fig. 2.2(a). The forward orbits of all points in the right half-plane above γ_1, are attracted to $z^{*(R)}$ until they cross Σ^- at a y-value greater than c_1, and become governed by the left half-system. Thus $y_1 = \mathcal{P}_R(y_0)$ is well-defined for any $y_0 \in \Sigma^+$ and $y_1 > c_1$, Fig. 2.2(b). Thus for any $y \ge 0$, $\mathcal{P}(y) < \mathcal{P}_L(c_1)$, Fig. 2.2(c). The point y^* is a fixed point of $\mathcal{P}$ if $h(y^*) \equiv \mathcal{P}(y^*) - y^* = 0$. We have $h(0) > 0$ and $h(\mathcal{P}_L(c_1)) < 0$ thus by the intermediate value theorem there exists y^* with $h(y^*) = y^*$. The point, $(0, y^*)$, lies on a periodic orbit of (2.1). The periodic orbit is unique since h is analytic and by Lemma 2.2 and is attracting.

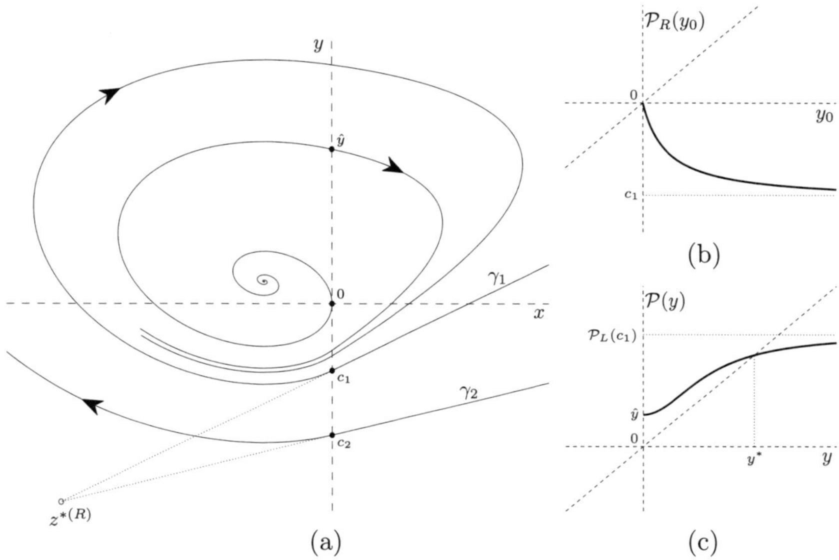

Fig. 2.2: Panel (a) shows a phase portrait of (2.1) when $z^{*(L)}$ is a repelling focus, $z^{*(R)}$ is an attracting node and $\mu < 0$. Panels (b) and (c) show sketches of $\mathcal{P}_R$ and $\mathcal{P}$, respectively. The Poincaré map, $\mathcal{P}$, has a unique fixed point corresponding to a periodic orbit of (2.1).

Now suppose $z^{*(R)}$ is a saddle, then $\delta_R < 0$. Since $\mu < 0$, $z^{*(R)}$ lies in the right half-plane, so is admissible. We have $\lambda_1 > 0 > \lambda_2$ and by (2.14), $c_2 > 0 > c_1$, see Fig. 2.3(a). Thus $\mathcal{P}_R(y_0) \to c_1$ as $y_0 \to c_2$ from below and $\mathcal{P}_R(y_0)$ is undefined for any $y_0 \geq c_2$, Fig. 2.3(b). A periodic orbit may or may not exist in this scenario. If $\mathcal{P}_L(c_1) < c_2$, as in Fig. 2.3(a), then $\mathcal{P}(y)$ must cross the 45° line at some point, y^*. As in the previous scenario, y^* is unique and corresponds to an attracting periodic orbit of (2.1). Alternatively, if $\mathcal{P}_L(c_1) > c_2$, there are three immediate possibilities. The function $\mathcal{P}(y)$ either intersects the 45° line at several points, grazes this line at one point, or fails to intersect this line anywhere. The first case is not permitted by Lemma 2.2. The grazing case also cannot occur, though this is not straight-forward to prove, see [Freire *et al.* (1998)]. Hence we have the third case, i.e. there are no periodic orbits.

Therefore the existence of a periodic orbit is determined by the sign of

$$q_\mu(\tau_L, \tau_R, \delta_L, \delta_R) \equiv c_2 - \mathcal{P}_L(c_1) \; . \tag{2.15}$$

If $q_\mu = 0$, there exists a homoclinic orbit. When we attempt to compute

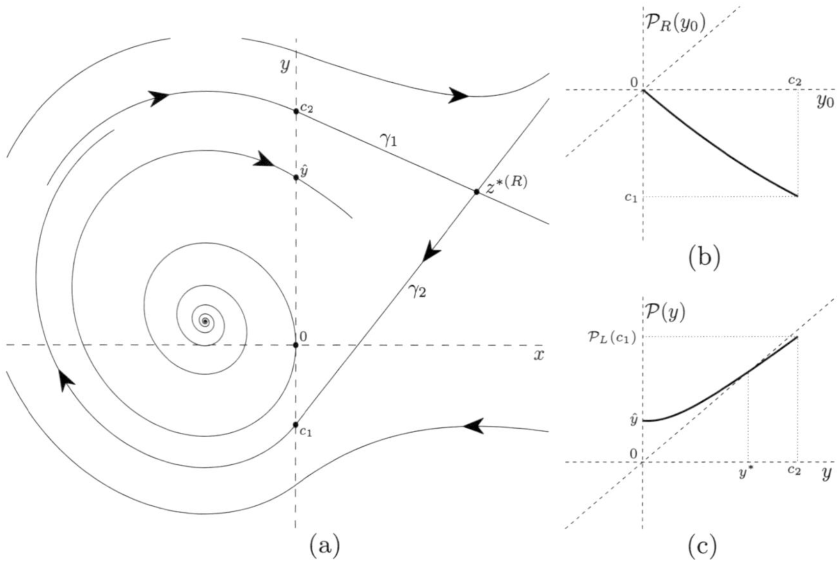

Fig. 2.3: Panel (a) shows a phase portrait of (2.1) when $z^{*(L)}$ is a repelling focus, $z^{*(R)}$ is a saddle and $\mu < 0$. Panels (b) and (c) show sketches of $\mathcal{P}_R$ and $\mathcal{P}$, respectively. For the particular parameter values chosen, the Poincaré map, $\mathcal{P}$, has a unique fixed point corresponding to a periodic orbit of (2.1).

an explicit expression for q_μ in terms of τ_L, τ_R, δ_L and δ_R, we encounter difficulties. The values c_1 and c_2 are given by (2.14). The map $\mathcal{P}_L$ is determined the left half-flow which is given by

$$\varphi_t^L(x,y;\mu) = \mathrm{e}^{\frac{\tau_L t}{2}} \begin{bmatrix} \cos(\Delta_L t) + \frac{\tau_L}{2\Delta_L}\sin(\Delta_L t) & \frac{1}{\Delta_L}\sin(\Delta_L t) \\ -\frac{\delta_L}{\Delta_L}\sin(\Delta_L t) & \cos(\Delta_L t) - \frac{\tau_L}{2\Delta_L}\sin(\Delta_L t) \end{bmatrix} \times \begin{bmatrix} x - \frac{\mu}{\delta_L} \\ y + \frac{\mu\tau_L}{\delta_L} \end{bmatrix} + \frac{\mu}{\delta_L}\begin{bmatrix} 1 \\ -\tau_L \end{bmatrix},$$

where $\Delta_L = \sqrt{\delta_L - \frac{1}{4}\tau_L^2}$. Then $\mathcal{P}_L(c_1) = e_2^\mathsf{T}\varphi_T^L(0,c_1;\mu)$ where T is the transition time obtained by solving $e_1^\mathsf{T}\varphi_T^L(0,c_1;\mu) = 0$. Unfortunately this last equation is transcendental (like $\mathrm{e}^X\cos(X) = 2$). Consequently q_μ has no known explicit form. This exemplifies a recurring theme regarding discontinuous bifurcations: *local* dynamics may be determined by *global* properties of a piecewise-linear system.

2.2 The Focus-Focus Case in Detail

This section studies discontinuous bifurcations of (1.10) in two dimensions in the case that the two equilibria, (1.21), are attracting and repelling foci for arbitrarily small values of μ. In this scenario, near the discontinuous bifurcation there exists a unique admissible equilibrium. The associated eigenvalues of the equilibrium are complex-valued and jump from one side of the imaginary axis to the other at the bifurcation. Consequently, the discontinuous bifurcation is akin to a Hopf bifurcation. In codimension-one situations a periodic orbit is generated at the bifurcation. If the orbit is stable, it encircles the repelling focus; this is known as the *supercritical* case. Conversely if the orbit is unstable, it encircles the attracting focus and this is known as the *subcritical* case. This section derives a condition governing the criticality of the bifurcation, extending the result of Freire *et al.* [Freire *et al.* (1997)] to piecewise-smooth systems.

As an example, consider the piecewise-linear, continuous system:

$$\begin{aligned} \dot{x} &= -x - |x| + y \,, \\ \dot{y} &= -3x + y + \mu \,. \end{aligned} \tag{2.16}$$

For any value of μ, (2.16) has a unique equilibrium, namely an attracting focus at $(\mu, 2\mu)$ when $\mu > 0$ (with eigenvalues, $-\frac{1}{2} \pm \frac{\sqrt{3}}{2} i$) and a repelling focus at $(\frac{\mu}{3}, 0)$ when $\mu < 0$ (with eigenvalues, $\frac{1}{2} \pm \frac{\sqrt{11}}{2} i$). The y-axis is a switching manifold and the equilibrium crosses this manifold at the origin when $\mu = 0$ and changes stability. Fig. 2.4 shows phase portraits of (2.16) for negative and positive values of μ. A stable periodic orbit exists for all negative values of μ and grows in size with $|\mu|$. This situation is analogous to a supercritical, Hopf bifurcation in a smooth system.

The main results of this section are summarized by the following theorem:

Theorem 2.1. *Suppose that the piecewise-C^k ($k \geq 1$), continuous ODE system (1.4) is two-dimensional and has an equilibrium, z^*, that transversely crosses a one-dimensional switching manifold when $\mu = \xi = 0$ at a point where the manifold is C^k. Suppose further that as $\mu \to 0_-$ the eigenvalues of the equilibrium approach $\nu_L \pm i\omega_L$ and as $\mu \to 0_+$ they approach $\nu_R \pm i\omega_R$, where $\nu_R < 0 < \nu_L$ and $\omega_L, \omega_R > 0$. Let*

$$\Lambda \equiv \frac{\nu_L}{\omega_L} + \frac{\nu_R}{\omega_R} \,, \tag{2.17}$$

denote the criticality parameter.

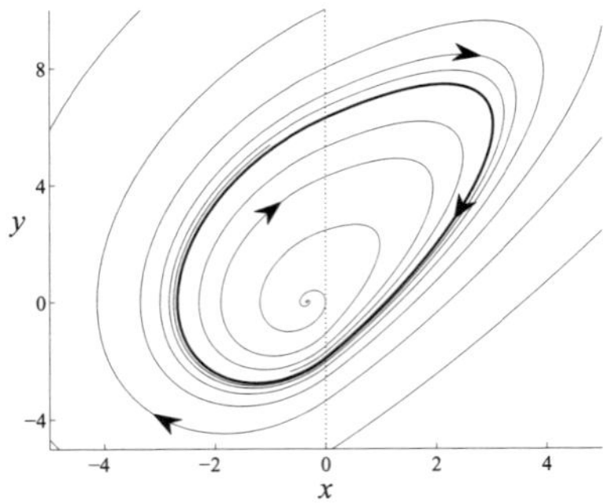

(a) $\mu = -1$

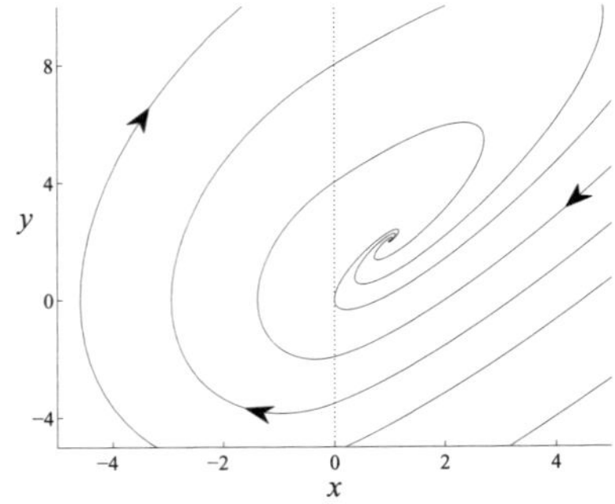

(b) $\mu = 1$

Fig. 2.4: Phase portraits of (2.16) for two different values of μ.

Then, if $\Lambda < 0$ there exists $\varepsilon > 0$ such that for all $-\varepsilon < \mu < 0$ there is a stable periodic orbit whose radius is $O(\mu)$ away from z^*, and for $0 < \mu < \varepsilon$ there are no periodic orbits near z^*.

If on the other hand $\Lambda > 0$, there exists $\varepsilon > 0$ such that for all $0 < \mu < \varepsilon$ there is an unstable periodic orbit whose radius is $O(\mu)$ away from z^*, and for all $-\varepsilon < \mu < 0$ there are no periodic orbits near z^*.

A proof of Theorem 2.1 is given in Appendix A. Sufficient non-degeneracy conditions are given in Theorem 2.1 such that the general system, (1.4), may be transformed to (1.10) with (1.11), as described in Sec. 1.2. Furthermore, as described in Sec. 2.1, the system (1.10) is observable, hence it suffices to analyze the observer canonical form, (1.10) with (1.34). See Sec. 3.3 for a detailed analysis of the codimension-two scenario, $\Lambda = 0$.

Theorem 2.1 predicts a local bifurcation diagram like Fig. 2.5 in the supercritical case. The curves y_+ and y_- denote the intercepts of the Hopf cycle with the y-axis. Unlike for a Hopf bifurcation in a smooth system, the Hopf cycle grows linearly in size with respect to the bifurcation parameter, to lowest order.

Theorem 2.1 implies that the criticality of the Hopf-like discontinuous bifurcation depends on the sign of Λ, (2.17). It is not difficult to understand why this makes sense geometrically. If (1.10) with (1.34) has a Hopf cycle for small μ, it must encircle the equilibrium and spend time in both the left and right half-planes. Suppose the equilibrium is repelling and lies in the left half-plane, see Fig. 2.6. Then, within the left half-plane, the Hopf cycle completes more than 180° around the repelling equilibrium of the left

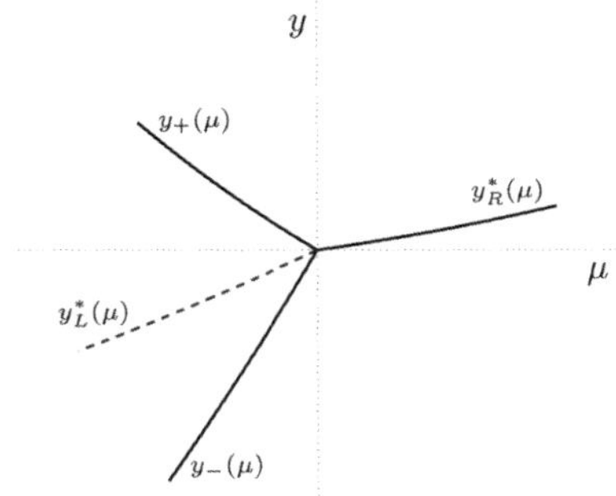

Fig. 2.5: A sketch of the local bifurcation diagram predicted by Theorem 2.1 when $\Lambda < 0$. The curves y_L^* and y_R^* denote the y-components of the equilibrium. The curves y_+ and y_- denote upper and lower intersections of a stable periodic orbit with the y-axis.

half-system. However, within the right half-plane it completes less than 180° around the virtual, attracting equilibrium of the right half-system. In order that the orbit be periodic it must, in some sense, spiral outward exactly as much as it spirals inwards. Consequently, the attracting nature of the attracting equilibrium must be stronger than the repelling nature of the repelling equilibrium. Since the time to move 180° is π/ω_j, this requirement is equivalent to $-\nu_R/\omega_R > \nu_L/\omega_L$, hence $\Lambda < 0$, in agreement with the theorem.

In smooth systems, Hopf cycles grow as ellipses and the period of an arbitrarily small Hopf cycle is easy to calculate [Kuznetsov (2004); Wiggins (2003)]. In order to calculate the period for our situation, we must first understand the shape of the Hopf cycle. As the Hopf cycle shrinks to a point, it becomes better approximated by the periodic orbit in the corresponding piecewise-linear system. A periodic orbit in a piecewise-linear system (like (2.16)), shrinks to zero in a self-similar manner due to the inherent scaling symmetry. Thus an arbitrarily small Hopf cycle generated in this discontinuous bifurcation takes the shape of the periodic orbit in the corresponding piecewise-linear system, hence the orbit consists of two spiral segments, see Fig. 2.6. Again, consider the case $\Lambda < 0$ where the equilibrium is repelling and lies in the left half-plane. The Hopf cycle completes, say, $\beta_L > \pi$ radians around the repelling equilibrium and $\beta_R < \pi$ radians around the virtual, attracting equilibrium. As $\mu \to 0_-$ the ratio β_L/β_R limits on a finite value strictly greater than one. However, even the knowledge of these angles does not give the period of the cycle since the time taken along each spiral segment is given by two different quantities, γ_L/ω_L and γ_R/ω_R, respectively. Here we observe that $0 < \gamma_R < \pi < \gamma_L < 2\pi$, but in general, $\beta_L \neq \gamma_L$ and $\beta_R \neq \gamma_R$. The period of the Hopf cycle is

$$T(\mu) = \left(\frac{\gamma_L}{\omega_L} + \frac{\gamma_R}{\omega_R}\right) + O(\mu)\,. \tag{2.18}$$

As a non-piecewise-linear example, consider the piecewise-smooth continuous system:

$$\begin{aligned} \dot{x} &= -x - |x| + y\,, \\ \dot{y} &= -3x + y + \mu + \frac{3}{100}y^2\,, \end{aligned} \tag{2.19}$$

which is identical to (2.16) except for the addition of a single nonlinear term. By Theorem 2.1, (2.19) will display the same local dynamical behavior about the origin for small μ as (2.16). Fig. 2.7 shows phase portraits of (2.19) at $\mu = \pm 1$, and for small values of x and y these plots indeed bear a strong resemblance to those of the piecewise-linear system, see Fig. 2.4.

A bifurcation diagram of (2.19) is illustrated in Fig. 2.8(a). Two equilibria are born in a saddle-node bifurcation at $\mu = \frac{25}{12} \approx 2.083$, and exist for all smaller values of μ. The attracting node becomes an attracting focus at $\mu = 100(\frac{7}{6}\sqrt{3} - 2) \approx 2.073$ and finally a repelling focus at $\mu = 0$ when its eigenvalues jump across the imaginary axis from $-\frac{1}{2} + \frac{\sqrt{3}}{2}\mathrm{i}$ to $\frac{1}{2} + \frac{\sqrt{11}}{2}\mathrm{i}$.

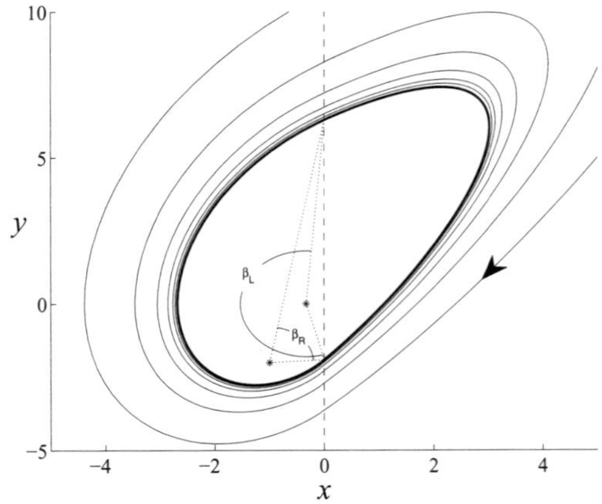

Fig. 2.6: A trajectory approaching a Hopf cycle of (2.16) when $\mu = -1$. The upper equilibrium is an admissible, repelling focus and the lower equilibrium is a virtual, attracting focus. The angles that the Hopf cycle subtends within each half-plane about each spiral center are labeled β_L and β_R.

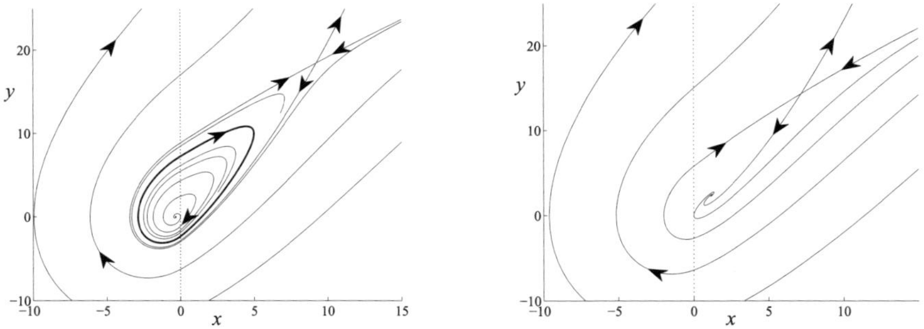

(a) $\mu = -1$ (b) $\mu = 1$

Fig. 2.7: Phase portraits of (2.19) for two different values of μ.

Here $\Lambda = -\frac{1}{\sqrt{3}} + \frac{1}{\sqrt{11}} < 0$ thus as predicted by Theorem 2.1, a stable periodic orbit is created at $\mu = 0$ and exists for small $\mu < 0$. The periodic orbit is destroyed in a collision with the saddle equilibrium in a homoclinic bifurcation at $\mu \approx -1.127$. The variation of the period with respect to μ is shown in Fig. 2.8(b). Note that the period at $\mu = 0$ is different from the nominal value, $T(0) \approx 5.522$, that it would have if $\gamma_L = \gamma_R = \pi$ in (2.18).

Codimension-two bifurcations can arise if the non-degeneracy condition $\Lambda \neq 0$ is not satisfied (see Sec. 3.3) or if the equilibrium does not cross the

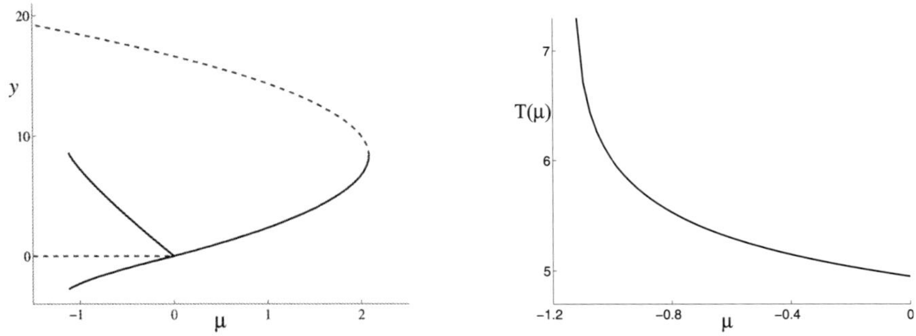

(a) (b)

Fig. 2.8: Panel (a) shows a bifurcation diagram of (2.19). The lower dotted curve denotes a repelling focus; the upper dotted curve denotes a saddle. The solid curve for $\mu > 0$ denotes a stable equilibrium. The solid curves for $\mu < 0$ denote the two locations at which the stable periodic orbit intersects the y-axis. Panel (b) is a plot of the period of the Hopf cycle generated at the discontinuous bifurcation.

switching manifold transversely at the bifurcation. In both cases higher order terms are necessary for a local analysis. For the special case that the equilibrium is fixed on the switching manifold, a periodic orbit is created as a parameter causes Λ to cross zero [Zou *et al.* (2006); Küpper and Moritz (2001)]. A further condition required for Theorem 2.1 is that the equilibrium intersects the switching manifold at a point where it is smooth. The scenario where this is not true could perhaps be understood by transforming the system so that the switching manifold lies on the positive halves of the x and y axes. A special case where the equilibrium remains at a corner was treated in [Zou and Küpper (2005)]. For this case, the parameter Λ is replaced by a sum of the ratios $\mathrm{Re}(\lambda_j)/\mathrm{Im}(\lambda_j)$ of the eigenvalues multiplied by the opening angle of each sector.

It would be useful to extend Theorem 2.1 to piecewise-smooth systems of an arbitrary number of dimensions. Since many complex phenomena have been observed for piecewise-linear systems in three dimensions [Carmona *et al.* (2005b); Llibre *et al.* (2007); Sparrow (1981)], this result may be more attainable if stringent conditions are imposed on the remaining $(N-2)$ eigenvalues of each limiting Jacobian matrix. For instance, a theoretical result may be within reach if one assumes that all the remaining eigenvalues have a large, negative real part, perhaps by using a multiple time-scales approach [Nayfeh (1973)]. Eigenvalues of this nature are observed at discontinuous bifurcations in the yeast growth model studied in Chapter 4.

2.3 Summary and Classification

This section summarizes and classifies codimension-one, discontinuous bifurcations of (1.10) with (1.11) when the system is two-dimensional. For an example of a physical system that exhibits many of the bifurcations discussed here see [Freire *et al.* (2004)]. All equilibria near a non-degenerate, discontinuous bifurcation were determined in Sec. 1.3, similarly all periodic orbits were determined in Sec. 2.1. There are five distinct situations:

(1) **NB - no bifurcation**
In this case, no bifurcation occurs. A single equilibrium persists and as it crosses the switching manifold neither of its associated eigenvalues cross the imaginary axis. This occurs, for instance, when the system is differentiable at the crossing point. Strictly speaking this is not a discontinuous bifurcation because it is not a

bifurcation.

(2) **DSN - discontinuous saddle-node bifurcation**
This is a nonsmooth analogue of the saddle-node bifurcation. The bifurcation is a nonsmooth fold, that is, two equilibria coexist and collide and annihilate at the discontinuous bifurcation. No other invariant sets are created at the bifurcation, see for example Fig. 1.5(a).

(3) **DHB - discontinuous Hopf bifurcation**
This is a nonsmooth analogue of the Hopf bifurcation. A single equilibrium persists; on one side of the bifurcation it is stable, on the other side it is unstable and on at least one side it is a focus. A periodic orbit is created at the discontinuous bifurcation and encircles a focus equilibrium of opposing stability, see for example Figs. 1.5(b) and 2.4.

(4) **DHBSN - discontinuous Hopf-saddle-node bifurcation**
This is a multiple-crossing bifurcation in the sense of Leine [Leine and Nijmeijer (2004)], see Sec. 1.5, and has no smooth analogue. It is so called because it exhibits both Hopf-like and saddle-node-like bifurcation properties. On one side of the bifurcation there are no invariant sets. On the other side there is a saddle, a focus, and a periodic orbit of opposite stability to the focus. An example is illustrated in Fig. 2.9.

(5) **SS - stability switching bifurcation**
This is also a multiple-crossing bifurcation. Here an attracting node changes to a repelling node. The absence of a focus negates the possibility of the creation of a periodic orbit. No bifurcation occurs in the sense of Seydel [Seydel (1994)] in that the number of invariant sets remains constant, however there is a topological change, thus a bifurcation in the usual sense. Fig. 2.10 illustrates an example. In panel (b) the eigenvectors of the virtual, repelling node form invariant, semi-infinite lines in the left half-plane. The orbit of a point on the upper line is a separatrix. The basin of attraction of the attracting node is the set of all points above this separatrix. In panel (a) there is a different separatrix that forms the boundary of repulsion of the repelling node. Interestingly, exactly at the bifurcation (when $\mu = 0$) there is an uncountable nested collection of homoclinic orbits, see Fig. B.1(b).

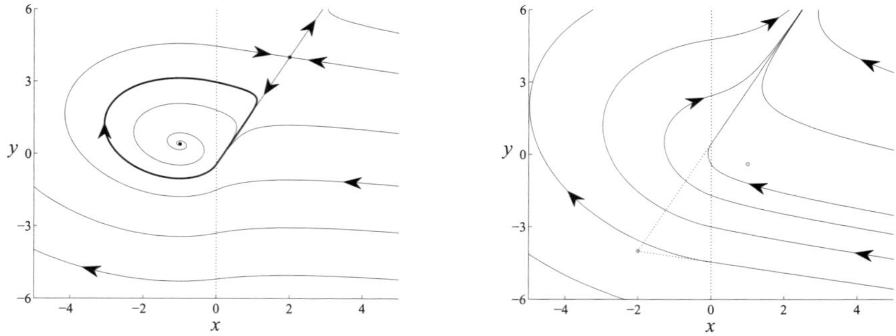

(a) $\mu = -1$ (b) $\mu = 1$

Fig. 2.9: Phase portraits of (2.1) when $(\tau_L, \tau_R, \delta_L, \delta_R) = (0.4, -2, 1, -0.5)$ for two different values of μ. The discontinuous bifurcation that occurs at $\mu = 0$ in this case is a discontinuous Hopf-saddle-node bifurcation. (A phase portrait when $\mu = 0$ is given in Fig. B.1(a).) In panel (b), both equilibria are virtual, indicated by small, unfilled circles.

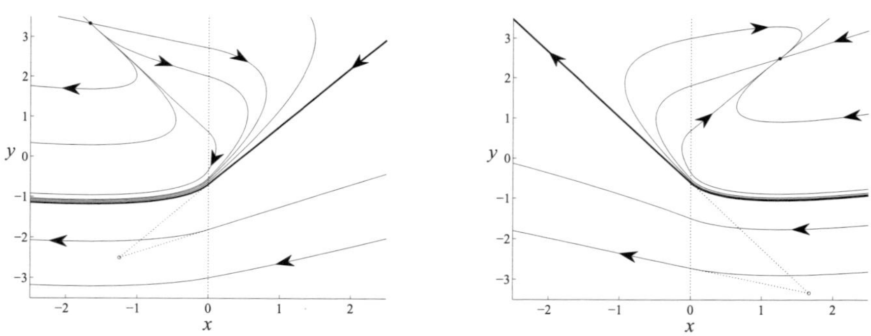

(a) $\mu = -1$ (b) $\mu = 1$

Fig. 2.10: Phase portraits of (2.1) when $(\tau_L, \tau_R, \delta_L, \delta_R) = (2, -2, 0.6, 0.8)$ for two different values of μ. The discontinuous bifurcation that occurs at $\mu = 0$ in this case is a stability switching bifurcation. Virtual equilibria are indicated by small, unfilled circles.

As argued in Sec. 1.4, the eigenvalues of the left and right equilibria of (1.10), $z^{*(L)}$ and $z^{*(R)}$, arbitrarily close to the discontinuous bifurcation determine the type of discontinuous bifurcation that occurs, in non-degenerate situations. In two dimensions any non-degenerate equilibrium fits one of the five following classifications:

(i) saddle (S), $\delta_i < 0$,
(ii) attracting node (AN), $\tau_i < 0$, $0 < \delta_i < \frac{1}{4}\tau_i^2$,
(iii) attracting focus (AF), $\tau_i < 0$, $\delta_i > \frac{1}{4}\tau_i^2$,
(iv) repelling focus (RF), $\tau_i > 0$, $\delta_i > \frac{1}{4}\tau_i^2$,
(v) repelling node (RN), $\tau_i > 0$, $0 < \delta_i < \frac{1}{4}\tau_i^2$.

The classification of $z^{*(L)}$ is independent to that of $z^{*(R)}$, hence there are 25 possible scenarios. The discontinuous bifurcation that results in each scenario is indicated in Fig. 2.11. In most cases the discontinuous bifurca-

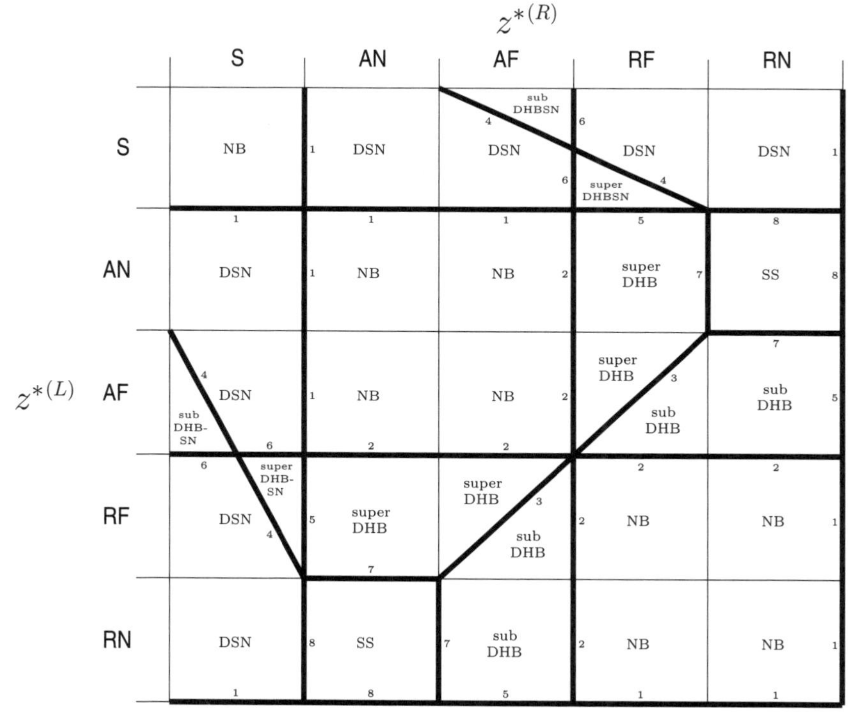

Fig. 2.11: Codimension-one, discontinuous bifurcations of (1.10) in two dimensions. Each row [column] corresponds to a different classification of $z^{*(L)}$ [$z^{*(R)}$] for an arbitrarily small value of the bifurcation parameter, μ, as described in the text. The abbreviations of the types of discontinuous bifurcations are also described in the text. Subcritical and supercritical (indicating the stability of the periodic orbit) are abbreviated to *sub* and *super* respectively. Boundaries of codimension-one, discontinuous bifurcations are numbered by type and correspond to codimension-two bifurcations.

tion is determined by the classification of $z^{*(L)}$ and $z^{*(R)}$. Ignoring symmetries there are only two exceptions. When the two equilibria are attracting and repelling foci, the discontinuous bifurcation is a discontinuous Hopf bifurcation with a criticality determined by the sign of Λ, (2.17). When the two equilibria are a focus and a saddle, the existence of a periodic orbit encircling the focus is determined by the sign of q_μ, (2.15).

The ordering of the five classifications of each $z^{*(i)}$ in Fig. 2.11 permits a classification of various codimension-two, discontinuous bifurcations. For example, suppose both equilibria are saddles. By Fig. 2.11, in this case there is no bifurcation at $\mu = 0$. Variation of the second independent parameter, η, in (1.10), may alter the classification of the equilibria. For example suppose $z^{*(L)}$ changes to an attracting node at $\eta = \eta^*$ beyond which at $\mu = 0$ there is a discontinuous saddle-node bifurcation. A codimension-two bifurcation occurs at $(\mu, \eta) = (0, \eta^*)$. This is indicated in Fig. 2.11 by the line segment joining the top-most two squares in the first column. This codimension-two bifurcation is labeled #1 and corresponds to the simultaneous occurrence of a saddle-node and a discontinuous bifurcation. A detailed unfolding of this bifurcation for an N-dimensional system is presented in Sec. 3.1.

Assuming η varies continuously, and the associated eigenvalues of the equilibria are never both zero (since this would generically produce a codimension-three discontinuous bifurcation), the classification of the equilibria may only change to one of the two classifications that are adjacent in Fig. 2.12. Thus, for instance, a saddle may change to a node, but not a focus. Consequently, the classification change of one equilibrium via variation of η corresponds to a movement in Fig. 2.11 from one square to an adjacent square. Whenever this corresponds to a change in the resulting bifurcation at $\mu = 0$, a codimension-two bifurcation occurs. The codimension-two bifurcations that arise in this manner may be classified by the two codimension-one scenarios on either side:

#1	NB $\leftrightarrow$ DSN
#2	NB $\leftrightarrow$ DHB
#3	DHB $\leftrightarrow$ DHB of opposite criticality
#4	DSN $\leftrightarrow$ DHBSN (and the stability of the focus is unchanged)
#5	DHB $\leftrightarrow$ DHBSN of the same criticality
#6	DSN $\leftrightarrow$ DHBSN (and the stability of the focus changes)
#7	DHB $\leftrightarrow$ SS
#8	DSN $\leftrightarrow$ SS

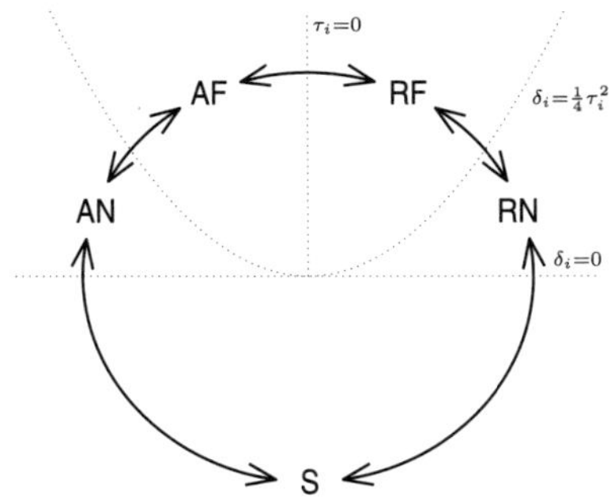

Fig. 2.12: A schematic of the classification of non-degenerate equilibria in planar systems.

Eight different codimension-two bifurcations are identified. Note that a saddle may change to a repelling node and vice versa, hence the left and right sides of Fig. 2.11 should be thought of as coincident, as should the top and bottom. Hence Fig. 2.11 is really a 2-torus! Codimension-two, discontinuous bifurcations are the focus of the next chapter.

Chapter 3

Codimension-Two, Discontinuous Bifurcations

As described in Sec. 1.4, non-degenerate discontinuous bifurcations are determined by the eigenvalues of the linearizations of the two locally defined half-systems at the bifurcation. That is, for the ODE system (1.10) with (1.11), generically, the eigenvalues of $A_L(0)$ and $A_R(0)$ determine the nature of all local dynamical phenomena. This attribute fails in special cases, for instance when one or more eigenvalues lie on the imaginary axis, or when a non-degeneracy condition (such as $\Lambda \neq 0$ (2.17), in the two-dimensional, focus-focus case, see Sec. 2.2) is not satisfied. In such situations nonlinear terms are important to local dynamical behavior. The present chapter details unfoldings of various codimension-two, discontinuous bifurcations seen the model of yeast growth discussed in Chapter 4.

Codimension-two bifurcations correspond to points in two-dimensional parameter space. Loci of codimension-one bifurcations emanate from such points. For instance, a codimension-two, discontinuous bifurcation point is generically intersected by a curve along which an equilibrium lies on the switching manifold. Throughout this book, such a curve is referred to as a *curve of discontinuity*. The reader might like to look ahead to Fig. 4.3 which shows three distinct curves of discontinuity passing through a variety of codimension-two points. Between codimension-two points a curve of discontinuity corresponds to a particular discontinuous bifurcation, or no bifurcation. At codimension-two points the curve of discontinuity changes type. As detailed below, loci of codimension-one, smooth bifurcations emanate from the codimension-two points.

All non-degenerate discontinuous bifurcations that may occur in planar systems were summarized in Fig. 2.11. The solid line boundaries in this figure correspond to codimension-two, discontinuous bifurcations. Eight distinct such bifurcations were identified (labeled #1-8). Generic unfoldings of the first six of these are shown in Fig. 3.1. Note, in most cases there are two distinct unfoldings. (For #7 and #8 all possible unfoldings are not known to the author. In these cases the loci of smooth bifurcations that emanate from the codimension-two point are dependent upon the nature of global properties of the separatrices of the stability switching bifurcation.) For each plot in Fig. 3.1, it is assumed that an equilibrium lies on the switching manifold when $\mu = 0$. In other words the curve of discontinuity coincides with the η-axis, where η is a second, suitable, unfolding parameter.

Along a curve of discontinuity in a planar system, the change from a discontinuous saddle-node bifurcation (DSN) to no bifurcation (NB) occurs at the codimension-two, discontinuous bifurcation, #1. Here, zero is one of the associated limiting eigenvalues. A change from DHBSN to DHB occurs at #5, and here also zero is a limiting eigenvalue. Section 3.1 presents a rigorous unfolding in N dimensions of the situation that one limiting eigenvalue is zero. As shown in Fig. 3.1 a locus of saddle-node bifurcations emanates from the codimension-two point and is tangent to the curve of discontinuity here. Consequently the codimension-two point is the simultaneous occurrence of a discontinuous bifurcation and a saddle-node bifurcation.

Section 3.2 analyzes in N dimensions the codimension-two scenario of a complex conjugate pair of limiting eigenvalues being purely imaginary. This situation corresponds to the simultaneous occurrence of a discontinuous bifurcation and an Andronov-Hopf bifurcation. A locus of Hopf bifurcations generically emanates from the codimension-two point and is non-tangent to the curve of discontinuity here. Associated Hopf cycles graze the switching manifold along a curve tangent to the Hopf locus at the codimension-two point. No bifurcation occurs at the grazing because the ODE system is continuous at the switching manifold.

In two dimensions, this scenario corresponds to codimension-two, discontinuous bifurcation #2 or #6. Furthermore, the criticality of the discontinuous Hopf bifurcation or discontinuous Hopf-saddle-node bifurcation bears no relation to the criticality of the smooth Hopf bifurcation because the criticality of the two bifurcations is determined by linear and nonlinear terms of the ODE system, respectively. As derived in Sec. 3.2, when the two criticalities are different the respective Hopf cycles are of opposing stability and collide and annihilate in a saddle-node bifurcation, indicated by the curve labeled SN_{orb} in Fig. 3.1. By way of rather nasty series expansions, later it will be shown that the corresponding locus of saddle-node bifurcations deviates only to order six from the grazing locus!

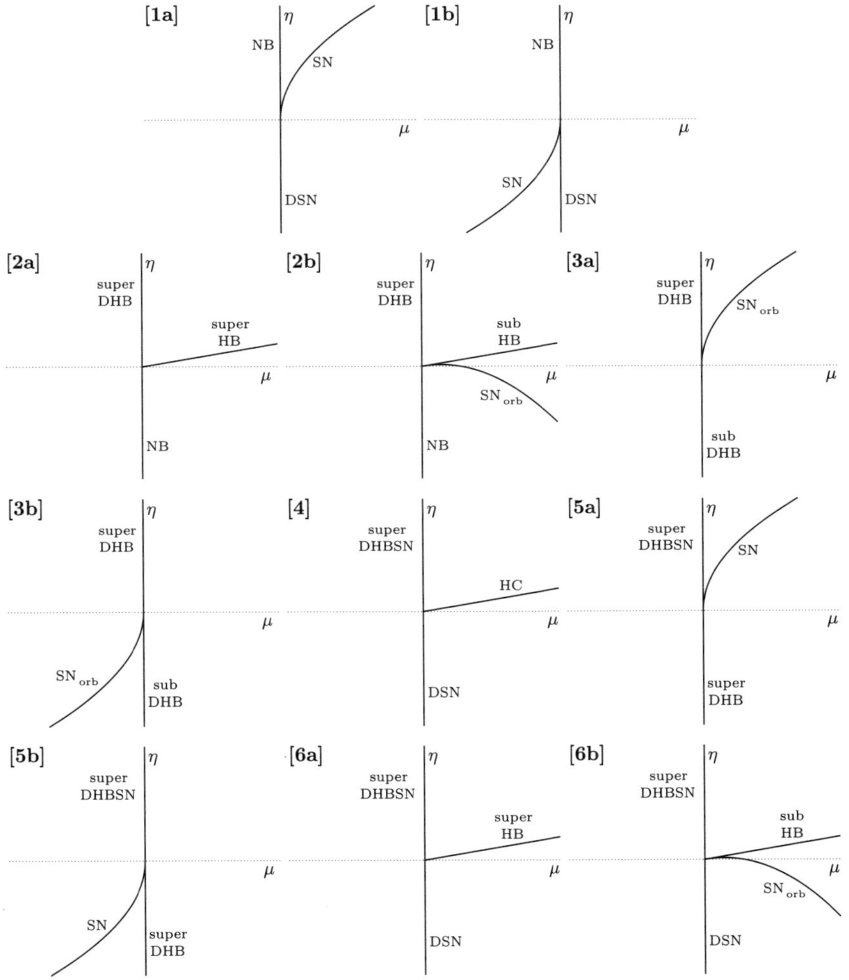

Fig. 3.1: Sketches of generic bifurcation sets local to the planar, codimension-two, discontinuous bifurcations labeled #1-6 in Fig. 2.11. SN [SN_{orb}] - saddle-node bifurcation of an equilibrium [periodic orbit]; HB - Hopf bifurcation; HC - homoclinic bifurcation. Meanings of the remaining abbreviations are given in Sec. 2.3.

As described in Sec. 2.2, in two dimensions, when the limiting left and right equilibria correspond to foci of opposite stabilities, a periodic orbit is generically created. The stability of the orbit and the criticality of the bifurcation is determined by the sign of Λ (2.17), which is a measure of the difference between the effective attracting and repelling strengths of the two foci. The codimension-two, discontinuous bifurcation, #3, is defined by $\Lambda = 0$. This scenario is unfolded in two dimensions in Sec. 3.3. As shown in Fig. 3.1, generically a locus of saddle-node bifurcations of periodic orbits emanates from the codimension-two point and is tangent to the curve of discontinuity here.

An unfolding of the codimension-two, discontinuous bifurcation, #4, is left for future work. Fig. 3.1 shows the hypothesized unfolding; namely that there exists a locus of orbits homoclinic to the saddle-type equilibrium emanating from the codimension-two point that is not tangent to the curve of discontinuity here. As $\mu \to 0$ the homoclinic orbit shrinks to a point. It seems plausible that a rigorous unfolding may be performed in two dimensions via a similar process to that in Sec. 3.3. However this situation has some unique difficulties. For instance, an explicit form for the condition, $q_\mu = 0$, (2.15), which defines the codimension-two situation, is not known. Also the Poincaré map, $\mathcal{P}$, is not well-defined in a neighborhood of the origin when $\mu = 0$. Other codimension-two, discontinuous bifurcations not discussed in this book include the occurrence of a discontinuous bifurcation at a non-differentiable point on the switching manifold.

3.1 A Nonsmooth, Saddle-Node Bifurcation

This section performs a rigorous unfolding of the simultaneous occurrence of a saddle-node bifurcation and a discontinuous bifurcation for the system (1.10) with (1.11), in N dimensions. At the bifurcation, one of A_L and A_R, say A_L, has a zero eigenvalue. Consequently, equilibria of the left half-system local to the bifurcation are determined by nonlinear terms. It is found that near the codimension-two bifurcation, equilibria in the left half-system undergo a saddle-node bifurcation.

The unfolding is summarized by the following theorem:

Theorem 3.1. *Consider the piecewise-C^k ($k \geq 4$), continuous ODE system (1.10) with (1.11). Suppose that near $(\mu, \eta) = (0, 0)$, $A_L(\mu, \eta)$ has an eigenvalue $\lambda(\mu, \eta) \in \mathbb{R}$ with an associated eigenvector, $v(\mu, \eta)$. In addition, suppose*

(i) $\lambda(0, 0) = 0$ *is of algebraic multiplicity 1 and is the only eigenvalue of* $A_L(0, 0)$ *with zero real part,*

(ii) $b_0 = \varrho^{\mathsf{T}}(0, 0)b(0, 0) \neq 0$,

where $\varrho^{\mathsf{T}} = \varrho_{\rm ode}^{\mathsf{T}}$ *is given by (1.22). Then* $v(0,0)$ *is not orthogonal to* e_1 *thus by scaling we may assume* $e_1^{\mathsf{T}} v(0,0) = 1$*. Finally suppose*

(iii) $\frac{\partial \lambda}{\partial \eta}(0,0) = 1$,

(iv) $a_0 = \varrho^{\mathsf{T}}((D_x^2 f^{(L)})(v,v))\big|_{(0,0,0)} \neq 0$.

Then

$$p_0 = \varrho^{\mathsf{T}}(0,0)v(0,0) \neq 0 \;, \tag{3.1}$$

and there exists a unique C^{k-2} *function* $h : \mathbb{R} \to \mathbb{R}$ *with* $h(0) = h'(0) = 0$ *and*

$$h''(0) = \frac{p_0^2}{a_0 b_0} \;, \tag{3.2}$$

such that in a neighborhood of $(\mu, \eta) = (0,0)$*, the curve* $\mu = h(\eta)$ *corresponds to a locus of saddle-node bifurcations of equilibria of (1.10) that are admissible when*

$$\mathrm{sgn}(\eta) = \mathrm{sgn}(a_0 p_0) \;. \tag{3.3}$$

A proof of Theorem 3.1 is given in Appendix A. Condition (i) is the defining singularity condition. Condition (ii) is the non-degeneracy condition, usually used to guarantee that the distance between the left and right equilibria and the switching manifold varies linearly with respect to μ, see (1.24). In Theorem 3.1 it ensures that the one-dimensional center manifold of the left half-system is not tangent to the switching manifold. The condition $\frac{\partial \lambda}{\partial \eta}(0,0) \neq 0$ is sufficient to ensure η is an appropriate unfolding parameter. Since η has no prior restrictions, it may be scaled such that $\frac{\partial \lambda}{\partial \eta}(0,0) = 1$, condition (iii). The constant a_0, given in condition (iv), is a function of quadratic coefficients of the left half-system. The nonlinear terms allow the bifurcation to be unfolded in the usual manner exactly when $a_0 \neq 0$. Note, the component, $D_x^2 f^{(L)}$, is a rank-three tensor that also is sometimes used in the statement of the saddle-node bifurcation theorem, see for instance [Guckenheimer and Holmes (1986)].

Put simply, Theorem 3.1 predicts that a locus of saddle-node bifurcations emanates from the codimension-two point and is tangent to the curve of discontinuity ($\mu = 0$) here. Furthermore, if $\det(A_R(0,0)) \neq 0$ generically the bifurcation set will resemble either Fig. 3.2(a) or Fig. 3.2(b) depending on the sign of a_0. In the first case (panel (a)), there exists a unique equilibrium except when $h(\eta) \leq \mu < 0$ and $\eta < 0$ where there are three coexisting equilibria and on the saddle-node bifurcation curve where there are two coexisting equilibria. In the second case (panel (b)), there exist two equilibria to the left of the solid curves and no equilibria to their right. Note that periodic orbits and other invariant sets may emanate from the discontinuous bifurcations but these are ignored in this section.

As an example, consider the one-dimensional, piecewise-smooth, continuous system

$$\dot{x} = \begin{cases} \mu + \eta x + \alpha x^2, & x \le 0 \\ \mu + x, & x \ge 0 \end{cases}, \tag{3.4}$$

where $\alpha \neq 0$. The equilibrium of the right half-system, $x^{*(R)} = -\mu$, is unstable and admissible when $\mu \le 0$. Equilibria of the left half-system are given by $x^{*(L)} = \frac{1}{2\alpha}(-\eta \pm \sqrt{\eta^2 - 4\alpha\mu})$ and exist when $\mu \le h(\eta) \equiv \frac{\eta^2}{4\alpha}$. Consequently, (3.4) exhibits the bifurcation diagrams shown in Fig. 3.3. Bifurcation sets for (3.4) are shown in Fig. 3.2. This is in full agreement with Theorem 3.1 because here $a_0 = \frac{\partial^2}{\partial^2 x^2}\alpha x^2|_{x=0} = 2\alpha$ and $\varrho = b = 1 \Rightarrow b_0 = p_0 = 1$ thus by (3.2), $h''(0) = \frac{1}{2\alpha}$ and by (3.3) the saddle-node bifurcations are admissible when η has the same sign as a_0.

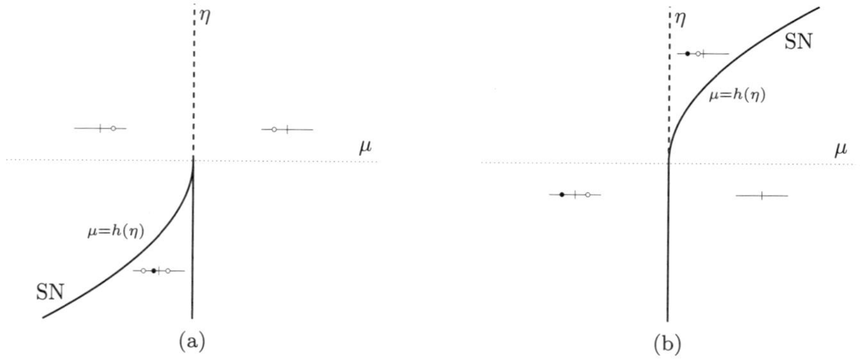

Fig. 3.2: This figure doubles as showing generic unfoldings predicted by Theorem 3.1 and bifurcation sets of the one-dimensional system (3.4) when $\alpha < 0$ in panel (a) and $\alpha > 0$ in panel (b). Dashed curves correspond to where an equilibrium lies on the switching manifold but no bifurcation results. Solid curves correspond to saddle-node bifurcations and nonsmooth folds. Within each region bounded by the curves, a representative phase portrait of the one-dimensional system (3.4) is given. Stable [unstable] equilibria are denoted by filled [unfilled] circles. Corresponding one-parameter bifurcation diagrams are shown in Fig. 3.3.

To illustrate Theorem 3.1 in higher dimensions consider the three-dimensional, piecewise-smooth, continuous system:

$$\begin{bmatrix} \dot{x} \\ \dot{y} \\ \dot{z} \end{bmatrix} = \begin{cases} \mu \begin{bmatrix} 1 \\ 0 \\ -1 \end{bmatrix} + \begin{bmatrix} -1 & 0 & 1 \\ \mu & 1 & 3 \\ -\eta - 2 & 0 & 2 \end{bmatrix} \begin{bmatrix} x \\ y \\ z \end{bmatrix} + \begin{bmatrix} 0 \\ 0 \\ 4x^2 \end{bmatrix}, & x \le 0 \\ \mu \begin{bmatrix} 1 \\ 0 \\ -1 \end{bmatrix} + \begin{bmatrix} 0 & 0 & 1 \\ 0 & 1 & 3 \\ 3 & 0 & 2 \end{bmatrix} \begin{bmatrix} x \\ y \\ z \end{bmatrix}, & x \ge 0 \end{cases} . \tag{3.5}$$

The left half-system of (3.5) has two equilibria with x-components given by

$$x^{*(L)} = \frac{1}{8}\eta \pm \frac{1}{8}\sqrt{\eta^2 + 48\mu} . \tag{3.6}$$

A saddle-node bifurcation occurs when the equilibria are equal, i.e. when $\mu = -\frac{1}{48}\eta^2 \equiv h(\eta)$ and at the saddle-node bifurcation $x^{*(L)} = \frac{1}{8}\eta$, hence the bifurcation is admissible when $\eta < 0$.

To compare this with Theorem 3.1, first note that $A_L(\mu, \eta)$ has an eigenvalue $\lambda(\mu, \eta) = \frac{1}{2} - \frac{1}{2}\sqrt{1 - 4\eta}$, thus indeed $\lambda(0,0) = 0$ and $\frac{\partial \lambda}{\partial \eta}(0,0) = 1$ as in the statement of the theorem. Furthermore, $b(0,0) = (1, 0, -1)^{\mathsf{T}}$, $v(0,0) = (1, -3, 1)^{\mathsf{T}}$ and $\varrho^{\mathsf{T}}(0,0) = (2, 0, -1)$. Consequently $(a_0, b_0, p_0) = (-8, 3, 1)$ thus Theorem 3.1 predicts saddle-node bifurcations along a curve $\mu = h(\eta)$ with $h(0) = h'(0) = 0$ and $h''(0) = -\frac{1}{24}$ that are admissible when η is negative, in agreement with the above directly computed results.

3.2 A Nonsmooth, Hopf Bifurcation

This section studies the codimension-two simultaneous occurrence of an Andronov-Hopf bifurcation and a discontinuous bifurcation. An unfolding is first obtained for a two-dimensional system, Theorem 3.2. Via careful construction of a Poincaré map, detailed results are derived. A locus of Hopf bifurcations emanates from the codimension-two point tangent to which is a locus along which the corresponding Hopf cycle grazes the switching manifold. No bifurcation occurs at the grazing because the system is continuous, however very shortly beyond the grazing the Hopf cycle may undergo a saddle-node bifurcation. A simple example is used to illustrate and verify Theorem 3.2. This example also provides a demonstration of the transformation to observer canonical form. In higher dimensions the same Poincaré map is considerably more difficult to compute. Such a computation is beyond the scope of this book. Theorem 3.3 states an unfolding in

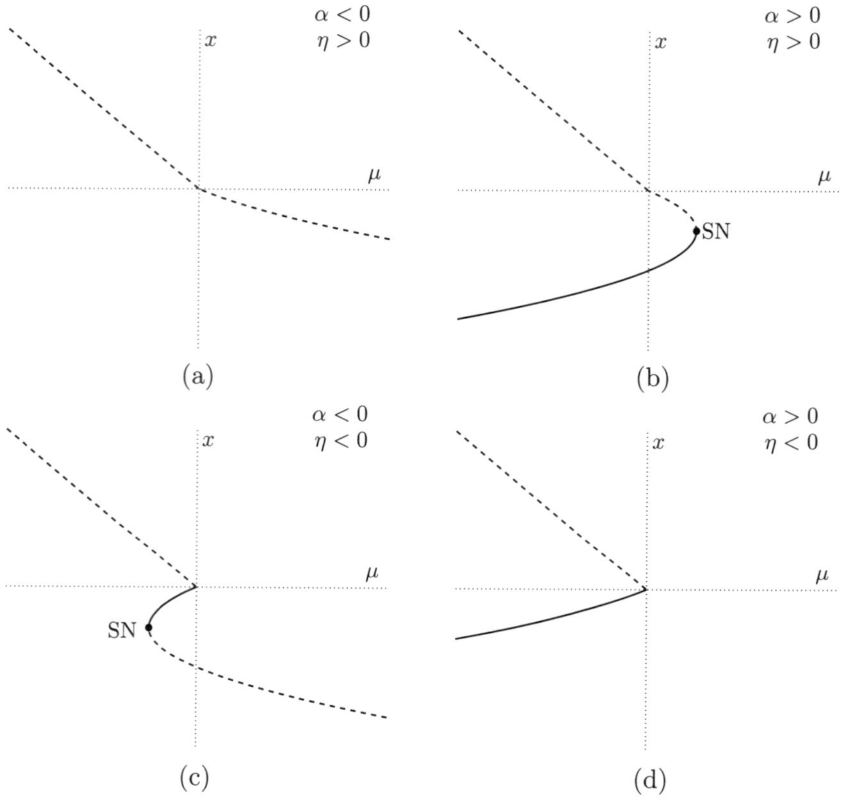

Fig. 3.3: Bifurcation diagrams of the one-dimensional system (3.4), for all four possible sign combinations of α and η. Stable [unstable] equilibria are indicated by a solid [dashed] curve. Saddle-node bifurcations appear in panels (b) and (c).

N dimensions. Here, the same loci of Hopf bifurcations and grazings are obtained, but a locus of saddle-node bifurcations of the Hopf cycle is not.

Preliminaries and Theorem Statement

Suppose the system (1.10) with (1.11) is two-dimensional and $A_L(0,0)$ has complex-valued eigenvalues. By Corollary (1.1) the system is observable when $\mu = \eta = 0$. By continuity $A_L(\mu, \eta)$ will have complex-valued eigenvalues for sufficiently small values of μ and η, thus (1.11) may be transformed to (1.34) for small μ and η. In two dimensions, the system (1.10) with

(1.34) may be written as

$$\begin{bmatrix} \dot{x} \\ \dot{y} \end{bmatrix} = \begin{cases} \begin{bmatrix} f^{(L)}(x,y;\mu,\eta) \\ g^{(L)}(x,y;\mu,\eta) \end{bmatrix}, & x \le 0 \\ \begin{bmatrix} f^{(R)}(x,y;\mu,\eta) \\ g^{(R)}(x,y;\mu,\eta) \end{bmatrix}, & x \ge 0 \end{cases}, \tag{3.7}$$

where

$$\begin{bmatrix} f^{(i)}(x,y;\mu,\eta) \\ g^{(i)}(x,y;\mu,\eta) \end{bmatrix} = \begin{bmatrix} 0 \\ \mu \end{bmatrix} + \begin{bmatrix} \tau^{(i)}(\mu,\eta) & 1 \\ -\delta^{(i)}(\mu,\eta) & 0 \end{bmatrix} \begin{bmatrix} x \\ y \end{bmatrix} + \begin{bmatrix} \tilde{f}^{(i)}(x,y;\mu,\eta) \\ \tilde{g}^{(i)}(x,y;\mu,\eta) \end{bmatrix} + o(k)\,, \tag{3.8}$$

where $\tilde{f}^{(i)}$, $\tilde{g}^{(i)}$ consist of terms that are nonlinear in the scalar variables x and y.

Theorem 3.2. *Consider the two-dimensional, piecewise-C^k ($k \ge 8$), continuous ODE system (3.7) with (3.8) and assume*

$$\tau^{(L)}(\mu,\eta) \equiv \eta\,. \tag{3.9}$$

Suppose,

(i) $\delta^{(L)}(0,0) = \omega^2$ *for* $\omega > 0$,

(ii) $a_0 \ne 0$, *where*

$$a_0 = \frac{1}{16}(f^{(L)}_{xxx} + g^{(L)}_{xxy} + \omega^2 f^{(L)}_{xyy} + \omega^2 g^{(L)}_{yyy}) - \frac{1}{16} f^{(L)}_{xy}(f^{(L)}_{xx} + \omega^2 f^{(L)}_{yy}) + \frac{1}{16} g^{(L)}_{xy}(\frac{1}{\omega^2} g^{(L)}_{xx} + g^{(L)}_{yy}) + \frac{1}{16}(\frac{1}{\omega^2} f^{(L)}_{xx} g^{(L)}_{xx} - \omega^2 f^{(L)}_{yy} g^{(L)}_{yy})\,, \tag{3.10}$$

evaluated at $(x,y;\mu,\eta) = (0,0;0,0)$, *and*

(iii) $\tau_R = \tau^{(R)}(0,0) \ne 0$.

Then, near $(x,y;\mu,\eta) = (0,0;0,0)$, *if* $\delta_R = \delta^{(R)}(0,0) \ne 0$ *when* $\delta_R \mu > 0$ *there is a unique equilibrium in the right half-plane given by* C^k *functions*

$$\begin{bmatrix} x^{*(R)}(\mu,\eta) \\ y^{*(R)}(\mu,\eta) \end{bmatrix} = \begin{bmatrix} \frac{1}{\delta_R} \\ -\frac{\tau_R}{\delta_R} \end{bmatrix} \mu + O(2)\,. \tag{3.11}$$

The equilibrium is repelling if $\tau_R, \delta_R > 0$, *attracting if* $-\tau_R, \delta_R > 0$ *and a saddle if* $\delta_R < 0$. *When* $\mu < 0$, *there is a unique equilibrium in the left half-plane given by* C^k *functions*

$$\begin{bmatrix} x^{*(L)}(\mu,\eta) \\ y^{*(L)}(\mu,\eta) \end{bmatrix} = \begin{bmatrix} \frac{1}{\omega^2} \\ 0 \end{bmatrix} \mu + O(2)\,. \tag{3.12}$$

Furthermore, there exist unique C^{k-1}, C^{k-1}, C^{k-2} functions $h_1, h_2, h_3 : \mathbb{R} \to \mathbb{R}$ respectively, with

$$h_1(\mu) = -\frac{1}{\omega^2}(f_{xx}^{(L)} + g_{xy}^{(L)})\big|_{(0,0;0,0)}\mu + O(\mu^2) , \tag{3.13}$$

$$h_2(\mu) = h_1(\mu) - \frac{2a_0}{\omega^4}\mu^2 + O(\mu^3) , \tag{3.14}$$

$$h_3(\mu) = h_2(\mu) - \frac{8\pi^2 a_0^3}{3\omega^{12}\tau_R^2}\mu^6 + o(\mu^6) , \tag{3.15}$$

such that when $\mu < 0$,

1. *(1) the curve $\eta = h_1(\mu)$ corresponds to a locus of Andronov-Hopf bifurcations of $(x^{*(L)}, y^{*(L)})^{\mathsf{T}}$ that are supercritical if $a_0 < 0$ and subcritical if $a_0 > 0$; this equilibrium is attracting if $\eta < h_1(\mu)$ and repelling if $\eta > h_1(\mu)$,*
2. *(2) the curve $\eta = h_2(\mu)$ corresponds to where the associated Hopf cycle grazes the y-axis,*
3. *(3) if $a_0\tau_R < 0$, the Hopf cycle exists for values of η between $h_1(\mu)$ and $h_3(\mu)$, a periodic orbit of opposing stability exists for $\eta < h_3(\mu)$ if $a_0 < 0$ and $\eta > h_3(\mu)$ if $a_0 > 0$ and the two orbits coincide at a locus of saddle-node bifurcations $\eta = h_3(\mu)$,*
4. *(4) if $a_0\tau_R > 0$, the Hopf cycle exists for $\eta > h_1(\mu)$ if $a_0 < 0$ and $\eta < h_1(\mu)$ if $a_0 > 0$.*

A proof of Theorem 3.2 is given in Appendix A; an outline of the proof is given later in this section. Equation (3.9) together with condition (i) define the codimension-two, singular situation. Condition (ii) is a non-degeneracy condition ensuring that the nonlinear terms unfold the bifurcation in the generic manner. Condition (iii) is the only restriction on the right half-system. A codimension-three bifurcation results if $\tau_R = 0$.

Theorem 3.2 predicts essentially two different unfolding scenarios. These are illustrated in Fig. 3.4. For each scenario, Fig. 3.5 shows a bifurcation diagram corresponding to fixing η at a small positive value, as indicated by the dash-dot line segments in Fig. 3.4.

Example

As an example consider the piecewise-C^∞, continuous system

$$\begin{aligned} \dot{u} &= -\alpha + \frac{2}{15}\beta + v + \frac{1}{5}u^2 + u^3 \,, \\ \dot{v} &= -\frac{5}{4}\alpha + \frac{1}{6}\beta - \frac{3}{8}u + \frac{1}{10}(\beta - 1)v + |\frac{1}{8}u - \frac{1}{10}v| \,. \end{aligned} \tag{3.16}$$

When $\alpha = \beta = 0$, the origin is an equilibrium on a switching manifold. Its two one-sided limiting associated eigenvalues are $\pm\frac{1}{\sqrt{2}}\mathrm{i}$ and $-\frac{1}{10} \pm \frac{\sqrt{6}}{5}\mathrm{i}$. This example exhibits the codimension-two scenario in which we are interested and satisfies the required non-degeneracy conditions.

A bifurcation set for (3.16) is shown in Fig. 3.6. For small values of α and β the bifurcation set is a smooth distortion of a mirror image of Fig. 3.4, panel (b). However for this example there also exists an unstable periodic orbit. The predictions of Theorem 3.2 break down away from $\alpha = \beta = 0$ where this orbit collides with the stable orbit. The resulting locus of saddle-node bifurcations of periodic orbits intersects the saddle-node locus anticipated by Theorem 3.2 at a cusp bifurcation at $(\alpha, \beta) \approx (0.019, -0.29)$.

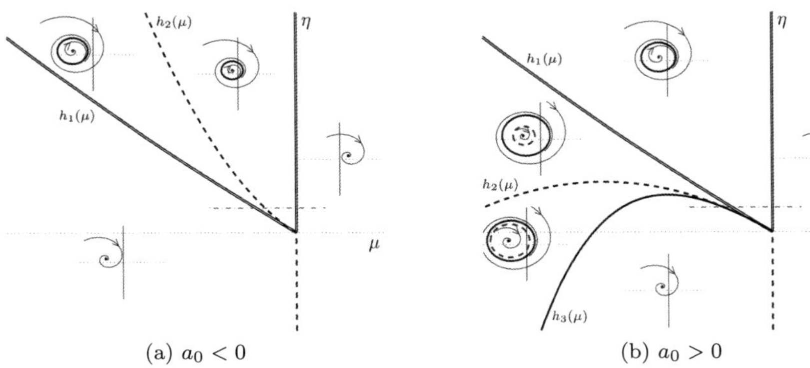

Fig. 3.4: Schematics showing unfoldings predicted by Theorem 3.2 when $\tau_R < 0$, $\delta_R > 0$. Double lines correspond to Hopf bifurcations or discontinuous bifurcations that create a periodic orbit. Dashed lines correspond to a collision of an equilibrium or periodic orbit with the switching manifold at which no bifurcation occurs. Solid lines correspond to saddle-node bifurcations of periodic orbits. Included are phase portraits showing local behavior. Stable orbits are solid; unstable orbits are dashed. Thin solid lines denote the switching manifold. Lastly, the horizontal dash-dot line segments approximate the μ and η values considered in Fig. 3.5.

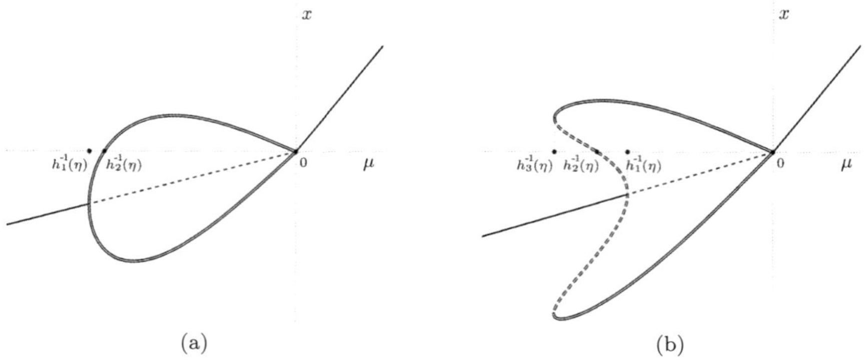

Fig. 3.5: Schematic bifurcation diagrams corresponding to the dash-dot line segments shown in Fig. 3.4. When stable [unstable] the equilibrium is indicated by a solid [dashed] line. The maximum and minimum x-values of periodic orbits are indicated by double lines; solid when stable and dashed when unstable.

In order to compare the predictions of Theorem 3.2 and in particular the scaling laws (3.13)-(3.15), we transform the system (3.16) to the form given in the theorem. The switching manifold is the line $u = \frac{4}{5}v$, therefore we let

$$x = u - \frac{4}{5}v \,. \tag{3.17}$$

This particular example has been chosen because the required transformation may be computed explicitly. Combining individual transformations discussed Sec. 1.2 and Sec. 1.4 produces

$$\begin{aligned} y &= -\frac{1}{10}(\beta - 4)u + v \,, \\ \mu &= \frac{1}{10}(\frac{33}{2} - \beta)(\alpha - \frac{2}{15}\beta) \,, \\ \eta &= \frac{1}{10}\beta \,. \end{aligned} \tag{3.18}$$

In the transformed coordinates the system is

$$\begin{bmatrix} \dot{x} \\ \dot{y} \end{bmatrix} = \begin{cases} \begin{bmatrix} 0 \\ \mu \end{bmatrix} + \begin{bmatrix} \eta & 1 \\ -\frac{1}{2} & 0 \end{bmatrix} \begin{bmatrix} x \\ y \end{bmatrix} + \begin{bmatrix} 1 \\ \frac{2}{5} - \eta \end{bmatrix} (\frac{1}{5}u^2 + u^3), & x \le 0 \\ \begin{bmatrix} 0 \\ \mu \end{bmatrix} + \begin{bmatrix} \eta - \frac{1}{5} & 1 \\ -\frac{1}{4} & 0 \end{bmatrix} \begin{bmatrix} x \\ y \end{bmatrix} + \begin{bmatrix} 1 \\ \frac{2}{5} - \eta \end{bmatrix} (\frac{1}{5}u^2 + u^3), & x \ge 0 \end{cases} , \tag{3.19}$$

where

$$u = \frac{25x + 20y}{33 - 20\eta} \,. \tag{3.20}$$

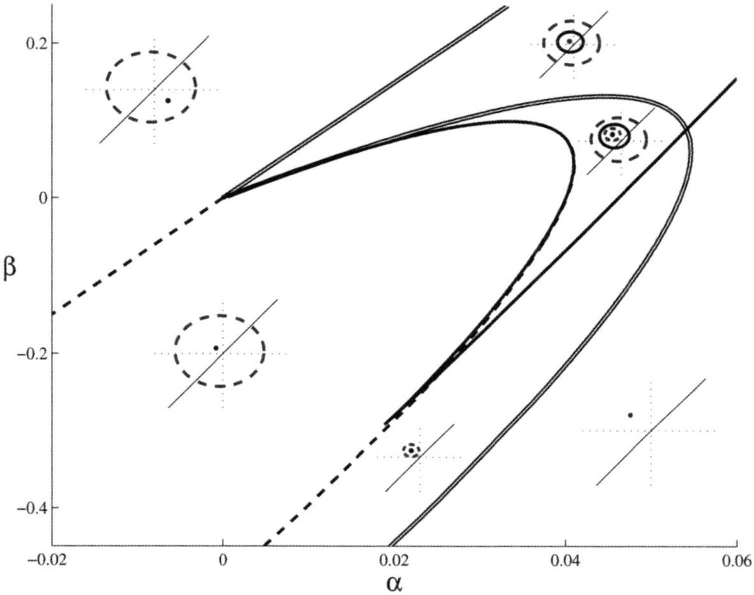

Fig. 3.6: A bifurcation set of the example system (3.16). Sketches of representative phase portraits are included though transient solutions are omitted for clarity. The line styles are the same as in Fig. 3.4.

The values of the important constants are

$$(a_0, \omega, \tau_R, \delta_R) = \left(\frac{25}{88}, \frac{1}{\sqrt{2}}, -\frac{1}{5}, \frac{1}{4}\right) ,$$
$$(f_{xx}^{(L)}, g_{xy}^{(L)})\big|_{(0,0;0,0)} = \left(\frac{250}{1089}, \frac{80}{1089}\right) . \tag{3.21}$$

Fig. 3.7 shows comparisons of numerical computations of the curves h_1, h_2 and h_3, with their predicted scalings, for the transformed example (3.19). The numerical results are in full agreement with Theorem 3.2. A calculation of error terms to quantitatively estimate the difference between the lowest order approximations and the true curves is beyond the scope of this book.

Proof Outline

The proof of Theorem 3.2, given in Appendix A, is now outlined. By assumption, when $\mu < 0$, there is an equilibrium, $(x^{*(L)}, y^{*(L)})^{\mathsf{T}}$, in the left half-plane. Close to this point there will exist an appropriate Poincaré section, $\hat{\Pi}$, in the left half-plane, see Fig. 3.8. The trajectory that begins from a point p_0 on $\hat{\Pi}$, spirals clockwise around $(x^{*(L)}, y^{*(L)})^{\mathsf{T}}$ and reintersects $\hat{\Pi}$ at some point p_5. We are interested in periodic orbits of the flow, thus when $p_5 = p_0$.

It will be more convenient to use a second Poincaré section, Π, that is a semi-infinite line intersecting the equilibrium and the origin. Artificially following the left half-flow from p_0, we arrive at a point p_1 on Π. If p_1 were in the left half-plane, by continuing from p_1 along the left half-flow we would arrive at p_5. However if p_1 is in the right half-plane, we must first calculate a correction, following the left half-flow from a different point on Π, p_4, in order to arrive at p_5. The correction results from the lack of smoothness at the switching manifold. The mapping between p_1 and p_4 is known as the *Poincaré-section discontinuity map* [Dankowicz and Nordmark (2000); di Bernardo *et al.* (2001b)], see Sec. 1.7.

To compute the discontinuity map, we follow the left half-flow back from p_1 until we arrive at a point, p_2, on the y-axis after a time $T_1 < 0$. We then follow the right half-flow from p_2 until the next intersection with the y-axis at p_3 after a time $T_2 > 0$. Finally we follow the left half-flow to p_4 on Π after a time $T_3 < 0$. Let $\mathcal{P}_1$, $\mathcal{P}_2$ and $\mathcal{P}_3$ denote maps relating to these three steps, respectively. The discontinuity map is then

$$\mathcal{P}_{\text{dm}} = \mathcal{P}_3 \circ \mathcal{P}_2 \circ \mathcal{P}_1 \,. \tag{3.22}$$

As will be shown, $\mathcal{P}_{\text{dm}}$ has a $\frac{3}{2}$-type singularity. This is because the system

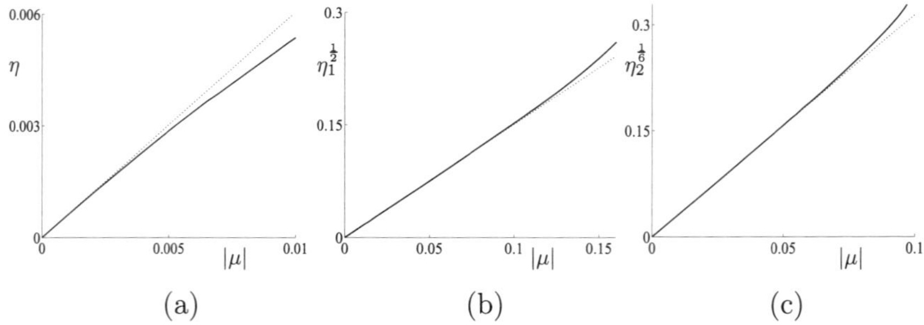

Fig. 3.7: Numerical verification of Theorem 3.2 for the transformed example system (3.19). Panel (a) shows $h_1(\eta)$. Panel (b) is a plot of the square root of the difference between $h_2(\eta)$ and $h_1(\eta)$. Panel (c) is a plot of the sixth root of the difference between $h_3(\eta)$ and $h_2(\eta)$. The dashed lines in the three panels are the lowest order scaling predictions (3.13)-(3.15), respectively.

is continuous on the switching manifold, see Sec. 1.7. When p_1 lies in the left half-plane the discontinuity map is taken to be the identity map.

A map from p_0 to p_5 may be derived by composing a map from $\hat{\Pi}$ to Π with $\mathcal{P}_{\rm dm}$ and with a map from Π to $\hat{\Pi}$. However, a map from p_1 to p_6 is simpler (where p_6 is the point on Π obtained by following the left half-flow from p_5) and equivalent. Let $\mathcal{P}_{\rm lhf}$ denote the map following the left half-flow from Π to itself and let T_4 denote the corresponding transition time. Then p_1 is mapped to p_6 via

$$\mathcal{P} = \mathcal{P}_{\rm lhf} \circ \mathcal{P}_{\rm dm} \,. \tag{3.23}$$

A fixed point of $\mathcal{P}$ corresponds to $p_6 = p_1$. Notice $p_6 = p_1$ exactly when $p_5 = p_0$.

The most intensive computations in the proof are in deriving the maps $\mathcal{P}_{\rm lhf}$, $\mathcal{P}_1$, $\mathcal{P}_2$ and $\mathcal{P}_3$ described above. Once this is accomplished, $\mathcal{P}$ is obtained by composition and it remains to derive its fixed points to find h_2

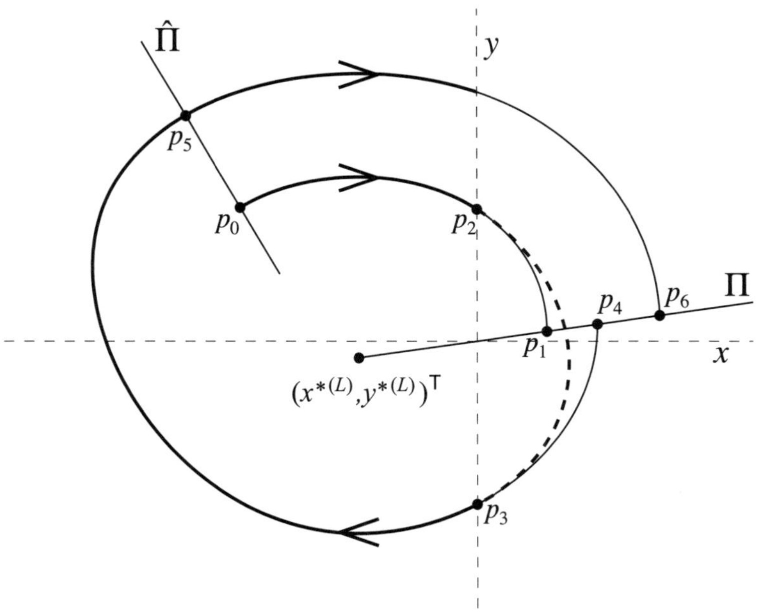

Fig. 3.8: A schematic illustrating a trajectory and Poincaré sections relating to Theorem 3.2 when $\mu < 0$. A solid [dotted] curve corresponds to following the left [right] half-flow. The true trajectory is the thick curve. The Poincaré maps constructed in the proof are $p_2 = \mathcal{P}_1(p_1)$, $p_3 = \mathcal{P}_2(p_2)$, $p_4 = \mathcal{P}_3(p_3)$, $p_6 = \mathcal{P}_{\rm lhf}(p_4)$.

and obtain the function h_3 by locating where $\mathcal{P}$ has a saddle-node bifurcation. In order to derive expressions, not for h_2 and h_3, but rather for $h_2 - h_1$ and $h_3 - h_2$, adjusted parameters η_1 and η_2, that represent the deviation from h_1 and h_2 respectively, are introduced.

Discussion

Heuristically, the curves $h_1(\mu)$ and $h_2(\mu)$ are tangent because, with respect to a linear change in parameter values, the distance between the equilibrium and the switching manifold increases linearly, whereas the amplitude of the Hopf cycle grows as the square root of the magnitude of the parameter change.

The mechanism behind the sixth order scaling of $h_3(\mu) - h_2(\mu)$ can be explained with a simple calculation. Omitting higher order terms, the Poincaré map described in the proof of Theorem 3.2, $\mathcal{P}$, (A.53), is essentially

$$\varepsilon' = \eta_2 + \Xi(\mu)\varepsilon + \gamma\varepsilon^{\frac{3}{2}} , \tag{3.24}$$

where ε is a measure of the distance from the origin, $\Xi(\mu)$ is the Floquet multiplier of the Hopf cycle and γ is a constant. Grazing occurs when $\eta_2 = 0$. It is easily determined that a saddle-node bifurcation of the fixed point of (3.24) occurs when

$$\eta_2^{\mathrm{SN}} = \frac{4}{27\gamma^2}(1 - \Xi)^3 . \tag{3.25}$$

The Floquet multiplier is unity at the Hopf bifurcation ($\eta_1 = 0$) and by assumption varies linearly with respect to η_1, i.e., $\Xi \approx 1 + \lambda\eta_1$, ($\lambda \neq 0$). The grazing occurs when $\eta_1 = O(\mu^2)$, thus $\Xi \approx 1 + \hat{\lambda}\mu^2$, ($\hat{\lambda} \neq 0$). Hence

$$\eta_2^{\mathrm{SN}} \approx -\frac{4\hat{\lambda}^3}{27\gamma^2}\mu^6 . \tag{3.26}$$

For simplicity, it was assumed that the switching manifold was infinitely differentiable. The analysis is unchanged if the switching manifold is only C^3. However, if the switching manifold were C^2 and not C^3, we would be unable to determine the same expression for the map $\mathcal{P}_3$, (A.48).

Generalization to Higher Dimensions

This section ends with a study of the simultaneous occurrence of a discontinuous bifurcation and Hopf bifurcation for the system (1.10) with (1.11) when $N \geq 2$. As above when $N = 2$, it is assumed A_L has purely imaginary eigenvalues at $\mu = \eta = 0$. In a similar fashion as for the N-dimensional system studied in Sec. 3.1, it is useful to derive ODEs describing the left half-flow restricted to the center manifold of equilibrium, $x^{*(L)}$, (1.21). The following theorem states that when certain non-degeneracy conditions are satisfied, such a system of ODEs may be put in the form (3.8), (where $i = L$), in preparation for a partial application of Theorem 3.2 as described below.

Theorem 3.3. *Consider the piecewise-C^k, continuous ODE system (1.10) with (1.11) and assume that $N > 2$ and $k \geq 6$. Suppose that near $(\mu, \eta) = (0, 0)$, $A_L(\mu, \eta)$ has complex eigenvalues $\lambda_\pm = \nu \pm \mathrm{i}\omega$ with associated eigenvectors, $z_\pm = u^{(1)} \pm \mathrm{i}u^{(2)}$. Suppose*

(i) $\nu(0, 0) = 0$, $\omega(0, 0) > 0$, *and $A_L(0, 0)$ has no other eigenvalues on the imaginary axis,*
(ii) $\varrho^{\mathsf{T}}(0, 0)b(0, 0) \neq 0$,
(iii) $\frac{\partial \nu}{\partial \eta}(0, 0) = 1$,
(iv) either $u_1^{(1)}$ or $u_1^{(2)}$ is nonzero.

Then, in the extended coordinate system (x, μ, η), there exists a C^{k-1} four-dimensional center manifold, W^c, for the left half-system that passes through the origin and is not tangent to the switching manifold at this point. Furthermore, $\exists j \neq 1$, such that in the coordinate system

$$\begin{bmatrix} \hat{x}_1 \\ \hat{x}_2 \end{bmatrix} = \begin{bmatrix} x_1 \\ x_j \end{bmatrix},$$

the left half-flow of (1.10) restricted to W^c is given by (1.11) with $i = L$ and $N = 2$ (with "hatted" variables). Moreover, this may then be transformed to (3.8) (with $i = L$) such that (3.9) and condition (i) of Theorem 3.2 are satisfied.

See Appendix A for a proof. Note that it is sufficient to assume only $k \geq 6$ since (3.15) is not be used here. Condition (i) is the defining singularity condition. Condition (ii) is a non-degeneracy condition necessary to apply Theorem 1.1 and obtain the observer canonical form. The requirement $\frac{\partial \nu}{\partial \eta}(0, 0) \neq 0$ is sufficient to ensure η is a suitable unfolding parameter. Here the derivative is set to one (condition (iii)) because η may be scaled. For the codimension-two situation to be generic, the center manifold of the left half-system must be non-tangent to the switching manifold. In Theorem 3.1 this requirement was given by condition (ii); for Theorem 3.3 it must be stated explicitly (condition (iv)).

Theorem 3.3 states that the restriction of the N-dimensional, left half-system to W^c is given by a two-dimensional system of the form (3.8). Theorem 3.2 may then be applied, but only partially because W^c does not extend into the right half-plane. That is, Theorem 3.3 does not describe an admissible, invariant manifold in a neighborhood of the origin. We may only extract the results of Theorem 3.2 that apply solely to the left half-system. Thus we have $\mu = h_1(\eta)$, (3.13): a locus of Hopf bifurcations for the N-dimensional system. If $a_0 \neq 0$, tangent to this is the locus $\mu = h_2(\eta)$, (3.14), along which the associated Hopf cycle grazes the switching manifold. By the analysis in this section, the curve $\mu = h_3(\eta)$ may only be speculated. It seems plausible that for the N-dimensional system there may exist a locus of saddle-node bifurcations of periodic orbits emanating from the codimension-two point and if so that this curve deviates only to order six from h_2. Such a locus remains to be derived in systems of any dimension.

3.3 A Codimension-Two, Discontinuous Hopf Bifurcation

This section studies the system (1.10) with (1.11) in two dimensions when the equilibria $x^{*(L)}$ and $x^{*(R)}$ are foci of opposing stabilities such that when $\eta = 0$, at the discontinuous bifurcation the effective strengths of attraction and repulsion of the foci are equal in the sense that $\Lambda = 0$, see (2.17). This is an analogue of the Bautin bifurcation [Kuznetsov (2004); Guckenheimer and Holmes (1986)] which occurs in smooth systems when an equilibrium has multipliers on the unit circle and the coefficient for a non-degenerate Hopf bifurcation vanishes. In a generic two-parameter picture in a neighborhood of a Bautin bifurcation, a locus of saddle-node bifurcations of the Hopf cycle emanates from a locus of Hopf bifurcations and the loci are tangent at the intersection point. The discontinuous analogue studied in this section exhibits essentially the same unfolding.

As in Sec. 3.2, since $A_L(0,0)$ has complex eigenvalues, the system (1.10) with (1.34) may be transformed to the observer canonical form in two dimensions, i.e. (3.7) with (3.8). The following theorem is the main result of this section.

Theorem 3.4. *Consider the two-dimensional, piecewise-C^k ($k \geq 3$), continuous ODE system (3.7) with (3.8).*
Let

$$\tau_i = \tau^{(i)}(0,0) \quad \text{and} \quad \delta_i = \delta^{(i)}(0,0) \quad \text{for } i = L, R\,, \tag{3.27}$$

and suppose

(i) $\delta_i > \frac{1}{4}\tau_i^2$, *for* $i = L, R$,
(ii) $\tau_R < 0 < \tau_L$.

Let

$$\Lambda = \frac{\tau^{(L)}}{\sqrt{4\delta^{(L)} - \tau^{(L)2}}} + \frac{\tau^{(R)}}{\sqrt{4\delta^{(R)} - \tau^{(R)2}}}\,, \tag{3.28}$$

and suppose

(iii) $\Lambda(0,0) = 0$,
(iv) $\Lambda_\eta(0,0) = 1$.

For small η, *let* $\gamma_L(\eta)$ *[*$\gamma_R(\eta)$*] be the positive* y*-value of the next [previous] intersection of the trajectory passing through the origin with the* y*-axis for the piecewise-linear approximation of (3.7) when* $\mu = -1$ *[*$\mu = 1$*]. Let*

$$K_i = \frac{1}{9\delta_i^2 - 2\delta_i\tau_i^2}\Big(-3\delta_i f_{xx}^{(i)} + 4\delta_i\tau_i f_{xy}^{(i)} - \delta_i(3\delta_i + 2\tau_i^2) f_{yy}^{(i)} - 2\tau_i g_{xx}^{(i)} + (-3\delta_i + 2\tau_i^2) g_{xy}^{(i)} + \tau_i(5\delta_i - 2\tau_i^2) g_{yy}^{(i)} \Big)\Big|_{(0,0;0,0)}\,, \tag{3.29}$$

for $i = L, R$, *and suppose*

(v) $K_L \neq K_R$.

Then, for small μ *and* η,

(1) *the positive and negative* η*-axes are loci of subcritical and supercritical discontinuous Hopf bifurcations (defined in Sec. 2.3), respectively,*
(2) *there exists a unique* C^{k-1} *function* $h : \mathbb{R} \to \mathbb{R}$ *with* $h(0) = h'(0) = 0$ *and*

$$h''(0) = \frac{\pi^2}{2\gamma_i(0)(K_L - K_R)\left(1 + e^{\frac{-\pi\tau_i}{\sqrt{4\delta_i - \tau_i^2}}}\right)} \tag{3.30}$$

where $i = L$ *if* $K_L < K_R$ *and* $i = R$ *if* $K_L > K_R$, *such that the curve* $\mu = h(\eta)$ *corresponds to saddle-node bifurcations of the associated Hopf cycles when* $\mathrm{sgn}(\eta) = -\mathrm{sgn}(K_L - K_R)$.

As with other theorems in this chapter, a proof is provided in Appendix A. Conditions (i) and (iii) define the codimension-two situation. Without loss of generality, $x^{*(L)}$ is chosen to be repelling and $x^{*(R)}$ chosen to be attracting (condition (ii)). Conditions (iv) and (v) are non-degeneracy conditions for the parameter η and the nonlinear terms of the system, respectively.

Theorem (3.4) predicts that a locus of saddle-node bifurcations of periodic orbits emanates from the codimension-two point and is tangent to the curve of discontinuity here, Fig. 3.9. However, the expression for the second derivative of this locus at the origin, (3.30), is not simple. In a similar fashion as for the Hopf bifurcation theorem [Guckenheimer and Holmes (1986); Kuznetsov (2004); Glendinning (1999)], and Theorem 3.2, it is necessary for a complicated function of the nonlinear terms, here $K_L - K_R$, to be nonzero. Also $h''(0)$ is dependent upon an intersection point of a trajectory, γ_i. As discussed in Sec. 2.2, γ_i is not known to be explicitly expressible in terms of coefficients of the ODE system. Note, the point γ_L is shown in Fig. 2.1, labeled $\hat{y}$.

To illustrate Theorem 3.4, consider the system

$$\begin{bmatrix} \dot{x} \\ \dot{y} \end{bmatrix} = \begin{cases} \begin{bmatrix} 0 \\ \mu \end{bmatrix} + \begin{bmatrix} \frac{3\sqrt{3}}{4}\eta + 1 & 1 \\ -1 & 0 \end{bmatrix} \begin{bmatrix} x \\ y \end{bmatrix}, & x \le 0 \\ \begin{bmatrix} 0 \\ \mu \end{bmatrix} + \begin{bmatrix} -2 & 1 \\ -4 & 0 \end{bmatrix} \begin{bmatrix} x \\ y \end{bmatrix} + \begin{bmatrix} \xi x^2 \\ 0 \end{bmatrix}, & x \ge 0 \end{cases} . \tag{3.31}$$

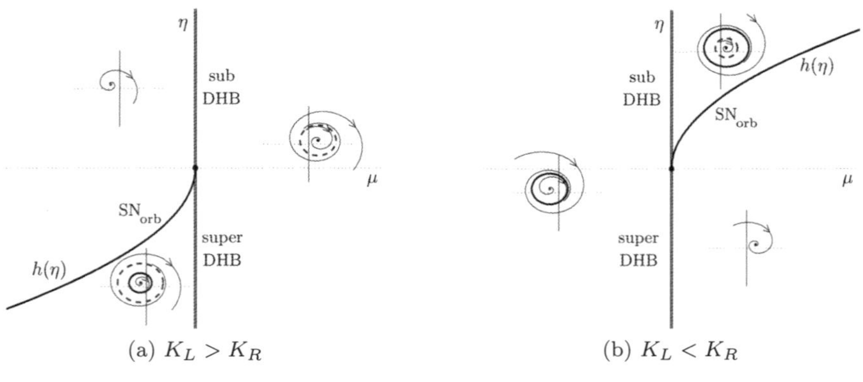

Fig. 3.9: Unfoldings predicted by Theorem 3.4. SN_{orb} - saddle-node bifurcation of a periodic orbit; super [sub] DHB - supercritical [subcritical] discontinuous Hopf bifurcation, see also Fig. 3.1. Included are phase portraits showing local behavior. Stable orbits are solid; unstable orbits are dashed. Thin solid lines denote the switching manifold.

Here $A_L(0,0)$ and $A_R(0,0)$ have eigenvalues $\frac{1}{2} \pm \frac{\sqrt{3}}{2}\mathrm{i}$ and $-1 \pm \sqrt{3}\mathrm{i}$, respectively. Also $\Lambda(0,0) = 0$, $\Lambda_\eta(0,0) = 1$, $K_L = 0$ and $K_R = \frac{-3(2\xi)}{9\delta_R - 2\tau_R^2} = -\frac{3\xi}{14}$. Numerically it has been found that $\gamma_L(0) \approx 12.1857$. Consequently, when for instance $\xi = 0.1$, $K_L > K_R$ thus Theorem 3.4 predicts the bifurcation set will resemble Fig. 3.9(a) and

$$h''(0) \approx \frac{\frac{70}{3}\pi^2}{12.1857(1 + \mathrm{e}^{\frac{-\pi}{\sqrt{3}}})} \approx 16.2492 \tag{3.32}$$

Fig. 3.10 shows that this prediction fits well with a numerical computation of $\mu = h(\eta)$.

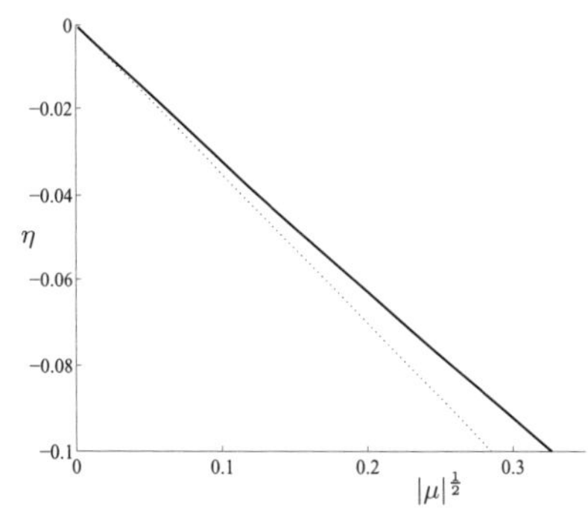

Fig. 3.10: The solid curve is a numerical computation of the saddle-node locus $\mu = h(\eta)$ for the example system (3.31). The dashed line is the predicted scaling (3.30).

Chapter 4

The Growth of Saccharomyces cerevisiae

Yeasts are single-celled fungi of which more than one thousand different species have been identified. The most commonly used yeast is *Saccharomyces cerevisiae* which has been utilized for the production of bread, wine and beer for thousands of years. Biologists in a wide variety of fields use *S. cerevisiae* as a model organism.

A common experimental method for observing biochemical processes involved in yeast growth is that of continuous cultivation in a chemostat [Najafpour (2007); Bailey and Ollis (1986)]. Fig. 4.1 shows the basic setup. Cell growth takes place in a vessel that is continuously stirred. A nutrient containing fluid is pumped into the vessel and cell culture flows out of the vessel at the same rate, ensuring that the volume of culture in the reaction vessel remains constant. The rate of flow in and out divided by culture volume is called the *dilution rate*, denoted D (in h^{-1}). Quantities such as concentrations of chemicals can be measured in a variety of ways, see [Porro *et al.* (1988); Satroutdinov *et al.* (1992)] for methods used in *S. cerevisiae* experiments.

As a continuous culture experiment is carried out, it is common for the system to reach a mathematical equilibrium (i.e. steady state, abbreviated to equilibrium throughout this chapter and different from an equilibrium in the biochemical sense). At equilibrium, the rate of cell division in the culture is equal to the dilution rate. However experimentalists in the late 1960's, [Fiechter (1968)], observed that instead of settling to an equilibrium, continuous culture experiments of *S. cerevisiae* could in some cases produce stable oscillations. Von Meyenburg, [von Meyenburg (1973)], discovered in subsequent experiments that these oscillations only occur in an intermediate range of values of the dilution rate (between about $0.08h^{-1}$ and $0.22h^{-1}$). Much work has since been done to understand the cause of such oscillations,

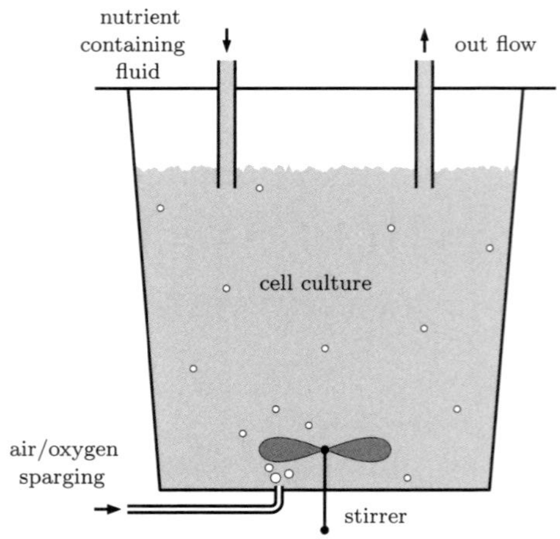

Fig. 4.1: A diagram illustrating continuous cultivation in a chemostat for *S. cerevisiae* growth. Nutrients pumped into the system allow for continuous yeast growth and the possibility of oscillatory behavior. A well-mixed solution exits the system such that the cell culture volume does not change with time. Air or oxygen bubbles enter the vessel from the bottom and are broken up by the stirrer.

see for instance [Richard (2003); Henson (2003); Zhang *et al.* (2002); Danø *et al.* (2001)].

S. cerevisiae has three metabolic pathways for glucose: fermentation, ethanol oxidation and glucose oxidation. The model of Jones & Kompala [Jones and Kompala (1999)] hypothesizes that the competing metabolic pathways of the growing yeast cells create feedback responses that produce stable oscillations. It assumes that micro-organisms will utilize the available substrates in a manner that maximizes their growth rate at all times. To enforce this optimization a "maximum function" is introduced in the model equations; as a result, the model is an example of a piecewise-smooth, continuous dynamical system.

In addition to the dilution rate, the speed of the stirrer is an easily controllable system parameter which affects qualitative changes to the yeast growth. Via a tube penetrating the bottom of the vessel, air or oxygen bubbles are pumped into the culture and are broken up by the stirrer. An increase in stirrer speed results in more bubble break up and therefore more surface area between the bubbles and the culture. This in turn leads to a faster mass transfer of oxygen into the culture liquid. The corresponding model parameter is the *dissolved oxygen mass transfer coefficient*, k_La (in h^{-1}).

This chapter studies the model of Jones & Kompala [Jones and Kompala (1999)] under variation of D and k_La. Section 4.1 states the model equations and provides some explanation as to their form. Emphasis is placed on the biochemical reasoning that leads to nonsmoothness. Section 4.2 briefly discusses some basic properties of the model including switching manifolds. A bifurcation analysis of the model is presented in Sec. 4.3. This section describes regions in parameter space where stable oscillations occur and uses the results of earlier sections to explain a variety of codimension-one and two discontinuous bifurcations. Stable oscillations are then studied in Sec. 4.4. This section gives a detailed description of the biochemical process that the model predicts induces stable oscillations and briefly discusses period-adding.

Lastly, the reader who wishes to numerically explore the dynamical behavior of the model themselves may do so without needing to code the equations by visiting the website [Simpson and Kompala (2008)]. The interface plots the time variation of several variables in response to user entered parameter values.

4.1 Mathematical Model

Jones & Kompala [Jones and Kompala (1999)] give the following model equations:

$$\frac{\mathrm{d}X}{\mathrm{d}t} = \left(\sum_i r_i v_i - D \right) X \,, \tag{4.1}$$

$$\frac{\mathrm{d}G}{\mathrm{d}t} = (G_0 - G)D - \left(\frac{r_1 v_1}{Y_1} + \frac{r_3 v_3}{Y_3} \right) X - \phi_4 \left(C \frac{\mathrm{d}X}{\mathrm{d}t} + X \frac{\mathrm{d}C}{\mathrm{d}t} \right) , \tag{4.2}$$

$$\frac{\mathrm{d}E}{\mathrm{d}t} = -DE + \left(\phi_1 \frac{r_1 v_1}{Y_1} - \frac{r_2 v_2}{Y_2} \right) X \,, \tag{4.3}$$

$$\frac{\mathrm{d}O}{\mathrm{d}t} = k_L a (O^* - O) - \left(\phi_2 \frac{r_2 v_2}{Y_2} + \phi_3 \frac{r_3 v_3}{Y_3} \right) X \,, \tag{4.4}$$

$$\frac{\mathrm{d}e_1}{\mathrm{d}t} = \alpha u_1 \frac{G}{K_1 + G} - \left(\sum_i r_i v_i + \beta \right) e_1 + \alpha^* \,, \tag{4.5}$$

$$\frac{\mathrm{d}e_2}{\mathrm{d}t} = \alpha u_2 \frac{E}{K_2 + E} - \left(\sum_i r_i v_i + \beta \right) e_2 + \alpha^* , \tag{4.6}$$

$$\frac{\mathrm{d}e_3}{\mathrm{d}t} = \alpha u_3 \frac{G}{K_3 + G} - \left(\sum_i r_i v_i + \beta \right) e_3 + \alpha^* , \tag{4.7}$$

$$\frac{\mathrm{d}C}{\mathrm{d}t} = \gamma_3 r_3 v_3 - (\gamma_1 r_1 v_1 + \gamma_2 r_2 v_2) C - \left(\sum_i r_i v_i \right) C , \tag{4.8}$$

where X, G, E and O represent the concentrations of cell mass, glucose, ethanol (in gL^{-1}) and dissolved oxygen (in mgL^{-1}) in the culture volume, respectively, C represents the intracellular carbohydrate mass fraction and each e_i represents the intracellular mass fraction of a key enzyme in the ith metabolic pathway. The subscripts 1, 2 and 3 correspond to the three pathways: fermentation, ethanol oxidation and glucose oxidation, respectively; r_i represents the growth rate on each pathway. Formulas for the growth rates and other functions are discussed later in this section. Values for all parameters except D and $k_L a$ (which are varied), are listed in Appendix D. Further details concerning the derivation of the model are found in [Jones and Kompala (1999); Kompala *et al.* (1986, 1984)] and in the M.S. thesis of Kenneth Jones [Jones (1995)].

As stated above, *S. cerevisiae* has the following three metabolic pathways for glucose:

(1) **Fermentation.**

$$C_6H_{12}O_6 \rightarrow 2C_2H_5OH + 2CO_2$$

Glucose is consumed by the well-known fermentation pathway to produce ethanol. This is the essential process in beer production. In continuous cultivation this pathway typically dominates at high dilution rates.

(2) **Ethanol oxidation.**

$$C_2H_5OH + 3O_2 \rightarrow 2CO_2 + 3H_2O$$

The presence of ethanol and oxygen allows for ethanol oxidation.

(3) **Glucose oxidation.**

$$C_6H_{12}O_6 + 6O_2 \rightarrow 6CO_2 + 6H_2O$$

Glucose is oxidized completely to carbon-dioxide and water. This is a fundamental process of bread making and is prevalent at low dilution rates.

As detailed below, the model assumes *S. cerevisiae* prefers different metabolic pathways at different times in a manner that maximizes its growth. This modeling approach replaces the thousands of rate equations necessary in a detailed metabolic model with a handful equations in terms of so-called *cybernetic variables*, u_i and v_i, that represent optimal growth strategies.

Many equations in the model of Jones & Kompala are based on the Michaelis-Menten rate equation or the similar Monod rate equation. The Michaelis-Menten rate equation concerns the conversion of a substrate into a product with the help of an enzyme giving the reaction rate, r, in terms of the substrate concentration, c_S, by

$$r = \frac{r_{\max} c_S}{K_S + c_S}, \tag{4.9}$$

where $r_{\max}$ and K_S are positive-valued constants [Lee (1992)]. By (4.9), when there is no substrate ($c_S = 0$), the reaction does not proceed. Furthermore, the reaction rate increases monotonically with substrate concentration but cannot exceed $r_{\max}$. The Monod rate equation has the same basic form as the Michaelis-Menten rate equation but applies to cell growth kinetics.

Jones & Kompala model the growth rates, r_i, for the three metabolic pathways by

$$r_1 = \mu_1 e_1 \frac{G}{K_1 + G}, \tag{4.10}$$

$$r_2 = \mu_2 e_2 \frac{E}{K_2 + E} \frac{O}{K_{O_2} + O}, \tag{4.11}$$

$$r_3 = \mu_3 e_3 \frac{G}{K_3 + G} \frac{O}{K_{O_3} + O}, \tag{4.12}$$

where each μ_i is a constant given in terms of other constants that are listed in Appendix D by

$$\mu_i = \mu_{i,\max} \frac{\mu_{i,\max} + \beta}{\alpha + \alpha^*}. \tag{4.13}$$

Each pathway is catalyzed by an enzyme of concentration, e_i, and in the absence of this enzyme, growth along the pathway does not proceed. The synthesis rate of the ith enzyme, r_{e_i}, is modeled by

$$r_{e_i} = \frac{\alpha S_i}{K_i + S_i},$$

where $S_1 = S_3 = G$, $S_2 = E$ and α is a constant. As in (4.5)-(4.7), the actual rate of synthesis of the ith enzyme is $r_{e_i} u_i$, where u_i is the fractional allocation of a critical resource for the synthesis of the ith enzyme. The u_i satisfy $0 \leq u_i \leq 1$ and $\sum_j u_j = 1$. The model of Jones & Kompala chooses these cybernetic variables to maximize the cellular growth rate of the yeast at all times. Via a simple application of Lagrange multipliers, see [Kompala *et al.* (1984)], cellular growth is maximized when the u_i's are in the same ratios as the r_i's. Therefore

$$u_i = \frac{r_i}{\sum_j r_j} \,. \tag{4.14}$$

Similarly, the actual growth rate of *S. cerevisiae* along the ith metabolic pathway is $r_i v_i$, where each v_i is a cybernetic variable satisfying $0 \leq v_i \leq 1$ that controls the activation and inhibition of the ith enzyme. Consequently the rate of change of cell mass, X, is given by (4.1). Maximal growth would occur when $v_i = 1$ for every i, but such growth is not observed experimentally. Inhibition occurs on the slower growth pathways; this is modeled by setting the corresponding v_i to values less than 1. As was first described in [Kompala *et al.* (1986)], see also [Ramkrishna *et al.* (1987)], the exact choice for v_i in this model is arrived at by assuming the degree of activation of the ith pathway is directly proportional to the magnitude of the growth rate of this pathway, i.e. $v_i \propto r_i$ or $v_i = \lambda r_i$, for some $\lambda \in \mathbb{R}$, for all i. Since $v_i = 1$ for some i, we have

$$v_i = \frac{r_i}{\max_j r_j} \,, \tag{4.15}$$

The presence of this maximum function causes the model to be piecewise-smooth, continuous. The nonsmooth choice (4.15) of v_i seems to be significant to the existence of stable oscillations over a large region of parameter space. Certainly if the denominator is replaced with an accurate smooth approximation such as the p-norm, $||r||_p = (r_1^p + r_2^p + r_3^p)^{\frac{1}{p}}$, on the vector $r = [r_1, \ldots, r_3]$ for a large value of p the model will still exhibit stable oscillations with the same qualitative properties. But numerical investigations for simple, smooth alternatives that have more physical meaning, such as $v_i = 1$ or $v_i = u_i$ for all i, have found that stable oscillations are less prevalent than with the nonsmooth choice for v_i. For instance if $v_i = u_i$ for all i, (equivalently, if the denominator in (4.15) is replaced with $||r||_1$), when $k_L a = 150$, stable oscillations have only been found for $0.09 \leq D \leq 0.096$, whereas for the piecewise-smooth model stable oscillations occur throughout the larger interval, $0.093 \leq D \leq 0.123$.

In many situations the model of Jones & Kompala fits well with experimental data, see for instance Fig. 4.2 (reproduced from [Jones and

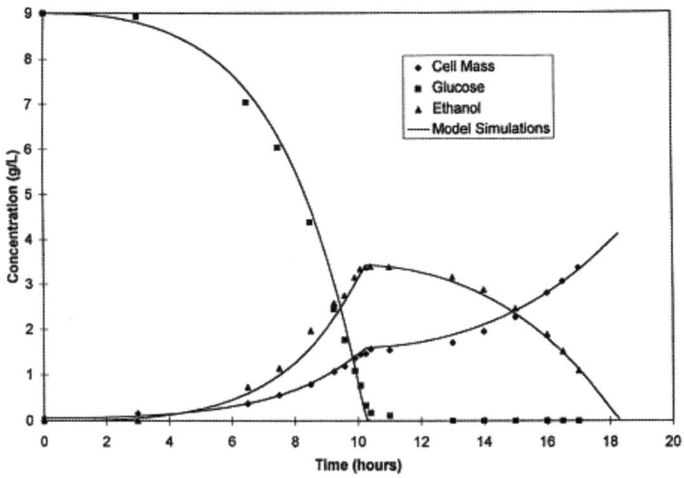

Fig. 4.2: This figure is taken from [Jones and Kompala (1999)]. It shows a comparison of the mathematical model of Jones & Kompala with experimental results of von Meyenburg in a batch process ($D = 0$). See [Jones and Kompala (1999)] or [Jones (1995)] for parameter values used. The yeast grows via fermentation until nearly all available glucose has been consumed after which the reaction proceeds along the ethanol oxidation pathway. This is known as *diauxic growth.*

Kompala (1999)]). In this particular *batch* experiment (in a batch process no nutrients are fed into the system, i.e. $D = 0$) the yeast initially prefers the fermentation pathway, creating cell mass by consuming glucose and producing ethanol. Throughout this stage the value of r_1 is greater than r_2 and r_3. After about 10 hours, once almost all the available glucose is consumed, the yeast switches, creating cell mass via the ethanol oxidation pathway. The rate of cell growth is low as the yeast switches pathways because some time is required to increase the level of the 2nd key enzyme to catalyze the ethanol oxidation pathway.

The related piecewise-smooth, continuous bacterial growth model of [Kompala *et al.* (1986)] is studied in detail in [Namjoshi and Ramkrishna (2001)]. Here the authors compute equilibria and describe discontinuous bifurcations (the term "catch-up bifurcation" is used in place of "nonsmooth fold"). Coexisting equilibria are found in a small region of parameter space and a cusp point, like that shown in Fig. 3.2(a), is located. A new model that extends [Kompala *et al.* (1986)] by including a "maintenance" term that summarizes non-growth associated cellular processes is formulated and analyzed in [Kumar *et al.* (2008)]. In this model stable oscillations are created via supercritical Hopf bifurcations.

4.2 Basic Mathematical Observations

A key property of (4.1)-(4.8) is that the positive hyper-octant (where the variables all have positive values) is forward invariant. That is, given any initial condition in the positive hyper-octant, any point on the solution curve at a later time will also be located in the positive hyper-octant. This is a common feature of mathematical models whose variables are not sensible when non-positive valued, such as the model studied here.

This property is easily proved by supposing one variable is zero and showing this implies the differential equation of that variable is non-negative, for all eight variables. For example if $G = 0$, then $r_1, v_1, r_3, v_3 = 0$ and $v_2 = 1$, thus $\frac{\mathrm{d}X}{\mathrm{d}t} = (r_2 - D)X$ and $\frac{\mathrm{d}C}{\mathrm{d}t} = -(\gamma_2 + 1)r_2 C$. Hence, upon further algebraic simplification, $\frac{\mathrm{d}G}{\mathrm{d}t} = G_0 D + \phi_4(D + r_2\gamma_2)CX$ which is non-negative because we are assuming $X, C, D \geq 0$. The bifurcation analysis in the next section is greatly simplified by a restriction to the positive hyper-octant. For instance several equilibria exist outside the positive hyper-octant; these are not studied.

Within the positive hyper-octant all trajectories are bounded forward in time. In other words, solutions always approach some attractor whether it is an equilibrium, a periodic orbit or a more complicated and possibly chaotic attractor. This is also an expected property and common in physical models.

As mentioned earlier, the model is piecewise-smooth, continuous. The system is non-differentiable on codimension-one switching manifolds. By (4.15), there are three distinct switching manifolds:

$$\begin{aligned} \Sigma_{1,2} &= \{ Z \in \mathbb{R}^8 \mid r_1 = r_2 \geq r_3 \} \,, \\ \Sigma_{1,3} &= \{ Z \in \mathbb{R}^8 \mid r_1 = r_3 \geq r_2 \} \,, \\ \Sigma_{2,3} &= \{ Z \in \mathbb{R}^8 \mid r_2 = r_3 \geq r_1 \} \,, \end{aligned} \tag{4.16}$$

where Z represents a point in phase space. At the intersection of any two switching manifolds, all the r_i are equal and therefore the third switching manifold intersects also. Except at these intersection points, the switching manifolds are smooth because each r_i is a smooth function of the system variables.

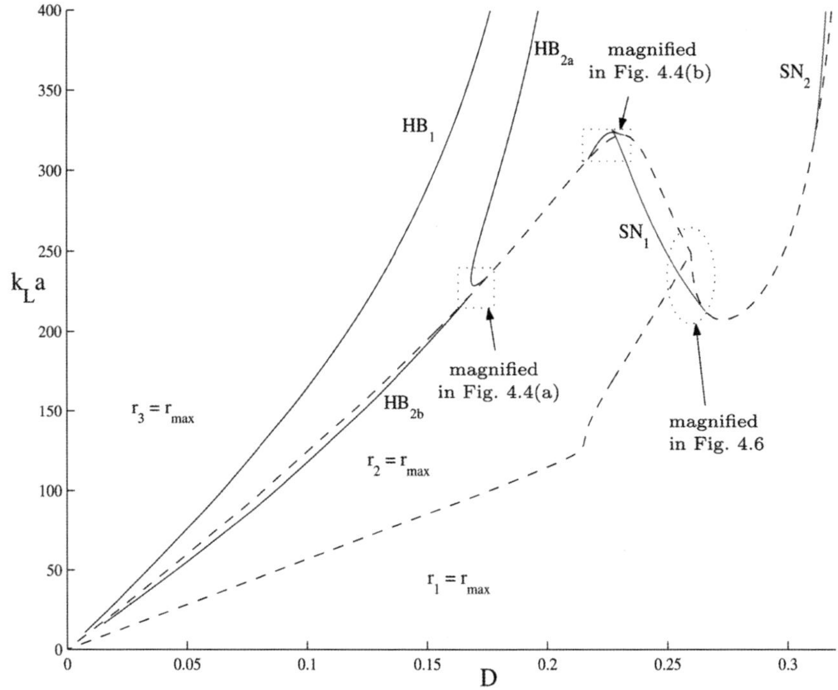

Fig. 4.3: A bifurcation set of the system (4.1)-(4.8). HB - Hopf bifurcation, SN - saddle-node bifurcation. Curves of discontinuity (where the equilibrium lies on a switching manifold, $\Sigma_{i,j}$ (4.16)) are shown dashed. In each region bounded by the curves of discontinuity, the growth rate that is maximal at equilibrium is indicated.

4.3 Bifurcation Structure

This section discusses the numerically computed bifurcation set for the system (4.1)-(4.8) shown in Fig. 4.3. Accurate numerical solutions were obtained by ensuring time steps were not taken across a switching manifold, i.e. whenever a trajectory crossed a switching manifold, a data point for the trajectory at the switching manifold was approximated, see [Gear and Østerby (1984)]. Numerical continuation was performed using the software AUTO [Doedel *et al.* (2007)].

The system (4.1)-(4.8) has a single, physically meaningful equilibrium (steady-state) except in small windows of parameter space between saddle-node or nonsmooth fold bifurcations. For small values of D, this equilibrium is stable. If we fix the value of k_La and increase D, the first bifurcation encountered is an Andronov-Hopf bifurcation (labeled HB_1 in Fig. 4.3). Slightly to the right of HB_1 the equilibrium is unstable and solutions approach a periodic orbit or complicated attractor (see Sec. 4.4). As the value of D is increased further a second Hopf bifurcation (HB_{2a} or HB_{2b}) is encountered that restores stability to the equilibrium (unless $k_La \approx 230$, see below).

Since the switching manifolds (4.16) are codimension-one, it is a codimension-one phenomenon for the equilibrium to lie precisely on a switching manifold. Though an analytical formula for the equilibrium is not known, one may numerically compute curves in two-dimensional parameter space along which this codimension-one phenomenon occurs (the dashed curves in Fig. 4.3). As mentioned in Chapter 3, throughout this book such curves are referred to as curves of discontinuity. The curves of discontinuity divide parameter space into three regions where one of the r_i is larger than the other two, at the equilibrium. They may also correspond to discontinuous bifurcations, as described below. Physically, crossing a curve of discontinuity corresponds to a change in the preferred metabolic pathway at equilibrium. At a codimension-two point near $(D, k_La) = (0.26, 250)$, the equilibrium lies at an intersection point of all three switching manifolds.

Fig. 4.4 shows enlargements of Fig. 4.3 near two points on the left-most curve of discontinuity along which the equilibrium lies on $\Sigma_{2,3}$, (4.16). In panel (a), along the curve of discontinuity, below the point (a) and above the point (c), there is no bifurcation. Between (a) and (c), numerical simulations show that a periodic orbit is created when the equilibrium crosses the switching manifold. Between (a) and (b), the orbit is unstable and emanates to the right of the curve of discontinuity. Between (b) and (c) the orbit is stable and emanates to the left.

Points (a) and (c) are codimension-two points studied from a general viewpoint in Sec. 3.2. At both points, on one side of $\Sigma_{2,3}$ there exists a complex conjugate pair of limiting associated eigenvalues of the equilibrium that are purely imaginary. As predicted by Theorem 3.3, a locus of Hopf bifurcations emanates non-tangentially from the curve of discontinuity. Grazing curves along which the corresponding Hopf cycle grazes $\Sigma_{2,3}$ emanate from (a) and (c) and are tangent to the Hopf loci here. Note that the grazing curve that emanates from (a) is so close to the Hopf locus, HB_{2b}, that in Fig. 4.4(a) the two curves are barely distinguishable.

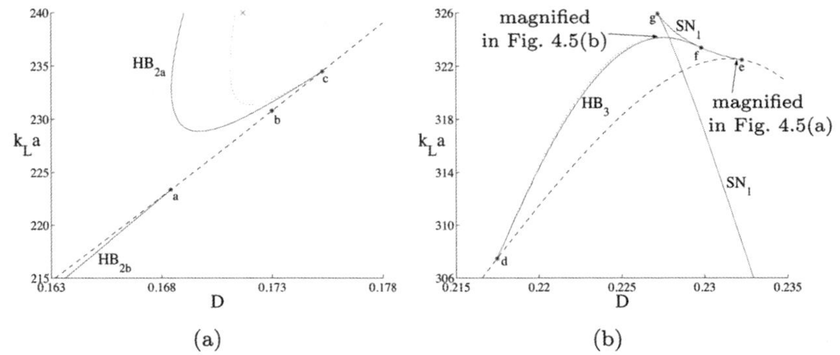

Fig. 4.4: Magnified views of Fig. 4.3. HB - Hopf bifurcation, SN - saddle-node bifurcation. The curves of discontinuity are shown dashed. The dotted curves correspond to the grazing of a Hopf cycle with a switching manifold. There are three such curves, these emanate from the points (a), (c) and (d) (note, the curve that emanates from (a) is barely distinguishable from HB_{2b}). At the top of panel (a), one point on a locus of saddle-node bifurcations of a Hopf cycle is shown with a cross.

Recall, Theorem 3.2 states that for two-dimensional systems when the stability of the periodic orbits created at the Hopf and Hopf-like bifurcations arbitrarily close to the codimension-two point are different, there generically exists a locus of saddle-node bifurcations of the periodic orbits and that this locus deviates from the grazing curve to only order six. Saddle-node loci obeying the same scaling laws are hypothesized to exist in higher dimensional systems, such as the yeast growth model. Near (a) and (c) the (smooth) Hopf bifurcations are subcritical. Since between (b) and (c) the discontinuous bifurcation generates a stable periodic orbit, a locus of saddle-node bifurcations is expected to emanate from (c) and lie extremely close to the grazing curve. No such locus is expected to emanate from (a) because along (a) and (b) unstable orbits are created (or more precisely orbits that have stable and unstable manifolds of the same dimensionality as those of Hopf cycles generated along HB_{2b}).

Unfortunately, periodic orbits of piecewise-smooth systems that lie in more than one smooth region of phase space are typically considerably more difficult to numerically compute than periodic orbits of a smooth system, or periodic orbits that lie in only one smooth component of a piecewise-smooth system. For the system (4.1)-(4.8), numerical difficulties also arise due to high dimensionality and stiffness. Periodic orbits at a saddle-node locus predicted by Theorem 3.2 lie in two smooth regions of phase space. Consequently, not more than a single point on a such a saddle-node locus has been computed numerically for the system (4.1)-(4.8), see Fig. 4.4(a). In contrast, grazing curves are relatively easy to compute because it suffices to solve only one smooth system component. Sophisticated algorithms are required for bifurcation continuation in piecewise-smooth systems. This is the subject of on-going research, see for instance [Batako *et al.* (2007); Piiroinen *et al.* (2004)].

The codimension-two point (b) is analogous to the scenario studied in two dimensions in Sec. 3.3 at which the discontinuous Hopf bifurcation is of indeterminable criticality. It is anticipated that as in the two-dimensional case summarized by Theorem 3.4, there exists a locus of saddle-node bifurcations of the Hopf cycles that emanates from (b) and is tangent to the curve of discontinuity here. Numerical analysis suggests that such a curve intersects (b) and exists below (b) and to the right of the curve of discontinuity and HB_{2b}. However, for the same reasons as above, an accurate computation of this proposed curve has not been accomplished.

Since periodic orbits generated at discontinuous bifurcations were not able to be numerically computed, the codimension-two point, (b), was located by studying piecewise-linear approximations of the form (1.18) via a procedure now summarized. For a given value of $k_L a$, first the value of D at the curve of discontinuity is computed. Then, by deriving a linear approximation to $\Sigma_{2,3}$, the system can be transformed such that locally the switching manifold is orthogonal to e_1. The matrices A_L and A_R are then determined by numerically computing Jacobians of the corresponding smooth components of the system at the equilibrium. Numerical simulations of the piecewise-linear approximation show the existence of a stable periodic orbit for relatively large values of $k_L a$. As $k_L a$ is decreased the amplitude of this orbit increases. The point at which the amplitude appears to limit on infinity is the point (b). Note, between (a) and (c), on both sides of the discontinuous bifurcation, the equilibrium has a complex pair of associated eigenvalues and six real-valued negative associated eigenvalues. At (b), Λ (2.17), as computed using the two complex eigenvalue pairs, is not equal to zero, which it would be if the system was two-dimensional, Sec. 3.3. This suggests that a condition (like $\Lambda = 0$) governing criticality in systems of arbitrary dimension may not have a simple form.

Fig. 4.4(b), shows a second magnification of Fig. 4.3 near $D = 0.225$ and $k_L a = 320$. Loci of Hopf bifurcations and saddle-node bifurcations of the equilibrium have endpoints at (d) and (e) that lie on a curve of discontinuity. The point (d) exhibits the unfolding predicted by Theorem 3.3. Near (d), HB_3 corresponds to Hopf bifurcations that are subcritical. Along the curve of discontinuity to the right of (d), a stable periodic orbit is created at the discontinuous bifurcation. Consequently, as for point (c), a locus of saddle-node bifurcations of periodic orbits is expected to exist very close to the grazing locus. The system exhibits stable oscillations in the region between the Hopf bifurcations and the curve of discontinuity, but oscillations in this parameter region have yet to be observed experimentally. Note that HB_{2a} and HB_3 are actually part of the same curve; the joining section is not seen because it corresponds to virtual solutions.

The point (e) lies at the tangential intersection of the curve of discontinuity with a locus of saddle-node bifurcations as predicted by Theorem 3.1. Along the curve of discontinuity to the right of (e), the discontinuous bifurcation is a nonsmooth fold. This is the scenario depicted in Fig. 3.2(b). A Takens-Bogdanov bifurcation occurs at (f) where the Hopf locus, HB_3, terminates at the saddle-node locus, and the point (g) corresponds to a cusp bifurcation [Kuznetsov (2004)].

Periodic orbits are created at the curve of discontinuity between (d) and a point to the left of (e). A precise computation of this point is left for

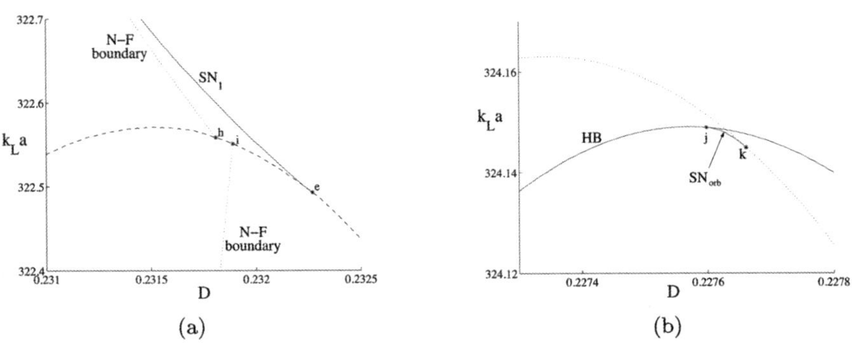

Fig. 4.5: Magnified views of Fig. 4.4(b). Along the dotted curves in panel (a) the admissible equilibrium has associated eigenvalues that are purely imaginary. Locally, to the left of the dotted curves the admissible equilibrium has a complex conjugate eigenvalue pair, whereas to the right all associated eigenvalues are real-valued.

future work but it is expected to lie near (h) and (i), see Fig. 4.5(a). Left of (h), complex-valued associated eigenvalues exist for the equilibrium on both sides of the switching manifold, thus here periodic orbits are likely to be created. Right of (i) all associated eigenvalues are real-valued and no periodic orbits are created. A complete analysis of all bifurcations in this region is left for future work.

Hopf bifurcations along HB_3 are subcritical between (d) and (j) and supercritical to the right of (j), see Fig. 4.5(b). A locus of saddle-node bifurcations of the associated Hopf cycle emanates from (j) as predicted by smooth bifurcation theory. This locus tangentially intersects the grazing curve at the point (k) beyond which the saddle-node locus is believed to persist. However, beyond (k) associated periodic orbits lie in two smooth regions of phase space and for this reason have not been computed.

Fig. 4.6 shows the bifurcation set superimposed on a sequence of bifurcation diagrams that vary with D. At the codimension-two point (l), at equilibrium all three metabolic pathways have the same growth rate. Equivalently, (l) lies at the intersection of three distinct curves of discontinuity.

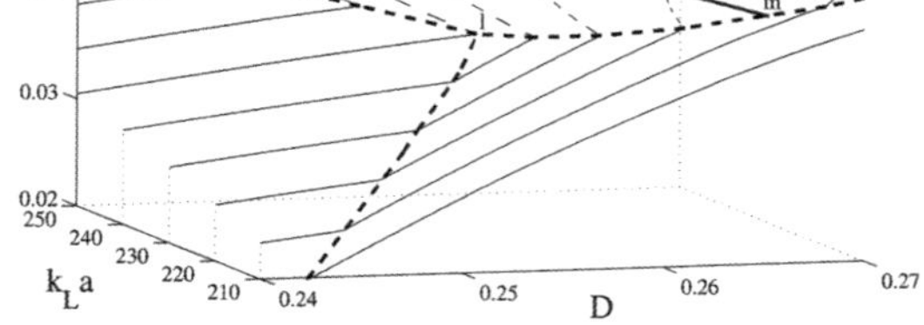

Fig. 4.6: A bifurcation set of the system (4.1)-(4.8) near the point, labeled (l), at which, at equilibrium, the growth rate along all three metabolic pathways is identical (i.e. $r_1 = r_2 = r_3$). The third axis shows the value of C (the variable C was chosen arbitrarily from the eight variables in the model). The locus of saddle-node bifurcations is shown as a thick, solid curve. Curves of discontinuity are shown dashed. For several fixed values of $k_L a$, the equilibrium is shown under variation of D. Equilibrium curves are shown solid [dashed] when the equilibrium is stable [unstable].

The point (m) lies at the intersection of SN_1 with a curve of discontinuity. The unfolding of (m) is in accordance with Theorem 3.1 and appears like Fig. 3.2(a).

Lastly, the intersection shown in Fig. 4.3 of SN_2 with a curve of discontinuity, occurs in the manner predicted by Theorem 3.1. In a neighborhood of the intersection the bifurcation set resembles Fig. 3.2(a). Three equilibria (two of which are stable) coexist in the thin region between SN_2 and the curve of discontinuity.

4.4 Simple and Complicated Stable Oscillations

Experimentally observed oscillations correspond to the region between HB_1 and HB_2 in Fig. 4.3. This section discusses the dynamics in this region in more detail.

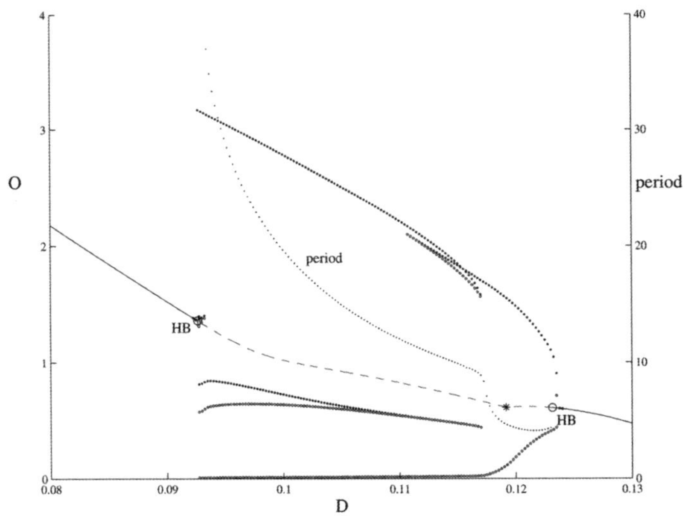

Fig. 4.7: A bifurcation diagram of the system (4.1)-(4.8) when $k_L a = 150$. The equilibrium is shown as a solid line when stable and a dashed line when unstable. Dots [small circles] correspond to local maximum [minimum] values of O that stable oscillations obtain. Two Hopf bifurcations are indicated by circles. The asterisk corresponds to where the equilibrium lies on the switching manifold, $\Sigma_{2,3}$, (4.16). The period of these oscillations is also indicated. See Fig. 4.10 for a magnification near the left-most Hopf bifurcation.

Fig. 4.7 shows a bifurcation diagram of the system (4.1)-(4.8) when the value of $k_L a$ is fixed at 150. For values of D between about 0.0927 and 0.1172, all three pathways are at some time preferred over one period of the solution. In particular, very soon after the preferred pathway changes from glucose oxidation to fermentation, the concentration of dissolved oxygen rebounds slightly before continuing to decrease, see Fig. 4.9. Thus local minima and maxima appear below the equilibrium value in Fig. 4.7. For larger values of D, still to the left of the rightmost Hopf bifurcation, fermentation is no longer a preferred pathway at any point on the stable solution, and the lower local maximum is lost. Also, the absolute maximum undergoes two cusp catastrophes [Ott (1993); Grebogi *et al.* (1983)] at $D \approx 0.111,\ 0.117$.

Different values of $k_L a$ yield similar bifurcation diagrams, a collection are shown in Fig. 4.8. As a general rule there is a rapid change from a stable equilibrium to a large amplitude orbit near the left-most Hopf bifurcation and as D is increased the amplitude and period of the orbit decrease.

A careful study of time series plots over one oscillation period can reveal the underlying biochemical processes that cause stable oscillations. Fig. 4.9

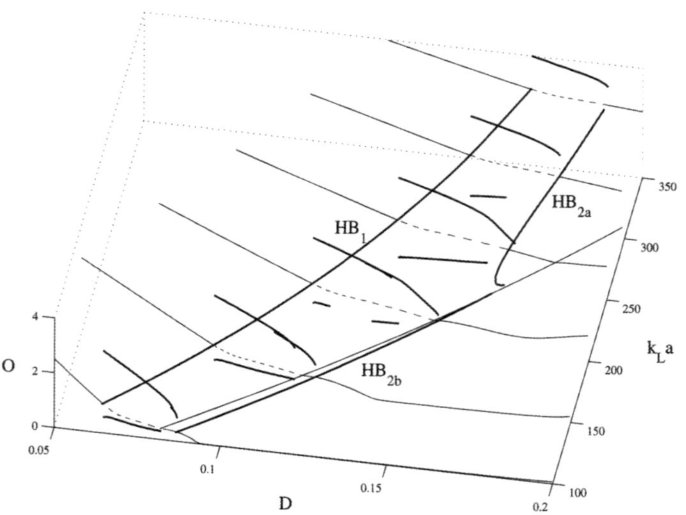

Fig. 4.8: Bifurcation diagrams of (4.1)-(4.8) at six different values of $k_L a$ with the same line style scheme as Fig. 4.7. Hopf loci and a curve of discontinuity (here indicated by a solid line) shown in Fig. 4.3 are also included.

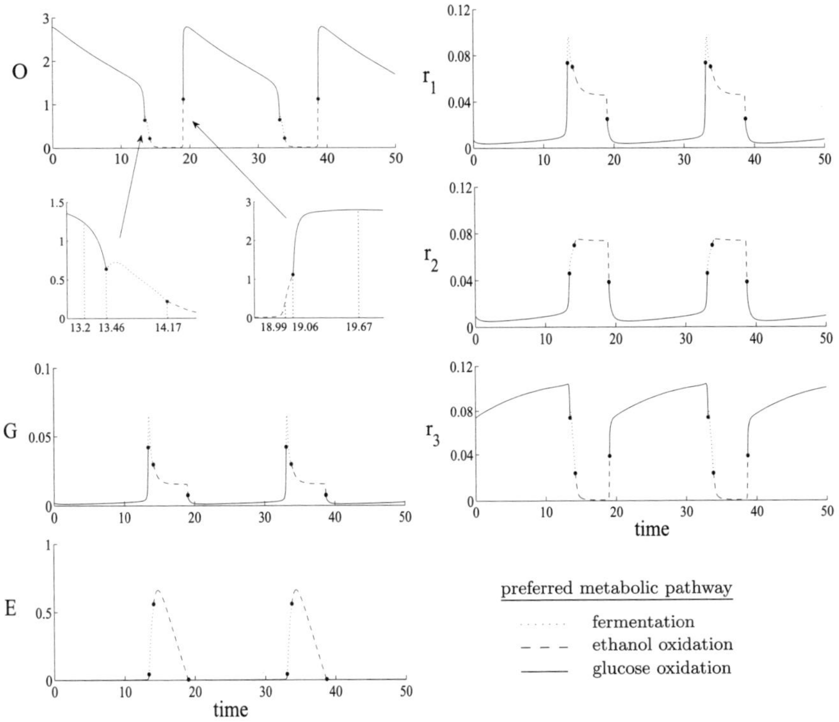

Fig. 4.9: Time series showing stable oscillations when $k_L a = 150$ and $D = 0.1$. Plots of the remaining five variables are shown in Fig. B.2. The point on the stable oscillation at which O is maximal has been taken as the initial condition. The preferred metabolic pathway changes with time and is indicated by the line style. Times discussed in the text at which an important change occurs are illustrated in the insets to the plot of dissolved oxygen.

shows the variation of dissolved oxygen, glucose, ethanol and the growth rates of the three metabolic pathways when $D = 0.1$ and $k_L a = 150$. Plots of the remaining variables are given in Fig. B.2. For ease of explanation, time is set to zero at a point where O is maximal.

Initially the concentration of dissolved oxygen is high, thus the yeast is expected to prefer one of the two oxidative pathways. The concentrations of glucose and ethanol are low thus since $K_3 \ll K_2$, we have $\frac{G}{K_3+G} \gg \frac{E}{K_2+E}$. The remaining terms in (4.11) and (4.12) are comparable thus r_3 is much greater than r_2. This is confirmed in Fig. 4.9: glucose oxidation dominates.

O decreases because the rate of oxygen consumption is greater than the rate at which oxygen is supplied.

At approximately $t = 13.2h$ (see the left inset of the time series plot of O in Fig. 4.9), the yeast begins to significantly oxidize ethanol in addition to glucose and as a result the rate of oxygen consumption increases sharply. The stimulus to ethanol oxidation at this point is not clear but may be due to a drop in the value of C. (No oscillations occur at these parameter values if the effect of C is removed from the model by setting $\phi_4 = 0$.) When $\frac{dC}{dt}$ becomes negative the last term of (4.2) becomes large and positive and G and consequently r_1 increase. This in turn causes E and r_2 to increase, by (4.3) and (4.11), thus we have ethanol oxidation.

At approximately $t = 13.46h$ the fermentation pathway becomes preferred due to the lack of oxygen. For a short time all three metabolic pathways are used to a significant degree. The most important feature of this "combined" stage is the rapid increase of ethanol concentration as a result of fermentation. Soon ethanol oxidation becomes the preferred pathway (at $t \approx 14.17h$). Fermentation still occurs significantly but glucose oxidation is minimal. This is because the concentration of dissolved oxygen is low, thus since $K_{O_2} \ll K_{O_3}$, we have $\frac{O}{K_{O_2}+O} \gg \frac{O}{K_{O_3}+O}$ and $r_2 \gg r_3$, see (4.11) and (4.12).

Throughout this stage of the yeast cycle, ethanol concentration decreases. At approximately $t = 18.99h$, the value of E drops below K_2 and r_2 decreases suddenly. Unlike before, the value of C is small and glucose concentration does not increase leading to fermentation. Instead, since neither oxidation pathway is being used greatly, the concentration of dissolved oxygen is able to increase. By $t \approx 19.67$, O has reached its maximum value, glucose oxidation dominates and the system has returned to its state at $t = 0$.

To summarize, glucose oxidation is the dominate metabolic pathway when the concentration of dissolved oxygen is high. Once available oxygen is consumed, the only pathway that does not require oxygen: fermentation, becomes preferred. As a result the level of ethanol increases; soon ethanol oxidation is preferred. Once the available ethanol is used up by this pathway the dissolved oxygen concentration rises and the system returns to glucose oxidation.

As mentioned above, when $D = 0.12$ (and $k_L a = 150$), fermentation is not a preferred pathway at any point during the oscillation. The basic feedback processes described above again cause oscillations in this case, however all three pathways are used to a significant degree throughout the oscillation and key metabolic changes are less obvious. Overall, chemical concentrations fluctuate less and the oscillation period is only about $4.33h$. Time series plots are given in Fig. B.3.

Hopf bifurcations provide the onset for stable oscillatory behavior, however, the behavior near the left-most Hopf bifurcation is actually quite complex, as indicated in Fig. 4.10, which is a magnification of Fig. 4.7. Here the Hopf bifurcation is supercritical giving rise to a stable orbit which then undergoes a period-doubling cascade to chaos over an extremely small interval. The first period-doubling occurs at $D \approx 0.092697$ and the solution appears chaotic by $D = 0.092700$. At $D \approx 0.092701$ the attractor suddenly explodes in size and the oscillation amplitude grows considerably. This sudden transition from a small amplitude orbit to a large amplitude orbit is reminiscent of the canard phenomenon [Krupa and Szmolyan (2001); Wechselberger (2005)].

As D decreases from the largest value shown in Fig. 4.10, a period-adding sequence is observed. As discussed in Sec. 1.8, period-adding sequences are characterized by successive jumps in the period in a manner that forms an approximately arithmetic sequence. Dynamical behavior

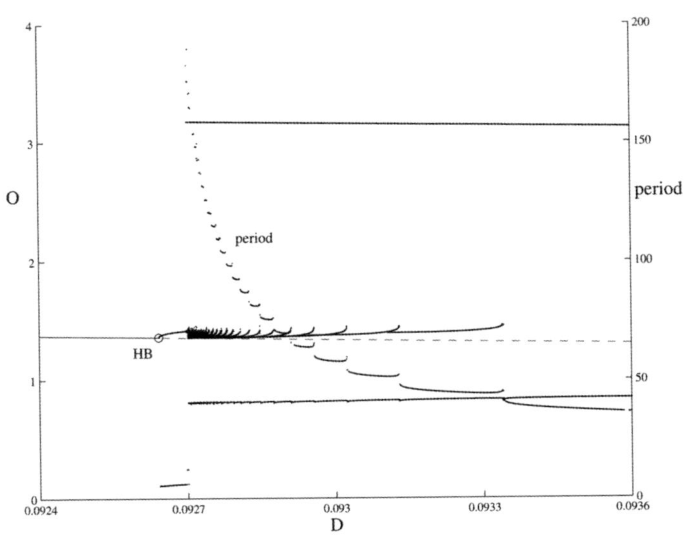

Fig. 4.10: A magnified view of Fig. 4.7, near the leftmost Hopf bifurcation. Local minimums are not shown. A different scale is used for the period axis.

between period-adding windows is determined by the types and order of various local bifurcations. In Fig. 4.10, the period appears to go to infinity in the period-adding sequence. Within the extremely small regions between windows, there exists period-doubling bifurcations and complicated attractors although these attractors deviate only slightly from the observed periodic orbits. The period adding observed here bears an especially strong resemblance to that in the neuronal Chay model [Yang *et al.* (2006)].

Chapter 5

Codimension-Two, Border-Collision Bifurcations

A border-collision bifurcation occurs for the piecewise-smooth, continuous map (1.14) with (1.11) at $\mu = 0$. At the bifurcation the origin is a fixed point that lies on the switching manifold. As the value of μ is varied from zero, one or more invariant sets such as fixed points, periodic cycles and invariant circles, may emanate from the bifurcation. As discussed in Sec. 1.6, the number and nature of these sets is typically determined by linear terms of the map. This chapter performs rigorous unfoldings of two special situations for which nonlinear terms affect local dynamical behavior.

In the literature there are a very limited number of investigations into piecewise-smooth maps for which nonlinear terms affect local dynamics. Perhaps this is because a complete understanding of piecewise-linear maps, even in only two dimensions, is unknown, see Chapter 7. Viktor Avrutin and coworkers have unfolded codimension-two and three bifurcations in one-dimensional, discontinuous maps [Avrutin and Schanz (2005a); Avrutin *et al.* (2007, 2006)]. They find a codimension-three point that is an organizing center for period-adding phenomena. Unfoldings of codimension-two, sliding bifurcations have been described [Kowalczyk and Piiroinen (2008); Nordmark and Kowalczyk (2006)], see also [Kowalczyk and di Bernardo (2005); Kowalczyk *et al.* (2006)]. Here a codimension-two point is the intersection of loci of various, distinct, codimension-one, sliding bifurcations. Codimension-two, grazing bifurcations have also been described, see [Zhao and Dankowicz (2006)].

When $\mu = 0$, the origin is a fixed point of both half-maps of (1.14). The multipliers associated with the origin are those of the matrix $A_L(0)$, for the left half-map, and $A_R(0)$ for the right half-map. If for one half-map, say the left half-map, the origin is non-hyperbolic (i.e., $A_L(0)$ has an multiplier on the unit circle), then a local smooth bifurcation may occur for the left half-map at $\mu = 0$ [Kuznetsov (2004); Robinson (2004)]. The nature of the smooth bifurcation is determined by nonlinear terms of the left half-map. Consequently these nonlinear terms affect dynamics of the full piecewise-smooth map local to the border-collision bifurcation at $\mu = 0$.

There are three codimension-two scenarios to consider. Namely that $A_L(0)$ has a multiplier of 1, a multiplier of -1, and a complex conjugate pair of multipliers with unit modulus. Cases involving a combination of these or $A_R(0)$ also having a multiplier on the unit circle, are of a codimension greater than two for the map (1.14) and for this reason are not considered in this book. To unfold codimension-two situations, as elsewhere in this book, a second independent parameter, η, is used. The codimension-two, border-collision bifurcations are assumed to occur at $\mu = \eta = 0$, and in each case a transversality condition is imposed on η to ensure its utility. A recent analysis of these situations in a general piecewise-smooth map is given in [Colombo and Dercole (unpublished)].

Section 5.1 studies the case that $A_L(0)$ has a multiplier of 1. This case is the discrete analogue of the situation studied in Sec. 3.1. As for a piecewise-smooth, ODE system, a locus of saddle-node bifurcations emanates from the codimension-two point and is tangent to the η-axis here. The case that $A_L(0)$ has a multiplier of -1 is more complicated because the existence and admissibility of 2-cycles must also be considered. Summarizing the analysis presented in Sec. 5.2, a locus of period-doubling bifurcations emanates non-tangentially from $\mu = 0$ and the generated 2-cycles collide with the switching manifold along a second tangential locus. Different 2-cycles are created at $\mu = 0$ for one sign of η; these collide with the switching manifold and coincide with the former 2-cycles along the second locus.

A rigorous unfolding of the case that $A_L(0)$ has complex multipliers on the unit circle, say $\lambda = \mathrm{e}^{\pm 2\pi i \omega}$, remains for future investigations, see however [Colombo and Dercole (unpublished)]. Resonance tongues may exist arbitrarily close to the codimension-two point and if ω is rational it seems reasonable that a resonance tongue boundary will emanate from this point.

5.1 A Nonsmooth, Saddle-Node Bifurcation

This section studies (1.14) with (1.11) when 1 is a multiplier of $A_L(0)$. This codimension-two situation corresponds to the simultaneous occurrence of a border-collision bifurcation with a saddle-node bifurcation. The generic unfolding is summarized by Theorem 5.1. Note that this theorem is very similar to Theorem 3.1 which unfolds the simultaneous occurrence of a discontinuous bifurcation with a saddle-node bifurcation.

Theorem 5.1. *Consider the piecewise-C^k ($k \geq 4$), continuous map (1.14) with (1.11). Suppose that near $(\mu, \eta) = (0,0)$, $A_L(\mu, \eta)$ has a multiplier $\lambda(\mu, \eta) \in \mathbb{R}$ with an associated eigenvector, $v(\mu, \eta)$. In addition, suppose*

(i) $\lambda(0,0) = 1$ is of algebraic multiplicity 1 and is the only multiplier of $A_L(0,0)$ on the unit circle,
(ii) $b_0 = \varrho^{\mathsf{T}}(0,0)b(0,0) \neq 0$,

where $\varrho^{\mathsf{T}} = \varrho^{\mathsf{T}}_{\mathrm{map}}$ is given by (1.26). Then $v(0,0)$ is not orthogonal to e_1 thus by scaling we may assume $e_1^{\mathsf{T}} v(0,0) = 1$. Finally suppose

(iii) $\frac{\partial \lambda}{\partial \eta}(0,0) = 1$,
(iv) $a_0 = \varrho^{\mathsf{T}}((D_x^2 f^{(L)})(v,v))\big|_{(0,0,0)} \neq 0$.

Then

$$p_0 = \varrho^{\mathsf{T}}(0,0)v(0,0) \neq 0 \,, \tag{5.1}$$

and there exists a unique C^{k-2} function $h : \mathbb{R} \to \mathbb{R}$ with $h(0) = h'(0) = 0$ and

$$h''(0) = \frac{p_0^2}{a_0 b_0} \,, \tag{5.2}$$

such that in a neighborhood of $(\mu, \eta) = (0,0)$, the curve $\mu = h(\eta)$ corresponds to a locus of saddle-node bifurcations of fixed points of (1.14) that are admissible when

$$\mathrm{sgn}(\eta) = \mathrm{sgn}(a_0 p_0) \,. \tag{5.3}$$

This theorem may be proved in the same manner as Theorem 3.1. See also the discussion immediately after Theorem 3.1. Basically Theorem 5.1 states that a locus of saddle-node bifurcations emanates from the codimension-two point and at this point is tangent to the η-axis (along which border-collision bifurcations of a fixed point occur).

To illustrate Theorem 5.1, consider the following two-dimensional, piecewise-C^∞, continuous map

$$\begin{aligned} x' &= 3\mu + \frac{1}{2}\eta x - 2y + \alpha x^2 \,, \\ y' &= \mu\eta - (\mu + 1)|x| + 3y \,, \end{aligned} \tag{5.4}$$

where $\alpha \neq 0$ is a constant. The map may easily be rewritten into the form (1.14) with (1.11). Then, here, $A_L(\mu, \eta) = \begin{bmatrix} \frac{1}{2}\eta & -2 \\ \mu + 1 & 3 \end{bmatrix}$ has an eigenvalue $\lambda(\mu, \eta) = 1 + 2\mu + \eta + O(2)$, thus conditions (i) and (ii) of Theorem 5.1 are satisfied. When $\mu = \eta = 0$, the associated eigenvector is $v(0,0) = [1, -\frac{1}{2}]^{\mathsf{T}}$. Also $\varrho^{\mathsf{T}}(0,0) = [2,2]$ and $b(0,0) = [3,0]^{\mathsf{T}}$. Consequently

$$(a_0, b_0, p_0) = (4\alpha, 6, 1) , \tag{5.5}$$

and thus Theorem 5.1 predicts saddle-node bifurcations along $\mu = h(\eta)$ where $h''(0) = \frac{1}{24\alpha}$, that are admissible when $\operatorname{sgn}(\eta) = \operatorname{sgn}(\alpha)$.

This prediction may be verified by a explicit computation of fixed points of (5.4). It is easily determined that the x-component of a fixed point in the left half-plane is given by

$$x^{*(L)}(\mu, \eta) = \frac{1}{2\alpha}\left(-\left(\frac{\eta}{2} + \mu\right) \pm \sqrt{\left(\frac{\eta}{2} + \mu\right)^2 - 4\alpha\mu(3 + \eta)}\right) . \tag{5.6}$$

Saddle-node bifurcations occur at values of μ and η for which the square root term of (5.6) is zero. Via a series expansion it follows that this term is zero when $\mu = h(\eta) = \frac{1}{48\alpha}\eta^2 + O(\eta^3)$ in agreement with the theorem. Furthermore, the saddle-node bifurcations are admissible when

$$x^{*(L)}(h(\eta), \eta) = -\frac{1}{4\alpha}\eta < 0 , \tag{5.7}$$

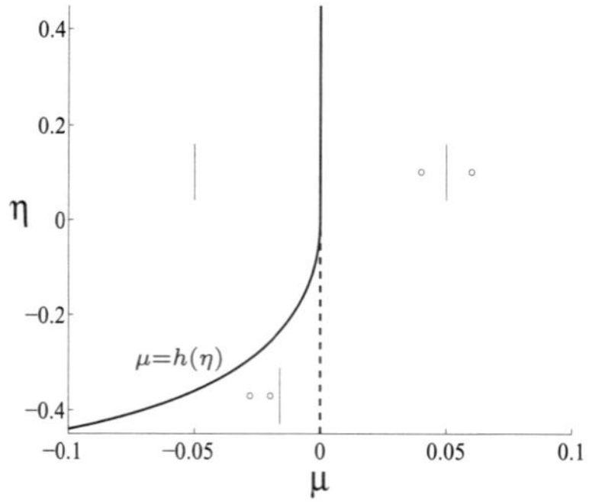

(a) $\alpha = -0.1$

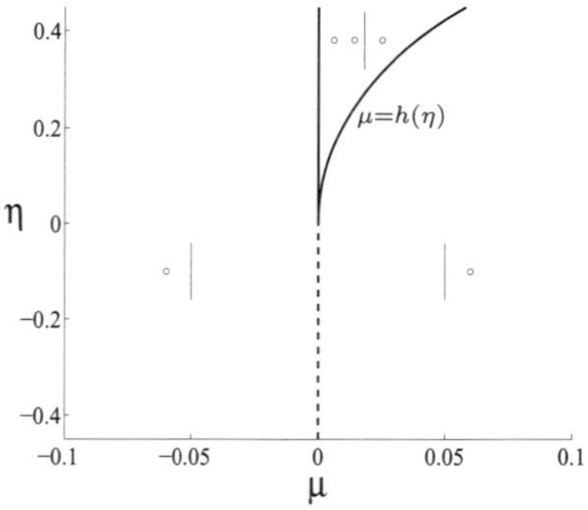

(b) $\alpha = 0.1$

Fig. 5.1: Bifurcation sets of the map (5.4) for two different values of α. Solid curves correspond to saddle-node bifurcations and nonsmooth folds. A single fixed point crosses the switching manifold along the dashed curves (persistence). Within each region bounded by the curves there are sketches showing the existence of admissible fixed points relative to the switching manifold. All fixed points shown are unstable.

i.e. when η has the same sign as α as predicted by the theorem. Fig. 5.1 shows bifurcation sets of the map (5.4).

5.2 A Nonsmooth, Period-Doubling Bifurcation

The coincidence of border-collision and period-doubling has been observed in real-world systems. Recall that a corner collision in a Filippov system provides an example of a border-collision bifurcation in a map [di Bernardo *et al.* (2001a)]. In [Angulo *et al.* (2005, 2008)] the authors describe the coincidence of a corner collision with period-doubling in a Filippov model of a DC/DC power converter. Grazing-sliding (which also corresponds to a piecewise-smooth, continuous Poincaré map, see [di Bernardo *et al.* (2002b)]) and period-doubling have been seen to occur simultaneously in a model of a forced, dry-friction oscillator [Kowalczyk *et al.* (2006)].

Here the map (1.14) with (1.11) is studied near the codimension-two scenario that $A_L(0)$ has a multiplier of -1. This situation corresponds to the coincidence of a border-collision bifurcation and a period-doubling bifurcation. As for the situation studied in Sec. 3.2, first a highly detailed unfolding is presented in the lowest possible dimension, in this case, in one dimension, see Theorem 5.2. In one dimension the border-collision bifurcations that occur at $\mu = 0$ when $\eta \neq 0$ are fully understood. The unfolding is then described in an arbitrary number of dimensions, see Theorem 5.3. This unfolding is accomplished by a restriction of the left half-map to the center manifold whereby results of the one-dimensional unfolding may be applied.

Theorem 5.2. *Consider the one-dimensional, piecewise-C^k $(k \geq 3)$, continuous map*

$$x' = \begin{cases} f^{(L)}(x;\mu,\eta), & x \leq 0 \\ f^{(R)}(x;\mu,\eta), & x \geq 0 \end{cases}, \tag{5.8}$$

where

$$\begin{aligned} f^{(L)}(x;\mu,\eta) &= \mu b(\mu,\eta) + a_L(\mu,\eta)x + p(\mu,\eta)x^2 + q(\mu,\eta)x^3 + o(x^3)\,, \\ f^{(R)}(x;\mu,\eta) &= \mu b(\mu,\eta) + a_R(\mu,\eta)x + O(x^2)\,. \end{aligned} \tag{5.9}$$

Suppose

(i) $a_L(0,0) = -1$,
(ii) $b(0,0) = 1$,

(iii) $\frac{\partial a_L}{\partial \eta}(0,0) = 1$.

Let

$$a_0^{(R)} = a_R(0,0) \,, \tag{5.10}$$

$$c_0 = p^2(0,0) + q(0,0) \,. \tag{5.11}$$

Then there exists $\delta > 0$ *such that the left half-map restricted to the neighborhood*

$$\mathcal{N} = \left\{ (x;\mu,\eta) \;\middle|\; |x|, |\mu|, |\eta| < \delta \right\} \tag{5.12}$$

has a unique fixed point given by a C^k *function,* $x^{*(L)}(\mu,\eta) = \frac{1}{2}\mu + O(2)$, *and there exist unique,* C^{k-1} *functions* $h_1, h_2 : \mathbb{R} \to \mathbb{R}$ *that satisfy*

$$h_1(\mu) = -\left(\frac{\partial a_L}{\partial \mu} + p \right)\bigg|_{\mu=\eta=0} \mu + O(\mu^2) \,, \tag{5.13}$$

$$h_2(\mu) = h_1(\mu) - \frac{c_0}{4}\mu^2 + o(\mu^2) \,, \tag{5.14}$$

such that

- *(1) when* $\mu < 0$, $x^{*(L)}(\mu, h_1(\mu))$ *has an associated multiplier of* -1, *and if* $c_0 \neq 0$ *the curve* $\eta = h_1(\mu)$ *corresponds to a locus of admissible, period-doubling bifurcations of* $x^{*(L)}$,
- *(2) the origin belongs to a period-two orbit of* $f^{(L)}$ *on* $\eta = h_2(\mu)$ *that is admissible when* $\mu < 0$,
- *(3) if* $a_0^{(R)} \neq \pm 1$ *and* $c_0 \neq 0$, *then fixed points and two-cycles of (5.8) exist in the sectors shown in Fig. 5.2,*
- *(4) if* $a_0^{(R)} < 1$ *or* $\mu > 0$ *or* $\eta > h_2(\mu)$, *(5.8) has no n-cycles for any* $n \geq 3$ *in* $\mathcal{N}$,
- *(5) if* $a_0^{(R)} > 1$ *and* $\mu < 0$ *and* $\eta < h_2(\mu)$, *then (5.8) exhibits chaos in* $\mathcal{N}$.

A proof is given is Appendix A. Condition (i) is the defining singularity condition. The border-collision bifurcation is degenerate if $b(0,0) = 0$, thus $b(0,0)$ must be nonzero. By scaling μ it may be assumed that $b(0,0) = 1$, condition (ii). Condition (iii) is the transversality condition and assumes an appropriate scaling of η. In view of extending Theorem 5.2 to higher dimensions, the theorem is written in a manner that yields a partial result when the non-degeneracy condition on the nonlinear terms, $c_0 \neq 0$, is not satisfied. The sign of c_0 determines the criticality of the period-doubling bifurcations.

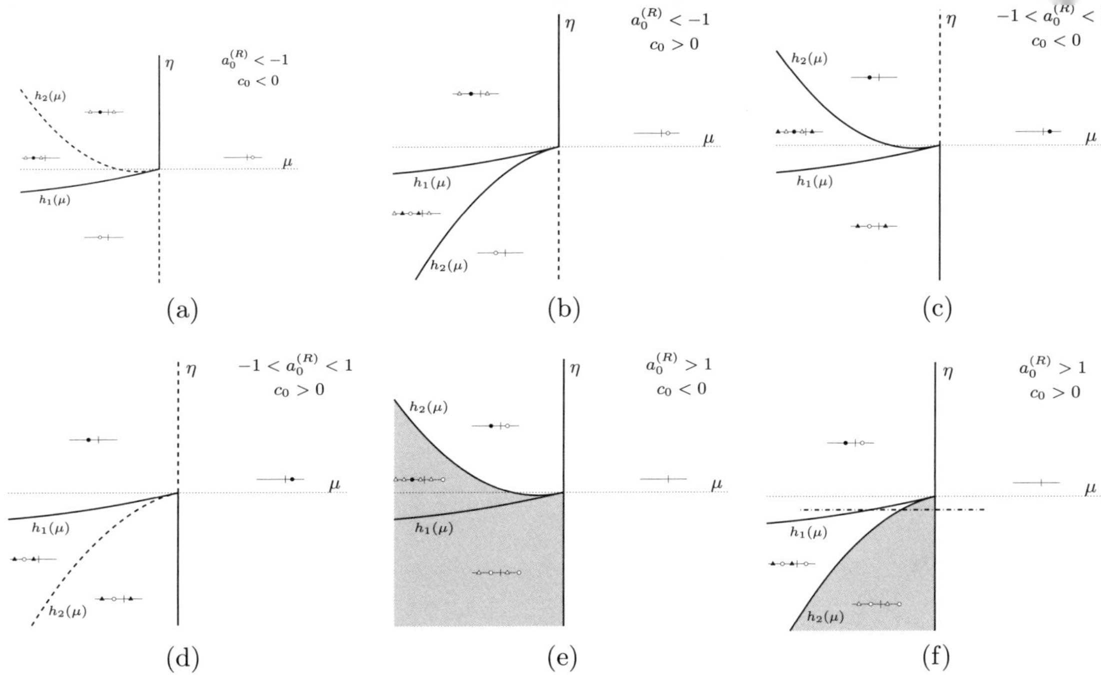

Fig. 5.2: Sketches of bifurcation sets of (5.8) for different values of $a_0^{(R)}$ and c_0, see Theorem 5.2. Each curve corresponds to either a locus of period-doubling bifurcations or a locus of border-collisions and is dashed when no topological change occurs at the border-collision. Insets show phase portraits with circles used to denote fixed points and triangles used to denote points on two-cycles. These are filled when the solutions are stable and unfilled otherwise. The small vertical lines denote the switching manifold. Chaotic dynamics occurs only for parameter values in the shaded regions; indeed, periodic solutions of a period higher than two only occur in these regions. The bifurcation diagram shown in Fig. 5.3 is taken along the dash-dot line segment of panel (f).

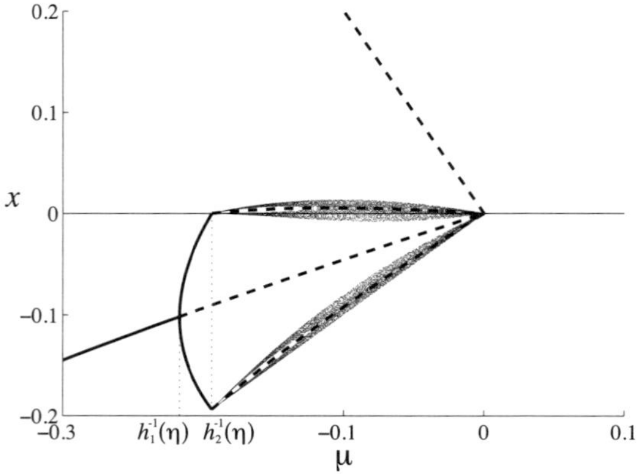

Fig. 5.3: A bifurcation diagram for (5.8) when $b = 1$, $a_L = \eta - 1$, $a_R = \frac{3}{2}$, $p = -1$, $q = \frac{3}{2}$ and there are no additional higher order terms. The value of η is fixed at -0.25. The parameters considered correspond to the dash-dot line segment shown in Fig. 5.2. Stable [unstable] fixed points and two-cycles are indicated by solid [dashed] curves. A supercritical period-doubling bifurcation occurs at $\mu = h_1^{-1}(\eta) \approx -0.2169$; a border-collision bifurcation of the period-doubled solution occurs at $\mu = h_2^{-1}(\eta) \approx -0.1937$. When $h_2^{-1}(\eta) < \mu < 0$ there exists a chaotic attracting set. When $\mu > 0$ there is no local attractor.

Theorem 5.2 predicts six different bifurcation sets depending on the value of $a_0^{(R)}$ and the sign of c_0, see Fig. 5.2. In each case the nature of the border-collision bifurcation at $\mu = 0$ for small $\eta \neq 0$ may be determined by referring to the discussion of the one-dimensional, piecewise-linear map (1.39) in Sec. 1.6, and Fig. 1.6. By condition (iii) of Theorem 5.2, if $\eta > 0$, then at $\mu = 0$, $a_L = -1 + \varepsilon$, and if $\eta < 0$, then at $\mu = 0$, $a_L = -1 - \varepsilon$, for some small $\varepsilon > 0$. For example, along the positive η-axis in Fig. 5.2(a), we have $a_L > -1$ and $a_R < -1$, thus here a single fixed point persists and a 2-cycle is created that coexists with the left fixed point and is stable.

In panels (e) and (f), along the negative η-axis we have $a_L < -1$ and $a_R > 1$. At points just to the left of the negative η-axis, both fixed points and a 2-cycle are admissible and unstable. Forward orbits are attracted to a chaotic solution created at the border-collision bifurcation at $\mu = 0$. Unstable high period orbits are also generated at the border-collision bifurcation; these are of even period. To further illustrate the unfolding, Fig. 5.3 shows a bifurcation diagram corresponding to a one-parameter slice of Fig. 5.2(f). The chaotic attracting set collapses to the origin at $\mu = 0$ and to a two-cycle at $\mu = h_2^{-1}(\eta)$.

Two-cycles created in border-collision bifurcations at $\mu = 0$ consist of one point in each half-plane. In contrast, 2-cycles created in period-doubling bifurcations along $\eta = h_1(\mu)$ lie entirely in the left half-plane. Along $\eta = h_2(\mu)$ the two 2-cycles coincide and are given by $\{0, \mu b(\mu, h_2(\mu))\}$. By computing the second iterate of (5.8), a map of the form (1.14) may be determined to describe the border-collision bifurcations along $\eta = h_2(\mu)$. This is achieved in the proof of Theorem 5.2, see (A.79). The two limiting multipliers are $1 - c_0\mu^2 + O(\mu^3)$ and $-a_0^{(R)} + O(\mu)$. Consequently 4-cycles arise from $\eta = h_2(\mu)$ exactly when $a_0^{(R)} > 1$, i.e. in panels (e) and (f) of Fig. 5.2, though these are not shown.

The following theorem unfolds the codimension-two scenario of interest in two or more dimensions.

Theorem 5.3. *Consider the piecewise-C^k, continuous map (1.14) with (1.11) and assume $N \geq 2$ and $k \geq 4$. Suppose that near $(\mu, \eta) = (0, 0)$, $A_L(\mu, \eta)$ has a multiplier $\lambda(\mu, \eta) \in \mathbb{R}$ with an associated eigenvector, $v(\mu, \eta)$. In addition, suppose*

(i) *$\lambda(0, 0) = -1$ is of algebraic multiplicity one and $A_L(0, 0)$ has no other multipliers on the unit circle,*
(ii) $\varrho^{\mathsf{T}}(0, 0)b(0, 0) \neq 0$,
(iii) $\frac{\partial \lambda}{\partial \eta}(0, 0) = 1$,
(iv) *the first element of $v(0, 0)$ is nonzero, thus by scaling it may be assumed that $e_1^{\mathsf{T}} v(0, 0) = 1$,*
(v) $\det(I - A_L(0, 0)A_R(0, 0)) \neq 0$.

Then, in the extended coordinate system (x, μ, η), there exists a C^{k-1} three-dimensional center manifold, W^c, for the left half-system that passes through the origin and is not tangent to the switching manifold at this point. Furthermore, the left half-map, $x' = f^{(L)}(x; \mu, \eta)$, restricted to W^c described in the coordinate system $(\hat{x}; \hat{\mu}, \hat{\eta}) = \left(e_1^{\mathsf{T}} x; \frac{2\varrho^{\mathsf{T}} b}{\det(I - A_L)}\Big|_{(0,0)} \mu, \eta\right)$ is given by the left half-map of (5.8) with "hatted variables" and the conditions (i)-(iii) of Theorem 5.2 will be satisfied. Moreover, a unique 2-cycle consisting of one point on each side of the switching manifold exists for small μ and η. This cycle is admissible exactly when $\hat{\mu} \leq 0$ and $\eta \leq \operatorname{sgn}\left(\frac{\det(I - A_L)\frac{\partial}{\partial \eta}\det(I + A_L)}{\det(I - A_L A_R)}\Big|_{(0,0)}\right) h_2(\hat{\mu})$.

A proof is given in Appendix A. Conditions (i)-(iii) are analogous to the first three conditions of Theorem 5.2. Condition (iv) is a non-degeneracy condition that ensures the period-doubled cycles collide with the switching manifold in the generic manner. If this condition is not satisfied, a higher codimension situation that may involve a linear separation between $h_1(\mu)$ and $h_2(\mu)$ will arise. Finally condition (v) is necessary to ensure that non-degenerate, 2-cycles are created at $\mu = 0$.

Theorem 5.3 essentially predicts the same basic bifurcation structure as Theorem 5.2. Since the restriction of the left half-map to the center manifold satisfies the conditions of Theorem 5.2 under an appropriate transformation, the curves h_1 and h_2 also exist for systems of two or more dimensions. The relative position of h_1 and h_2 is determined by the sign of c_0. In one dimension, c_0 is a simple function of coefficients of nonlinear terms of the left half-map (5.11). To obtain the value of c_0 in higher dimensions, one may derive an expression for the restriction of the left half-map to the center manifold, then apply the one dimensional result. This is done for an example below. An explicit expression for c_0 in N dimensions is not known but is probably too complicated to be of practical use.

To illustrate Theorem 5.3 consider the following map

$$\begin{aligned} x' &= -\frac{1}{2}\mu - \frac{1}{2}x + y \,, \\ y' &= -\frac{1}{2}|x| - \frac{3}{2}\eta x + \frac{1}{4}x^2 \,. \end{aligned} \tag{5.15}$$

As with the example in the previous section, (5.15) is easily rewritten in the piecewise form (1.14). Fixed points of (5.15) are given by

$$\begin{bmatrix} x^{*(L)}(\mu,\eta) \\ y^{*(L)}(\mu,\eta) \end{bmatrix} = \begin{bmatrix} -\frac{1}{2} \\ -\frac{1}{4} \end{bmatrix} \mu + O(2) \,, \tag{5.16}$$

$$\begin{bmatrix} x^{*(R)}(\mu,\eta) \\ y^{*(R)}(\mu,\eta) \end{bmatrix} = \begin{bmatrix} -\frac{1}{4} \\ \frac{1}{8} \end{bmatrix} \mu + O(2) \,. \tag{5.17}$$

Thus the left fixed point is admissible when $\mu \geq 0$ and the right fixed point is admissible when $\mu \leq 0$.

A 2-cycle of the left half-map (obtained by replacing $|x|$ with $-x$ in (5.15)) consists of two points. Of these two points, the one that lies closer to the switching manifold is given explicitly by

$$\begin{aligned} x^{*(LL)}(\mu,\eta) &= 3\eta + \sqrt{2\mu + 3\eta(4+3\eta)} \,, \\ y^{*(LL)}(\mu,\eta) &= \frac{1}{2}\mu + \frac{9}{2}\eta + \frac{1}{2}\sqrt{2\mu + 3\eta(4+3\eta)} \,. \end{aligned} \tag{5.18}$$

Period-doubling bifurcations occur when the square root term is zero, i.e. when $\mu = -6\eta - \frac{9}{2}\eta^2$. Rearranging to make η the subject of the equation yields

$$\eta = h_1(\mu) = -\frac{1}{6}\mu - \frac{1}{48}\mu^2 + O(\mu^3)\,. \tag{5.19}$$

To determine h_2, we solve $x^{*(LL)}(\mu,\eta) = 0$ for η. This produces

$$\eta = h_2(\mu) = -\frac{1}{6}\mu\,. \tag{5.20}$$

Now consider 2-cycles of (5.15) that are comprised of one point on each side of the switching manifold. The multipliers of the product $A_L(0,0)A_R(0,0)$ are $\frac{1\pm\sqrt{17}}{8}$, thus since these do not lie on the unit circle, the 2-cycles exist (but are not necessarily admissible) in a neighborhood of $(\mu,\eta) = (0,0)$, see Sec. 6.2. Feigin's results [di Bernardo *et al.* (1999)], see Sec. 1.6, may be used to determine the sign of η for which admissible 2-cycles are created at $\mu = 0$. The multipliers of $A_L(0,0)$ and $A_R(0,0)$ are $-1, \frac{1}{2}$ and $-\frac{1}{4} \pm \frac{\sqrt{7}}{4}\mathrm{i}$, respectively. Therefore, for small $\eta < 0$, when $\mu = 0$, A_L has one multiplier less than -1 and A_R has no multipliers less than -1, thus by Feigin's results the 2-cycle is created here. Since the 2-cycle is admissible along $\eta = h_2(\mu)$ for small $\mu > 0$, the 2-cycle is admissible between the negative η-axis and $\eta = h_2(\mu)$. The stability of the fixed points and 2-cycles is easily determined and the map (5.15) has the bifurcation set shown in Fig. 5.4.

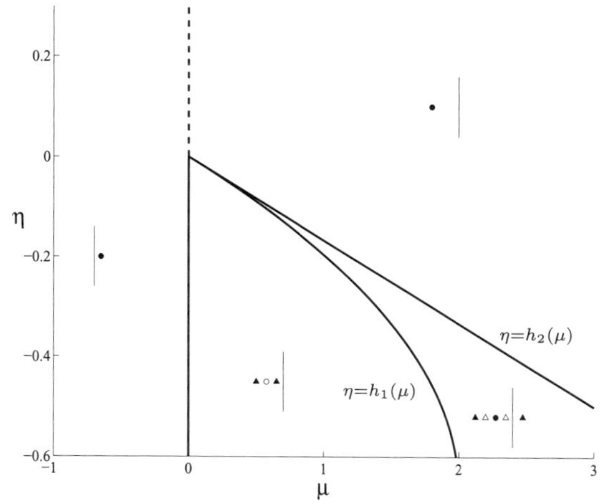

Fig. 5.4: A bifurcation set of the two-dimensional, piecewise-C^∞ map (5.15) near $\mu = \eta = 0$. Period-doubling and border-collision bifurcations occur along $\eta = h_1(\mu)$ and $\eta = h_2(\mu)$, respectively, as predicted by Theorem 5.3. As in Fig. 5.2, schematics showing fixed points and 2-cycles are included.

Now consider the implications of Theorem 5.3 to the map (5.15). Here we have $\varrho^{\mathsf{T}}(0,0) = [1,1]$, $b(0,0) = [-\frac{1}{2},0]^{\mathsf{T}}$ and $\det(I - A_L(0,0)) = 1$, thus $\hat{\mu} = -\mu$ and therefore $h_1(\mu)$ corresponds to admissible solutions when $\mu > 0$, in agreement with Fig. 5.4. Also, $\frac{\partial}{\partial \eta} \det(I + A_L)\Big|_{(0,0)} = \frac{3}{2}$ and $\det(I - A_L(0,0)A_R(0,0)) = \frac{1}{2}$, hence by the last statement of Theorem 5.3, 2-cycles with points on each side of the switching manifold are admissible below h_2, which again is confirmed by Fig. 5.4.

Via a series expansion (see also the proof of Theorem 5.3), one finds that the left half-map of (5.15) restricted to the center manifold is given by

$$x' = \hat{\mu} - x + \frac{1}{2}x^2 - \frac{2}{3}\hat{\mu}x + \eta x + \frac{1}{3}\hat{\mu}^2 - 2\hat{\mu}\eta - \frac{1}{3}x^3 + \cdots \qquad (5.21)$$

Here the term $-(\frac{\partial a_L}{\partial \hat{\mu}} + p)\Big|_{(0,0)}$, given in Theorem 5.2, is equal to $-\frac{1}{6}$. Also $c_0 = -\frac{1}{12}$. Thus $h_1(\mu) = -\frac{1}{6}\mu + O(\mu^2)$ and $h_2(\mu) - h_1(\mu) = \frac{1}{48}\mu^2 + O(\mu^3)$ matching the explicitly derived results above.

Chapter 6

Periodic Solutions and Resonance Tongues

Recall from Chapter 1, a bifurcation resulting from the collision of a fixed point with a switching manifold in a piecewise-smooth, continuous map is known as a border-collision bifurcation. As illustrated for the one-dimensional, piecewise-smooth, continuous map (1.39), a variety of invariant sets may be created at such bifurcations. Fixed points are the simplest invariant sets. The behavior of fixed points near border-collision bifurcations was described in Sec. 1.6. This chapter studies the existence, admissibility and bifurcations of periodic solutions near border-collision bifurcations. Some discussion of more complex invariant sets may be found in Chapter 7.

Dynamical behavior local to a border-collision bifurcation is described by a map of the form (1.14) with (1.11). As argued in Sec. 1.3, nonlinear terms of (1.14) do not influence structurally stable local dynamics. In particular, except in special situations, periodic solutions that emanate from border-collision bifurcations are determined by linear terms of (1.14). This will become clear in Sec. 6.2, where periodic solutions are computed explicitly. Consequently, this chapter studies (1.14) in the absence of nonlinear terms (except in Sec. 6.3 where some special situations are considered).

It is useful to denote the first component of a point in phase space, x, by a new variable:

$$s = e_1^{\mathsf{T}} x \, . \tag{6.1}$$

The map under investigation is then

$$x' = f(x;\mu) = \begin{cases} f^{(L)}(x;\mu), & s \le 0 \\ f^{(R)}(x;\mu), & s \ge 0 \end{cases} , \tag{6.2}$$

where

$$f^{(i)}(x;\mu) = \mu b + A_i x \, , \tag{6.3}$$

for $i = L, R$. The vector b and the matrices A_L and A_R (which by continuity of (6.2) differ in only possibly the first column) are assumed to be smooth functions of parameters $\mu \in \mathbb{R}$ and $\eta \in \mathbb{R}^{M-1}$. A border-collision bifurcation occurs at $\mu = 0$.

For fixed $\mu \neq 0$, under variation of η a bifurcation may result when one point of an n-cycle (periodic solution of period n) of (6.2) collides with the switching manifold. Behavior of n-cycles near this bifurcation may be analyzed by studying fixed points of the appropriate n^{th} iterate map of (6.2). As shown in Sec. 6.2, remarkably the nth iterate map will be of the form (6.2) with (6.3). Therefore this bifurcation is a border-collision bifurcation and Feigin's results [di Bernardo *et al.* (1999)] (see Sec. 1.6) may be used to determine whether one n-cycle persists or two n-cycles collide and annihilate in a nonsmooth fold. The map (6.2) may be also applied to border-collision bifurcations of periodic solutions that are not local to a single switching manifold.

Resonance (or Arnold) tongues are regions in parameter space within which there exists a particular stable periodic solution. Since all bounded invariant sets of (6.2) collapse to a point as $\mu \to 0$, in this chapter resonance tongues of (6.2) will be considered within η-space, for fixed $\mu \neq 0$. As was first described in a 1987 paper by Yang and Hao [Yang and Hao (1987)], resonance tongues of piecewise-smooth, continuous maps commonly exhibit a distinctive lens-chain (or sausage) geometry. Alternative nonsmooth scenarios may yield resonance tongues with other interesting geometries, see for instance [Maistrenko *et al.* (1995); Griffiths and Chou (1986)].

Boundaries of resonance tongues near shrinking points correspond to nonsmooth folds of the stable n-cycle. At such a boundary, one point of the n-cycle lies on the switching manifold. Consequently, a shrinking point is defined geometrically by the occurrence of an n-cycle with two points on the switching manifold. However, it will be seen that there are two cases that must be treated separately: terminating and non-terminating shrinking points. Non-terminating shrinking points are defined algebraically by the singularity of certain matrices (the "border collision" matrices). Terminating shrinking points correspond to the case that the n-cycle has only one point in one half plane and are also defined algebraically but by the occurrence of certain multipliers. It will be shown that the algebraic shrinking point conditions imply the geometrical condition of having two points on the switching manifold. Furthermore, at a shrinking point, the map has an invariant polygon (that is typically nonplanar) on which the dynamics are conjugate to a rigid rotation.

Since a shrinking point is defined by two conditions (two points on the switching manifold), the unfolding of a shrinking point bifurcation requires the variation of two parameters. In the generic two-parameter picture, a (non-terminating) shrinking point appears to lie at the intersection of two smooth curves, see Fig. 6.1. However, really there are four distinct curves corresponding to four different nonsmooth folds. Each of the four boundaries is quadratically tangent to one of the others and appears on only one side of the shrinking point. The extension of these boundaries through the shrinking point corresponds to virtual solutions. At each boundary one point of a particular n-cycle lies on the switching manifold. There are curves in parameter space along which each point of this cycle lies on the switching manifold, see Fig. 6.1; however except for the four that form the resonance tongue boundaries, these correspond to virtual solutions.

The remainder of this chapter is organized as follows. Symbolic dynamics are introduced in Sec. 6.1 to describe periodic solutions. The periodic orbit corresponding to a particular symbol sequence may be found by solving a linear system, see Sec. 6.2. Section 6.3 discusses resonance tongue boundaries which correspond to bifurcations of stable periodic solutions. The new concept of a "rotational symbol sequence" is introduced in Sec. 6.4 and using symbolic dynamics, periodic solutions may be classified as either rotational or non-rotational. The motivation is that rotational periodic solutions correspond to lens-chain resonance regions whereas non-rotational periodic solutions may not. In Sec. 6.5 the number of distinct rotational periodic solutions is determined. Formal definitions for terminating and non-terminating shrinking points are provided in Sec. 6.6. This section describes the singular nature of shrinking points and the construction of invariant, nonplanar polygons. This sets up Sec. 6.7 that gives the unfolding of the shrinking point bifurcation.

6.1 Symbolic Dynamics

Consider a map

$$g : \mathcal{M} \to \mathcal{M} \,. \tag{6.4}$$

Sometimes it is not the precise values of points in $\mathcal{M}$ that are of interest, but rather their location relative to one or more key points or lines in $\mathcal{M}$. With this in mind, one may partition the phase space $\mathcal{M}$ into a small handful of physically meaningful regions, $R_1, \ldots, R_r$. Then each point, $x \in \mathcal{M}$, is assigned one of r distinct symbols as determined by the region, R_i, in which

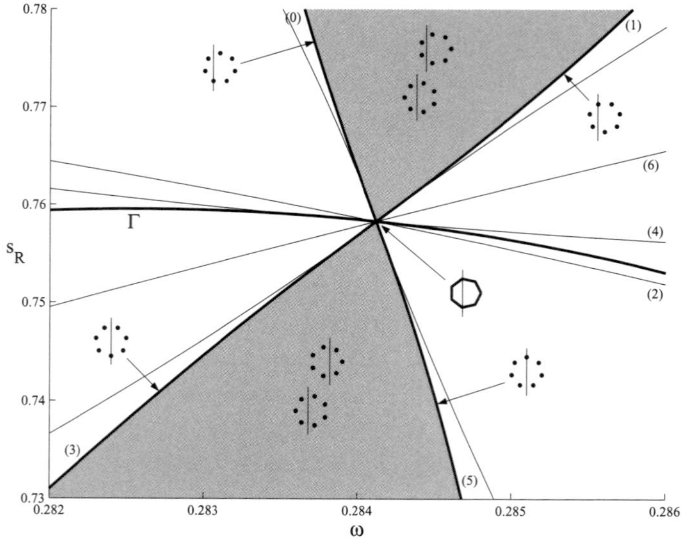

Fig. 6.1: A magnified view of Fig. 7.6 showing part of a resonance tongue corresponding to stable 7-cycles for the two-dimensional, piecewise-linear, continuous map (7.5) investigated in Chapter 7. Sketches of four 7-cycles are shown. The orbit with $l = 3$ (where l denotes the number of points in the left half-plane) is admissible in both gray regions. In the upper region this orbit is stable; in the lower region it is a saddle. Let $\{x_i\}$ denote the points of this orbit. On the curves labeled (i) the point x_i lies on the switching manifold. By Theorem 6.1, the curves (0), (1), (3) and (5) form tongue boundaries. A saddle 7-cycle with $l = 2$ is admissible in the upper region and coincides with $\{x_i\}$ on the boundaries (0) and (1). Similarly, a stable 7-cycle with $l = 4$ is admissible in the lower region and coincides with $\{x_i\}$ on the boundaries (3) and (5). At the shrinking point there is an invariant heptagon. On the curve Γ, the linear solution system (6.9) is singular.

x is located. Thus any forward orbit, $\{x_0, x_1, x_2, x_3, x_4, \ldots\}$ is described by a symbol sequence, e.g. $\{ABCBC\ldots\}$. A study of the symbol sequences of g may lead to a useful understanding of the map. This is known as *symbolic dynamics*.

Symbolic dynamics seems to have been first developed for one-dimensional, unimodal maps (such as the logistic map and the tent map), see [Metropolis *et al.* (1973)]. Points to the left of the single critical point of the map are assigned the symbol L; similarly points to the right are labeled R. The critical point is labeled C. Corresponding symbol sequences are found to obey certain ordering and composition rules [Hao and Zheng (1998)]. For maps that have multiple critical points more symbols are used. Symbol sequences that begin at a critical point are known as *kneading sequences*. The overall dynamical behavior of a map is often well determined by its kneading sequences [Milnor and Thurston (1988)].

Symbolic dynamics are commonly applied to circle maps, i.e. with $\mathcal{M} = \mathbb{S}^1$ in (6.4), see [Hao and Zheng (1998); Dullin *et al.* (2005)]. If g is a homeomorphism, it is useful to assign the symbol L to any $\theta \in \mathbb{S}^1$ for which $g(\theta) > \theta$, and R otherwise. This formulization leads to an association between symbol sequences and rotation numbers and consequently a useful symbolic representation of the Farey tree [Hao and Zheng (1998)].

Piecewise-smooth maps have a natural division of phase space by switching manifolds thus are well-suited to be analyzed with symbolic dynamics. Recent studies of piecewise-smooth, continuous maps that utilize symbolic dynamics include [Zhusubaliyev and Mosekilde (2006b); Sushko *et al.* (2005); Chang and Juang (2004); Belghith (2000)]. Symbolic dynamics applied to piecewise-smooth, discontinuous maps may be found in [Avrutin and Schanz (2008); Fu and Ashwin (2003); Tse *et al.* (2005)].

For the remainder of this book, a *symbol sequence* refers to any sequence, $\mathcal{S}$, that has elements taken from the alphabet, $\{L, R\}$. Symbolic dynamics will be used to describe periodic solutions. For this reason all symbol sequences will be of a finite length, n. The elements of $\mathcal{S}$ are indexed from 0 to $n-1$. Arithmetic on the indices of $\mathcal{S}$ is usually modulo n. For clarity, "mod n" will be omitted where it is clear modulo arithmetic is being used.

Given $n \in \mathbb{N}$, the collection of all symbol sequences of length n is $\{L, R\}^n \equiv \Sigma_2^n$. The $i^{\rm th}$ left cyclic permutation is an operator, $\sigma_i : \Sigma_2^n \to \Sigma_2^n$, defined by $(\sigma_i \mathcal{S})_j = \mathcal{S}_{i+j}$ and the alternative notation, $\mathcal{S}^{(i)} \equiv \sigma_i \mathcal{S}$ will also be used. The $i^{\rm th}$ flip permutation is an operator, $\chi_i : \Sigma_2^n \to \Sigma_2^n$, that flips the $i^{\rm th}$ element of $\mathcal{S}$ (i.e. $L \to R$ and $R \to L$) and leaves all other elements unchanged. The notation, $\mathcal{S}^{\bar{i}} \equiv \chi_i \mathcal{S}$ will be used. For example if $\mathcal{S} = LRLRR$ then $\mathcal{S}^{\bar{3}} = LRLLR$ and $\mathcal{S}^{(2)} = LRRLR$. In general $\mathcal{S}^{\bar{i}(j)} \equiv (\mathcal{S}^{\bar{i}})^{(j)} \neq (\mathcal{S}^{(j)})^{\bar{i}} \equiv \mathcal{S}^{(j)\bar{i}}$ because with the same example $\mathcal{S}^{\bar{3}(2)} = LLRLR$ and $\mathcal{S}^{(2)\bar{3}} = LRRRR$.

The $i^{\rm th}$ multiplication permutation is an operator, $\pi_i : \Sigma_2^n \to \Sigma_2^n$, defined by $(\pi_i \mathcal{S})_j = \mathcal{S}_{ij}$. Notice $(\pi_i \pi_j \mathcal{S})_k = (\pi_j \mathcal{S})_{ik} = \mathcal{S}_{ijk} = (\pi_{ij} \mathcal{S})_k$ thus

$$\pi_i \pi_j = \pi_{ij} \, . \tag{6.5}$$

Consequently π_i is an invertible operator if and only if $\gcd(i, n) = 1$ and the inverse of π_i is $\pi_{i^{-1}}$ (where i^{-1} is the multiplicative inverse of i modulo n). Also notice $(\sigma_i \pi_j \mathcal{S})_k = (\pi_j \mathcal{S})_{i+k} = \mathcal{S}_{ij+jk} = (\sigma_{ij} \mathcal{S})_{jk} = (\pi_j \sigma_{ij} \mathcal{S})_k$. Hence

$$\sigma_i \pi_j = \pi_j \sigma_{ij} \ . \tag{6.6}$$

Let $\mathcal{S}\hat{\mathcal{S}}$ denote the concatenation of the symbol sequences $\mathcal{S}$ and $\hat{\mathcal{S}}$ and let $\mathcal{S}^k \in \Sigma_2^{kn}$ denote the symbol sequence formed by the concatenation of k copies of $\mathcal{S}$. A symbol sequence is called *primitive* if it cannot be written as a power, $\mathcal{S}^k$, for any $k > 1$. It is straight-forward to show that $\mathcal{S}$ is primitive if and only if $\mathcal{S} \neq \mathcal{S}^{(i)}$ for all $i \neq 0$.

6.2 Describing and Locating Periodic Solutions

Each orbit of (6.2) can be coded by a symbol sequence that gives its itinerary relative to the switching manifold. However, instead of defining symbol sequences for orbits, here it preferable to do the reverse. Given a point $x = x_0 \in \mathbb{R}^N$, let x_i denote the ith iterate of x under the maps $f^{(L)}$ and $f^{(R)}$ in the order determined by $\mathcal{S} \in \Sigma_2^n$:

$$x_{i+1} = f^{(\mathcal{S}_i)}(x_i; \mu) \ . \tag{6.7}$$

In general this is different from iterating x under the map (6.2). However, if the sequence $\{x_i\}$ satisfies the *admissibility condition*:

$$\mathcal{S}_i = \begin{cases} L, \text{ whenever } s_i < 0 \\ R, \text{ whenever } s_i > 0 \end{cases} \tag{6.8}$$

for every i, then $\{x_i\}$ coincides with the forward orbit of x under (6.2). When (6.8) holds for every i, $\{x_i\}$ is admissible, otherwise it is virtual.

For a given symbol sequence $\mathcal{S}$, it is of interest to find $x_0 \in \mathbb{R}^N$ such that $x_0 = x_n$, because then $\{x_0, x_1, \ldots, x_{n-1}\}$ is an n-cycle. This particular orbit will be referred to as an $\mathcal{S}$*-cycle*. $\mathcal{S}$-cycles are determined by the linear system

$$\begin{aligned} x_1 &= A_{\mathcal{S}_0} x_0 + \mu b \ , \\ x_2 &= A_{\mathcal{S}_1} x_1 + \mu b \ , \\ &\vdots \\ x_0 &= A_{\mathcal{S}_{n-1}} x_{n-1} + \mu b \ . \end{aligned}$$

Elimination of the points $x_1, \ldots, x_{n-1}$, gives

$$(I - M_{\mathcal{S}}) x_0 = \mu P_{\mathcal{S}} b \ . \tag{6.9}$$

where

$$M_{\mathcal{S}} = A_{\mathcal{S}_{n-1}} \cdots A_{\mathcal{S}_0} \,, \tag{6.10}$$

$$P_{\mathcal{S}} = I + A_{\mathcal{S}_{n-1}} + A_{\mathcal{S}_{n-1}} A_{\mathcal{S}_{n-2}} + \cdots + A_{\mathcal{S}_{n-1}} \cdots A_{\mathcal{S}_1} \,. \tag{6.11}$$

The linear system (6.9) will be referred to as the *n-cycle solution system* of $\mathcal{S}$. If $(I - M_{\mathcal{S}})$ is nonsingular, then (6.9) has the unique solution

$$x_0 = (I - M_{\mathcal{S}})^{-1} P_{\mathcal{S}} b \mu \,. \tag{6.12}$$

(Note, if (6.2) were to include nonlinear terms, then we would have $x_0(\mu) = (I - M_{\mathcal{S}})^{-1} P_{\mathcal{S}} b \Big|_{\mu=0} \mu + O(\mu^2)$. Thus as long as $(I - M_{\mathcal{S}})$ is nonsingular and $e_1^{\mathsf{T}}(I - M_{\mathcal{S}})^{-1} P_{\mathcal{S}} b \neq 0$, the point x_0 is not qualitatively affected by nonlinear terms. If $e_1^{\mathsf{T}}(I - M_{\mathcal{S}^{(i)}})^{-1} P_{\mathcal{S}^{(i)}} b \neq 0$ for all i, as is typically the case, then by Lemma 6.7 nonlinear terms are unimportant to the entire n-cycle.)

Stability of the $\mathcal{S}$-cycle is determined by $M_{\mathcal{S}}$ and for this reason $M_{\mathcal{S}}$ will be called the *stability matrix* of $\mathcal{S}$. In view of Corollary 6.1 below, $P_{\mathcal{S}}$ will be called the *border-collision matrix* of $\mathcal{S}$. Notice $P_{\mathcal{S}}$ is independent of $\mathcal{S}_0$, thus

$$P_{\mathcal{S}} = P_{\mathcal{S}^{\overline{0}}} \,. \tag{6.13}$$

A relationship between $M_{\mathcal{S}}$ and $M_{\mathcal{S}^{\overline{0}}}$ is deduced from the following lemma.

Lemma 6.1. *For any $N \times N$ matrix, X, $XA_R = XA_L + \xi e_1^{\mathsf{T}}$, for some $\xi \in \mathbb{R}^N$.*

Proof. Since A_L and A_R are identical in all but possibly their first columns, we may write $A_R = A_L + \zeta e_1^{\mathsf{T}}$ for some $\zeta \in \mathbb{R}^N$. This proves the result with $\xi = X\zeta$. □

By putting $X = A_{\mathcal{S}_{n-1}} \cdots A_{\mathcal{S}_1}$ into Lemma 6.1 it follows that $M_{\mathcal{S}}$ and $M_{\mathcal{S}^{\overline{0}}}$ differ in only their first column. Consequently, the map that describes the nth iterate of x_0 under either $\mathcal{S}$ or $\mathcal{S}^{\overline{0}}$:

$$x_n = \begin{cases} \mu P_{\mathcal{S}} b + M_{\mathcal{S}} x, & s \le 0 \\ \mu P_{\mathcal{S}} b + M_{\mathcal{S}^{\overline{0}}} x, & s \ge 0 \end{cases} , \tag{6.14}$$

is piecewise-smooth, continuous and has the same form as (6.2). For this reason (6.2) may be used to investigate dynamical behavior local to border-collision bifurcations of periodic solutions.

By (6.12), the first component of the solution to the n-cycle solution system of $\mathcal{S}$ is

$$s_0 = e_1^{\mathsf{T}}(I - M_{\mathcal{S}})^{-1} P_{\mathcal{S}} b \mu \,.$$

If we let

$$\varrho_{\mathcal{S}}^{\mathsf{T}} = e_1^{\mathsf{T}} \mathrm{adj}(I - M_{\mathcal{S}}) \, , \tag{6.15}$$

(where $\mathrm{adj}(X)$ denotes the adjugate of X, see Appendix C), then

$$s_0 = \frac{\varrho_{\mathcal{S}}^{\mathsf{T}} P_{\mathcal{S}} b}{\det(I - M_{\mathcal{S}})} \mu \, . \tag{6.16}$$

An alternative route by which to arrive at (6.16) is to apply the fixed point formula, (1.25), to the piecewise-smooth, continuous map (6.14).

The following two lemmas lead to a quite remarkable third formula for s_0.

Lemma 6.2. *The matrices $P_{\mathcal{S}}(I - A_{\mathcal{S}_0})$ and $(I - M_{\mathcal{S}})$ differ in only possibly their first columns.*

Proof.

$$\begin{aligned}
P_{\mathcal{S}}(I - A_{\mathcal{S}_0}) &= (I + A_{\mathcal{S}_{n-1}} + A_{\mathcal{S}_{n-1}} A_{\mathcal{S}_{n-2}} + \cdots + A_{\mathcal{S}_{n-1}} \ldots A_{\mathcal{S}_1})(I - A_{\mathcal{S}_0}) \\
&\quad \text{expand and group terms differently:} \\
&= I - M_{\mathcal{S}} + (A_{\mathcal{S}_{n-1}} - A_{\mathcal{S}_0}) + A_{\mathcal{S}_{n-1}}(A_{\mathcal{S}_{n-2}} - A_{\mathcal{S}_0}) + \cdots \\
&\qquad + A_{\mathcal{S}_{n-1}} \ldots A_{\mathcal{S}_2}(A_{\mathcal{S}_1} - A_{\mathcal{S}_0}) \\
&\quad \text{apply Lemma 6.1:} \\
&= I - M_{\mathcal{S}} + \xi_{n-1} e_1^{\mathsf{T}} + \xi_{n-2} e_1^{\mathsf{T}} + \cdots + \xi_1 e_1^{\mathsf{T}} \\
&\quad \text{where each } \xi_{\mathrm{i}} \in \mathbb{R}^{\mathrm{N}}, \\
&= I - M_{\mathcal{S}} + \left(\sum_{i=1}^{n-1} \xi_i \right) e_1^{\mathsf{T}} \, .
\end{aligned}$$

□

Lemma 6.3. $\varrho_{\mathcal{S}}^{\mathsf{T}} P_{\mathcal{S}} = \det(P_{\mathcal{S}}) \varrho^{\mathsf{T}}$*, where $\varrho_{\mathcal{S}}^{\mathsf{T}}$ is given by (6.15) and $\varrho^{\mathsf{T}} = \varrho_{\mathrm{map}}^{\mathsf{T}}$ is given by (1.26).*

Proof. By Lemma 6.2 and (C.3) we have

$$\begin{aligned}
e_1^{\mathsf{T}} \mathrm{adj}(P_{\mathcal{S}}(I - A_{\mathcal{S}_0})) &= e_1^{\mathsf{T}} \mathrm{adj}(I - M_{\mathcal{S}}) = \varrho_{\mathcal{S}}^{\mathsf{T}} \\
\Rightarrow \quad e_1^{\mathsf{T}} \mathrm{adj}(I - A_{\mathcal{S}_0}) \mathrm{adj}(P_{\mathcal{S}}) &= \varrho_{\mathcal{S}}^{\mathsf{T}} \quad \text{by (C.2)} \\
\Rightarrow \quad \varrho^{\mathsf{T}} \mathrm{adj}(P_{\mathcal{S}}) P_{\mathcal{S}} &= \varrho_{\mathcal{S}}^{\mathsf{T}} P_{\mathcal{S}} \quad \text{by (1.26)} \\
\Rightarrow \quad \det(P_{\mathcal{S}}) \varrho^{\mathsf{T}} &= \varrho_{\mathcal{S}}^{\mathsf{T}} P_{\mathcal{S}} \quad \text{by (C.1)}
\end{aligned}$$

□

By Lemma 6.3, if $(I - M_{\mathcal{S}})$ is nonsingular, then by (6.16),

$$s_0 = \frac{\det(P_{\mathcal{S}})}{\det(I - M_{\mathcal{S}})} \varrho^{\mathsf{T}} b \mu \,. \tag{6.17}$$

This elegant expression relates s_0 simply in terms of the key quantities $\varrho^{\mathsf{T}} b$ and μ and the determinants of $P_{\mathcal{S}}$ and $(I - M_{\mathcal{S}})$. (Feigin's result concerning the existence of 2-cycles (see Sec. 1.6) can be seen by substituting $\mathcal{S} = LR$ and $\mathcal{S} = RL$ into (6.17).) The next result follows immediately from (6.17).

Corollary 6.1. *Suppose $(I - M_{\mathcal{S}})$ is nonsingular, $\mu \neq 0$ and $\varrho^{\mathsf{T}} b \neq 0$. Then the point x_0, given by (6.12), lies on the switching manifold if and only if $P_{\mathcal{S}}$ is singular.*

The following lemma has many symmetries with Corollary 6.1.

Lemma 6.4. *Suppose $P_{\mathcal{S}}$ is nonsingular, $\mu \neq 0$ and $\varrho^{\mathsf{T}} b \neq 0$. Then the n-cycle solution system (6.9) has a solution if and only if $(I - M_{\mathcal{S}})$ is nonsingular.*

Proof. Clearly if $(I - M_{\mathcal{S}})$ is nonsingular, (6.9) has the unique solution (6.12). To prove the converse, suppose that $(I - M_{\mathcal{S}})$ is singular and for a contradiction suppose (6.9) has a solution, x_0. That is we have

$$(I - M_{\mathcal{S}}) x_0 = \mu P_{\mathcal{S}} b \,.$$

Multiplication of this by $\varrho_{\mathcal{S}}^{\mathsf{T}}$ (6.15) on the left yields

$$\det(I - M_{\mathcal{S}}) s_0 = \mu \varrho^{\mathsf{T}} b \,,$$

where we have also used Lemma 6.3. This provides a contradition because the left hand side of the previous equation is zero, whereas by assumption the right hand side is nonzero. □

An interpretation of Corollary 6.1 and Lemma 6.4 is presented in Table 6.1. The situation $\det(I - M_{\mathcal{S}}) = \det(P_{\mathcal{S}}) = 0$ is generically codimension-two. When appropriate non-degeneracy conditions are satisfied it is equivalent to a shrinking point, see Sec. 6.6.

Lastly, three basic lemmas relating to n-cycle solution systems that will be used in Sec. 6.6 and Sec. 6.7 are stated and proved.

Lemma 6.5. *Suppose x solves the n-cycle solution system (6.9) of $\mathcal{S}$ and $s_i = 0$. Then x also solves the n-cycle solution system of $\mathcal{S}^{\bar{i}}$.*

Proof. By continuity: $A_L x_i = A_R x_i$, hence there is no restriction on the ith element of $\mathcal{S}$. □

Table 6.1: A grid summarizing the nature of solutions to (6.9) when $\mu \neq 0$ and $\varrho^{\mathsf{T}} b \neq 0$ as determined by Corollary 6.1 and Lemma 6.4. $M_{\mathcal{S}}$ is the stability matrix of $\mathcal{S}$, (6.10), and $P_{\mathcal{S}}$ is the border-collision matrix of $\mathcal{S}$, (6.11).

	$\det(I - M_{\mathcal{S}}) \neq 0$	$\det(I - M_{\mathcal{S}}) = 0$
$\det(P_{\mathcal{S}}) \neq 0$	unique solution and $s_0 \neq 0$	no solution
$\det(P_{\mathcal{S}}) = 0$	unique solution and $s_0 = 0$	possibly uncountably many solutions

Lemma 6.6. *Suppose x and $\hat{x}$ solve the n-cycle solution systems of $\mathcal{S}$ and $\mathcal{S}^{\overline{0}}$ respectively. Then* $\det(I - M_{\mathcal{S}})s = \det(I - M_{\mathcal{S}^{\overline{0}}})\hat{s}$.

Proof. By (6.9) and (6.13), we have $(I - M_{\mathcal{S}})x = (I - M_{\mathcal{S}^{\overline{0}}})\hat{x}$. Since $(I - M_{\mathcal{S}})$ and $(I - M_{\mathcal{S}^{\overline{0}}})$ are identical except in the first column, by (C.3), the first row of their adjugates are identical. Multiplication on the left by this row to the previous equation yields the desired result. □

Lemma 6.7. *For any i,* $\det(I - M_{\mathcal{S}^{(i)}}) = \det(I - M_{\mathcal{S}})$.

Proof. Suppose w.l.o.g., $\mathcal{S}_0 = L$. If A_L is nonsingular, then $(I - M_{\mathcal{S}^{(1)}}) = A_L(I - M_{\mathcal{S}})A_L^{-1}$ which verifies the result for $i = 1$. Since nonsingular matrices are dense in the set of all matrices and the determinant of a matrix is a continuous function of its elements, the result for $i = 1$ is also true even when A_L is singular. Repetition of this argument completes the result for any i. □

6.3 Resonance Tongue Boundaries

A resonance tongue is a region in parameter space within which a particular periodic solution exists and is attracting, see Fig. 6.2 for an example. For a smooth map, codimension-one boundaries correspond to familiar smooth bifurcations such as saddle-node and period-doubling bifurcations. For a piecewise-smooth system, tongue boundaries may be border-collision bifurcations. This section looks at the bifurcations that form tongue boundaries for the piecewise-linear, continuous map (6.2). Some discussion concerning the effects of nonlinear terms is also presented.

Let R be a resonance tongue for the map (6.2). Let $\{x_i\}_{i=0}^{n-1}$ be the associated attracting periodic solution and $\mathcal{S}$ its symbol sequence. The

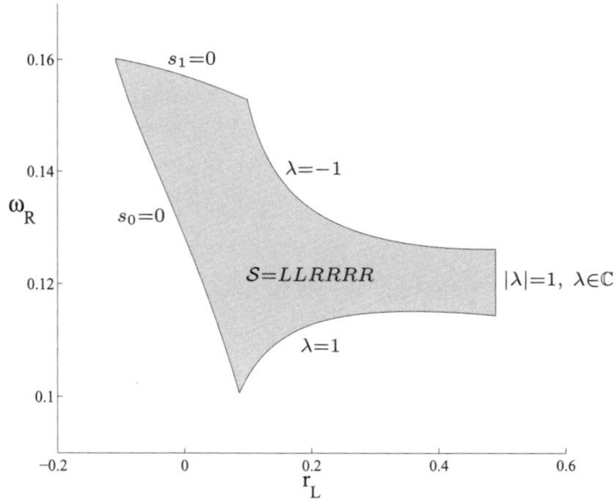

Fig. 6.2: A resonance tongue corresponding to an attracting 6-cycle, $\{x_i\}$, with symbol sequence $\mathcal{S} = LLRRRR$, for the piecewise-linear map (7.5) when $\omega_L = 0.45$, $s_R = 0.7$ and $\mu = 1$. The left-most and upper-most tongue boundaries correspond to a collision of x_0 and x_1 (in the notation of this chapter) with the switching manifold, respectively. Each of the remaining three boundaries corresponds to a multiplier of the associated stability matrix, $M_{\mathcal{S}}$, crossing the unit circle.

following two types of codimension-one boundaries arise commonly:

(I) $s_j = 0$, for exactly one index j.
(II) $M_{\mathcal{S}}$, given by (6.10), has a multiplier on the unit circle.

A third scenario is the simultaneous occurrence of several x_i on the switching manifold. When this occurs in a codimension-one fashion the boundary is a type I boundary of some lower period solution (or a fixed point) which may or not be stable.

Border-collision bifurcation tongue boundaries (type I)

For a tongue boundary of type I, there is an index, j, for which x_j collides with the switching manifold at the boundary. As shown in Sec. 6.2, this bifurcation is a border-collision bifurcation because local dynamics are unfolded by the map (6.14), which, to lowest order, has the same form as the original map (6.2). Consequently, Feigin's results [di Bernardo *et al.* (1999)] (see Sec. 1.6) may be used to classify the bifurcation as either persistence or a nonsmooth fold. The latter case is more frequently encountered and is sometimes called a border-collision fold bifurcation or a border-collision pair bifurcation. In this book it will be referred to as a nonsmooth fold of the periodic solution, or simply a nonsmooth fold.

At a nonsmooth fold of $\{x_i\}$, the periodic solution collides with another periodic solution, say $\{y_i\}$, that has the same period, at the bifurcation. The orbit $\{y_i\}$ will be unstable and its symbol sequence will be $\mathcal{S}^{\bar{j}}$. The two periodic solutions will coexist in R near the tongue boundary. This is reminiscent of resonance tongues of circle maps [Boyland (1986)] where stable and unstable periodic solutions coexist throughout tongues and collide and annihilate in saddle-node bifurcations along tongue boundaries.

Fig. 6.3 illustrates the other type of border-collision bifurcation: persistence, for a 9-cycle. Note that in both panels only three of the nine points of the periodic solution are shown. Note also that the stable periodic solutions are non-rotational and are discussed more in Sec. 7.4. In panel (a), two distinct, stable 9-cycles are admissible on opposite sides of the bifurcation. In panel (b), a stable 18-cycle is created. The effect of period-doubling is not due to a multiplier of -1, but is rather an artifact of the particular border-collision bifurcation. In general, any solutions with

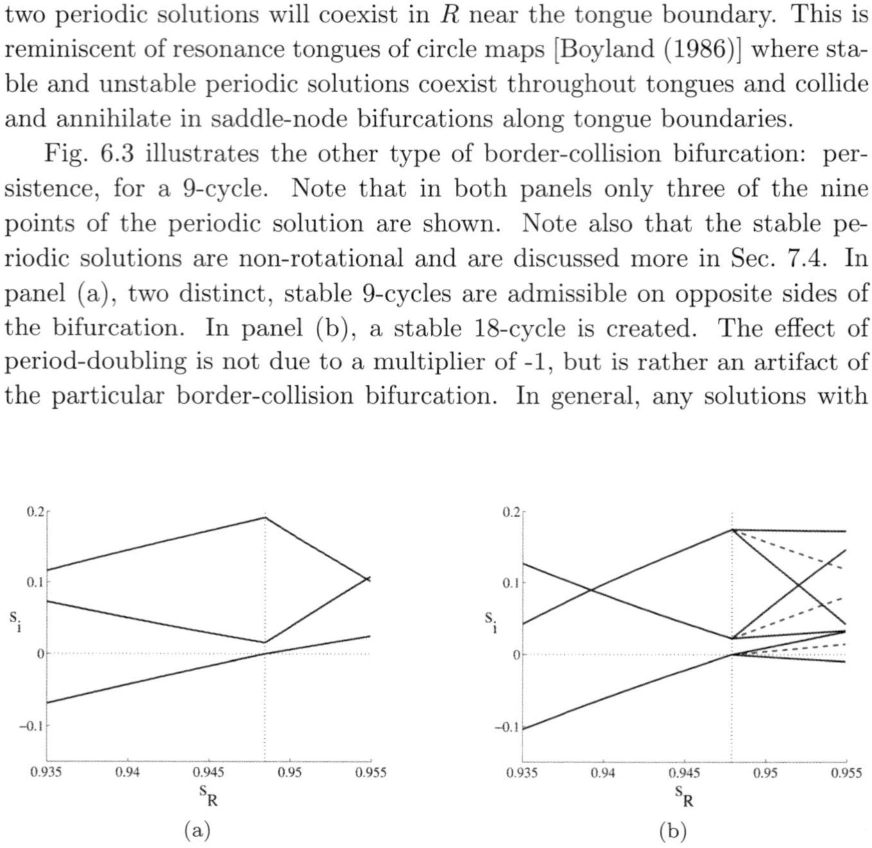

Fig. 6.3: Bifurcation diagrams showing persistence of a 9-cycle of the piecewise-linear map (7.5) when $r_L = 0.475$, $\omega_L = 0.09$, $\mu = 1$ and $\omega_R = 0.38$ in panel (a), and $\omega_R = 0.375$ in panel (b). For all values of s_R shown there exists a 9-cycle; drawn solid when stable and dashed when unstable. For clarity, only the three points of the 9-cycle closest to the switching manifold are shown. In panel (b) for s_R greater than approximately 0.9479 there exists a stable 18-cycle. Associated resonance tongues are shown in Fig. 7.16.

a period that is a multiple of the period of $\{x_i\}$, in this case nine, may conceivably be created.

Except in special cases, the inclusion of nonlinear terms to the map will not affect local dynamics near a border-collision bifurcation tongue boundary. This is because the nth-iterate map, (6.14), has the same piecewise form as (1.14) and the discussion concerning nonlinear terms and structurally stable dynamics given in Sec. 1.2 applies. In contrast, nonlinear terms are important to boundaries of the second type.

Stability loss tongue boundaries (type II)

The second tongue boundary type listed above corresponds to a loss of stability of $\{x_i\}$. At such a boundary, one or more multipliers of the stability matrix, $M_{\mathcal{S}}$ (6.10), escape the unit circle. In codimension-one situations this may happen in three ways: a real multiplier may pass through 1 or -1 or a complex conjugate multiplier pair may cross the unit circle.

Interestingly, for the piecewise-linear map (6.2), the first case does not resemble a saddle-node bifurcation in a smooth map. This is because if 1 is a multiplier of $M_{\mathcal{S}}$, then $(I - M_{\mathcal{S}})$ is singular. Therefore by Lemma 6.4, the n-cycle solution system of $\mathcal{S}$, (6.9), has no solution (provided the non-degeneracy assumptions of the lemma are satisfied). Consequently,

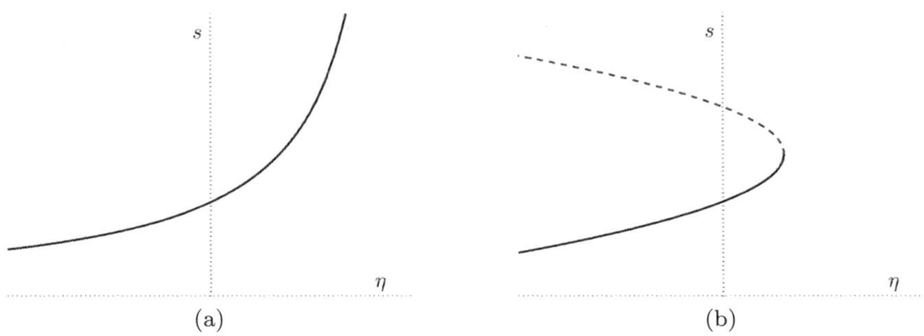

Fig. 6.4: Schematic bifurcation diagrams showing the s-component of one point of an $\mathcal{S}$-cycle near a parameter value at which $M_{\mathcal{S}}$ has a multiplier of 1. Panel (a) is a sketch for the piecewise-linear map (6.2). Here the amplitude of the $\mathcal{S}$-cycle approaches infinity as the critical multiplier of $M_{\mathcal{S}}$ approaches 1. Panel (b) is a diagram for a piecewise-smooth map with the same linearization as that for panel (a). Here a saddle-node bifurcation occurs as a result of the presence of nonlinear terms.

as a multiplier of $M_{\mathcal{S}}$ approaches 1, the $\mathcal{S}$-cycle becomes unbounded, see Fig. 6.4(a). Of course this dynamical behavior is somewhat artificial: a result of the absence of nonlinear terms in (6.2). With the addition of nonlinear terms the bifurcation diagram may appear like that shown in Fig. 6.4(b). As $\mu \to 0$, the value of η at which the saddle-node bifurcation occurs will approach the bifurcation value in panel (a).

The case that a multiplier of $M_{\mathcal{S}}$ passes through -1 is now described. Since the resulting bifurcation appears often in Chapter 7, it is useful to assign it a name: *nonsmooth period-doubling bifurcation.*

Assume $(I - M_{\mathcal{S}})$ is nonsingular and let x_0 be given by (6.12). Further, assume that the $\mathcal{S}$-cycle is admissible and has no points that lie exactly on the switching manifold. Then for any point w_0 near x_0, the first few iterates of w_0 will follow the sequence $\mathcal{S}$. If w_0 is sufficiently close to x_0, the first n iterates of w_0 will follow the sequence $\mathcal{S}$ and consequently $w_n = \mu P_{\mathcal{S}} b + M_{\mathcal{S}} w_0$.

Near the bifurcation, $M_{\mathcal{S}}$ has a multiplier near -1, call it λ_{PD}. Let $v_{PD}^{(0)}$ denote the eigenvector of $M_{\mathcal{S}}$ at x_0 associated with λ_{PD}. Since $v_{PD}^{(0)}$ corresponds to the dynamically slow direction of the periodic solution, and since we are assuming the periodic solution is stable on one side of the bifurcation, the remaining multipliers lie inside the unit circle and orbits of (6.2) that start sufficiently close to the periodic orbit approach the image vectors $v_{PD}^{(i)}$ associated with λ_{PD} for each x_i. This is illustrated in Fig. 6.5.

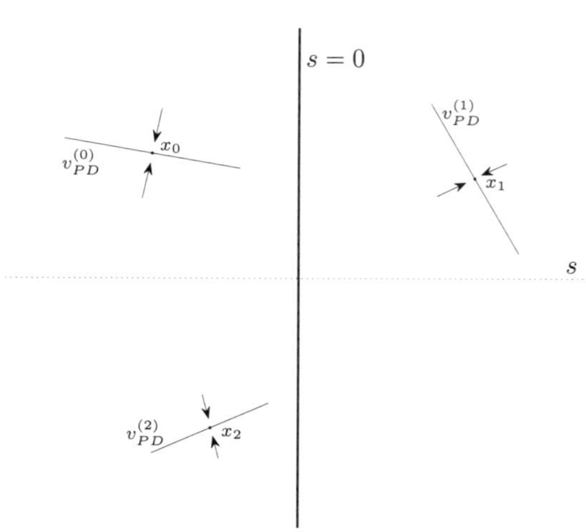

Fig. 6.5: A schematic showing a 3-cycle and its associated slow eigenvectors near a nonsmooth period-doubling bifurcation.

In the absence of nonlinear terms, when $\lambda_{PD} = -1$ points on the slow eigenvectors that are sufficiently close to the n-cycle return to themselves after $2n$ iterations; therefore, there exists a segment of $2n$-cycles with symbol sequence $\mathcal{S}^2$ (the concatenation of $\mathcal{S}$ with itself). This family of $2n$-cycles has a "first" point of intersection with the switching manifold. Let $q_\delta = x_0 + \delta v_{PD}^{(0)}$, then for $\lambda_{PD} = -1$ the orbits of q_δ have the symbol sequence $\mathcal{S}^2$ providing $\delta \in [0, \delta_{\max}]$ for some $\delta_{\max} > 0$. Moreover, a single point on the $2n$-cycle will generically touch the switching manifold at $\delta_{\max}$. Without loss of generality, it may be assumed that a permutation of the orbit is selected so that the point that hits the switching manifold first is $q_{\delta_{\max}}$. Thus, when $\lambda_{PD} = -1$ there is a $2n$-cycle with symbol sequence $\mathcal{S}^2$ that has its first point on the switching manifold.

Since the first symbol is ambiguous, we can also declare that this orbit has the symbol sequence $(\mathcal{S}^2)^{\overline{0}}$. The first component of this solution will typically change and will be admissible for one sign of $\lambda_{PD} + 1$. This $2n$-cycle will coexist with either the $\mathcal{S}$-cycle when it is unstable as in Fig. 6.6(a), or when it is stable. A major difference between the nonsmooth period-doubling bifurcation and its smooth counterpart is that the period-doubled solution appears non-local to $\{x_i\}$.

With the addition of nonlinear terms the bifurcation of Fig. 6.6(a) may appear like in panel (b). Here a smooth period-doubling bifurcation occurs. The generated period-doubled solution quickly collides with the switching manifold in a border-collision bifurcation. Period-doubling-like behavior

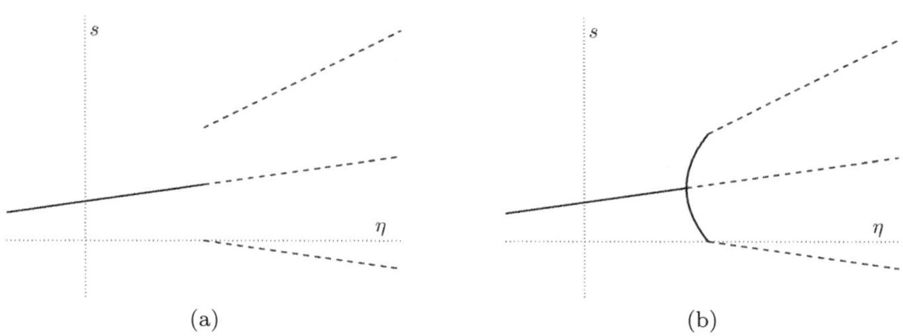

Fig. 6.6: Schematic bifurcation diagrams showing one point of an $\mathcal{S}$-cycle near where $M_{\mathcal{S}}$ has a multiplier of -1. Panel (a) corresponds to the piecewise-linear map (6.2). The inclusion of nonlinear terms will lead to a period-doubling bifurcation and possibly a bifurcation diagram like that shown in panel (b).

has also been observed in piecewise-smooth, discontinuous maps [Avrutin and Schanz (2005b)].

A detailed study of the case that $M_{\mathcal{S}}$ has complex multipliers that cross the unit circle remains for future work. Some numerical results regarding this situation are presented towards the end of Sec. 7.4. It is anticipated that the bifurcation will have many similarities with the nonsmooth period-doubling bifurcation described above. Additional complications arise from the effects of strong and weak resonances.

6.4 Rotational Symbol Sequences

Of particular interest is the situation that the map (6.2) exhibits an invariant, topological circle that crosses the switching manifold at two points. When the restriction of the map to this circle is homeomorphic to a monotone increasing circle map, periodic solutions on the circle can only have certain symbol sequences. Here such sequences are given the name rotational symbol sequences. They differ from, but are related to, *Sturmian sequences* [Morse and Hedlund (1940); Fogg (2002)], see also [Siegel *et al.* (1992)].

Definition 6.1. Let $l, m, n \in \mathbb{N}$, with $l, m < n$ and $\gcd(m, n) = 1$. Let $\mathcal{S} = \mathcal{S}[l, m, n]$ be the symbol sequence of length n defined by

$$\mathcal{S}_{id} = \begin{cases} L, & i = 0, \ldots, l-1 \\ R, & i = l, \ldots, n-1 \end{cases} \tag{6.18}$$

where d is the multiplicative inverse of m modulo n, i.e. $dm = 1 \bmod n$. Then $\mathcal{S}$, and any cyclic permutation of $\mathcal{S}$, is said to be a *rotational symbol sequence.*

Notice d always exists and is unique because m/n is an irreducible fraction [Gallian (1998)]. For example if $(l, m, n) = (3, 2, 7)$ then $d = 4$, hence $\mathcal{S}_0 = \mathcal{S}_4 = \mathcal{S}_{8 \bmod 7 = 1} = L$, thus $\mathcal{S}[3, 2, 7] = LLRRLRR$. Alternatively d may be computed via the Farey tree [Lagarias and Tresser (1995)]. If m_1/n_1 and m_2/n_2 are the Farey neighbors of m/n ($m_1 + m_2 = m$ and $n_1 + n_2 = n$), then $mn_i \bmod n = \pm 1$, for $i = 1, 2$, thus d is either n_1 or n_2.

A pictorial method for computing $\mathcal{S}$ in terms of l, m and n is to select n points on a circle, then draw a vertical line through the circle such that l points lie to the left of the line, see Fig. 6.7. Label the first point to the left of the lower intersection of the circle and line, point 0. Move m

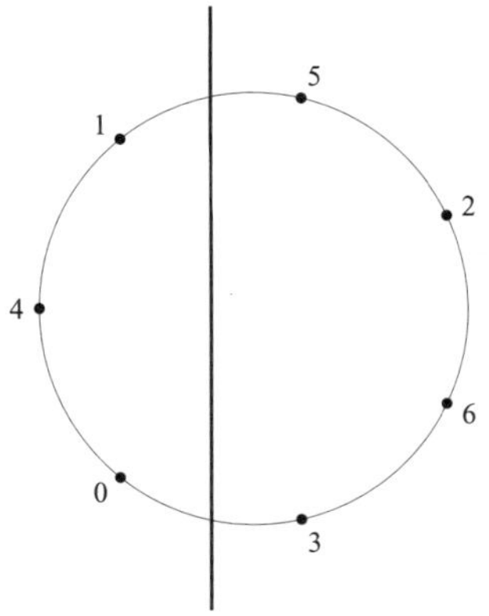

Fig. 6.7: An illustration of the pictorial method for determining $\mathcal{S}[3,2,7] = LLRRLRR$.

points clockwise from 0 and label this point 1. Continue stepping clockwise labeling every m^{th} point with a number that is one greater than the previous number until all points are labeled. Then $\mathcal{S}_i = L$ if the point i lies to the left of the vertical line and R otherwise.

The circle represents an invariant circle of (6.2) and the vertical line represents the switching manifold. Taking m steps clockwise corresponds to evaluating the map (6.2) once. Thus $\mathcal{S}[l,m,n]$ is the symbol sequence of an n-cycle of (6.2) that has l points to the left of the switching manifold and with *rotation number* m/n.

The following lemma states some basic properties of rotational symbol sequences.

Lemma 6.8. *Let $\mathcal{S}[l,m,n]$ be a rotational symbol sequence.*

- *(a)* $\mathcal{S}[l,m,n] = \pi_m \mathcal{S}[l,1,n]$.
- *(b) If* $\mathcal{S}[l,m,n]^{(i)}_0 = L$ *and* $\mathcal{S}[l,m,n]^{(i)}_{-d} = R$ *then* $i = 0$.
- *(c)* $\mathcal{S}[l,m,n]$ *is primitive.*
- *(d)* $\mathcal{S}[l,m,n]^{((l-1)d)} = \mathcal{S}[l,n-m,n]$.
- *(e)* $\mathcal{S}[l,m,n]^{((l-1)d)\overline{0}} = \mathcal{S}[l,m,n]^{\overline{0}(ld)}$.
- *(f) If* $0 < m_1, m_2 < \frac{n}{2}$ *are distinct integers coprime to* n *and* $l \neq 1, n-1$, *then* $\mathcal{S}[l,m_1,n]$ *is not a cyclic permutation of* $\mathcal{S}[l,m_2,n]$.

Proof. Recall the notation $\sigma_i \mathcal{S} = \mathcal{S}^{(i)}$, for the ith left cyclic permutation and $\chi_i \mathcal{S} = \mathcal{S}^{\bar{i}}$ for the flip of the ith element.

(a) Since $dm = 1$,

$$(\pi_m \mathcal{S}[l,1,n])_{id} = \mathcal{S}[l,1,n]_i = L \text{ if and only if } i = 0,\ldots,l-1 \,,$$

matching the definition of $\mathcal{S}[l,m,n]$, (6.18).

(b) Let $j = im$ (equivalently, $i = jd$). Then $\mathcal{S}[l,m,n]_0^{(i)} = \mathcal{S}[l,m,n]_{jd} = L$ implies $0 \le j \le l-1$. Similarly, $\mathcal{S}[l,m,n]_{-d}^{(i)} = \mathcal{S}[l,m,n]_{(j-1)d} = R$ implies $l \le j-1 \le n-1$. The only value of j that satisfies both inequalities is zero, hence $i = 0$.

(c) From part (b), $\mathcal{S}[l,m,n]$ differs from any non-trivial cyclic permutation of itself in either the $0^{\rm th}$ or the $(-d)^{\rm th}$ element. Therefore, $\mathcal{S}[l,m,n]$ is primitive.

(d) By definition,

$$\mathcal{S}[l,m,n]_{id} = L \text{ if and only if } i = 0,\ldots,l-1 \,, \quad (6.19)$$

$$\text{and } \mathcal{S}[l,n-m,n]_{i(n-d)} = L \text{ if and only if } i = 0,\ldots,l-1 \,, \quad (6.20)$$

since the multiplicative inverse of $(n-m)$ is $(n-d)$. From (6.19), by letting $j = i-l+1$ we obtain

$$\mathcal{S}[l,m,n]_{jd}^{((l-1)d)} = L \text{ if and only if } j = -l+1,\ldots,0 \,, \quad (6.21)$$

$$\text{thus } \mathcal{S}[l,m,n]_{j(n-d)}^{((l-1)d)} = L \text{ if and only if } j = 0,\ldots,l-1 \,,$$

which matches (6.20).

(e) From (6.21) we obtain

$$\mathcal{S}[l,m,n]_{id}^{((l-1)d)\overline{0}} = L \quad \text{if and only if} \quad i = -l+1,\ldots,-1 \,. \quad (6.22)$$

Also

$$\mathcal{S}[l,m,n]_{id}^{\overline{0}} = L \text{ if and only if } i = 1,\ldots,l-1 \,,$$

$$\text{thus } \mathcal{S}[l,m,n]_{jd}^{\overline{0}(ld)} = L \text{ if and only if } j = -l+1,\ldots,-1 \,, \quad (6.23)$$

where we have set $j = i-l$. Equation (6.23) matches (6.22) proving the result.

(f) Let d_1 and d_2 denote the multiplicative inverses of m_1 and m_2 modulo n, respectively. Let

$$\check{\mathcal{S}} = \pi_{d_1} \mathcal{S}[l,m_1,n] \,,$$
$$\hat{\mathcal{S}} = \pi_{d_1} (\mathcal{S}[l,m_2,n]^{(k)}) \,,$$

where $k \in \mathbb{Z}$. Since π_{d_1} is an invertible operator, it remains to show that $\check{\mathcal{S}} \ne \hat{\mathcal{S}}$ for any $k \in \mathbb{Z}$. Using (6.5) and part (a) we find

$$\check{\mathcal{S}} = \mathcal{S}[l,1,n] \,. \quad (6.24)$$

Re-expressing $\hat{\mathcal{S}}$ in the form we desire is a little more complicated but requires no more that the basic known properties concerning multiplicative permutations, π.

$$\begin{aligned}\hat{\mathcal{S}} &= (\pi_{d_1}\mathcal{S}[l,m_2,n])^{(km_1)} , \quad \text{by (6.6)}\\ &= (\pi_{d_1,n}\pi_{m_2,n}\mathcal{S}[l,1,n])^{(km_1)} , \quad \text{by part (a)}\\ &= (\pi_{\hat{m},n}\mathcal{S}[l,1,n])^{(km_1)} , \quad \text{where } \hat{m} = d_1m_2 , \text{ by (6.5)}\\ &= \mathcal{S}[l,\hat{m},n]^{(km_1)} , \quad \text{by part (a).} \end{aligned} \tag{6.25}$$

Notice $\hat{m} \neq 1$ since $m_1 \neq m_2$. Also $\hat{m} \neq n-1$ since, otherwise, $d_1m_2 = -1 \Rightarrow (n-d_1)m_2 = 1 \Rightarrow m_2 = n - m_1$ (since the multiplicative inverse of $(n-d_1)$ is $(n-m_1)$), thus $m_1 + m_2 = n$ which is a contradiction since $m_1, m_2 < \frac{n}{2}$ by assumption.

Using (6.25), by the definition of a rotational symbol sequence (6.18), if $\hat{\mathcal{S}}_i = L$, then $\hat{\mathcal{S}}_{i+\hat{d}} = L$ for all but one value of $i \in \{0,\ldots,n-1\}$ (where $\hat{d}\hat{m} = 1$). We now show this property of $\hat{\mathcal{S}}$ is not exhibited by $\check{\mathcal{S}}$ and hence $\check{\mathcal{S}}$ and $\hat{\mathcal{S}}$ cannot be equal.

Using (6.24) and remembering $\hat{d}, l \neq 1, n-1$, if $\hat{d} \leq l$, then $\check{\mathcal{S}}_{l-\hat{d}} = \check{\mathcal{S}}_{l-\hat{d}+1} = L$ and $\check{\mathcal{S}}_l = \check{\mathcal{S}}_{l+1} = R$. Similarly if $\hat{d} > l$, we have $\check{\mathcal{S}}_0 = \check{\mathcal{S}}_1 = L$ and $\check{\mathcal{S}}_{\hat{d}} = \check{\mathcal{S}}_{\hat{d}+1} = R$. In either case

$$\check{\mathcal{S}} \neq \hat{\mathcal{S}} ,$$

for any $k \in \mathbb{Z}$. □

6.5 Cardinality of Symbol Sequences

Let $\mathcal{N}_n$ $[\mathcal{N}_n^{\mathrm{rot}}]$ denote the number of primitive symbol sequences [primitive rotational symbol sequences] of length n that are distinct up to cyclic permutation. To use combinatorics terminology, $\mathcal{N}_n$ is the number of n-bead necklaces of two colors with primitive period n, and it is the number of binary *Lyndon words* of length n [Lothaire (1983); Charalambides (2002); van Lint and Wilson (2001)]. The formulas below for $\mathcal{N}_n$ and $\mathcal{N}_n^{\mathrm{rot}}$ use the Möbius function:

$$\mu(n) = \begin{cases} 1 & \text{if } n = 1 \text{ or } n \text{ is the product of an even number of primes,} \\ -1 & \text{if } n \text{ is the product of an odd number of primes,} \\ 0 & \text{otherwise, i.e. } n \text{ is not square free,} \end{cases}$$

and Euler's totient function:

$\varphi(n) =$ the number of positive integers less than n and coprime to n, where $\varphi(1) = 1$. The first few values of $\mu(n)$ and $\varphi(n)$ are:

n	1	2	3	4	5	6	7	8	9	10	11	12
$\mu(n)$	1	-1	-1	0	-1	1	-1	0	0	1	-1	0
$\varphi(n)$	1	1	2	2	4	2	6	4	6	4	10	4

Then

$$\mathcal{N}_n = \frac{1}{n} \sum_{a|n} \mu(\frac{n}{a}) 2^a \ , \tag{6.26}$$

$$\mathcal{N}_n^{\rm rot} = 2 + \frac{n-3}{2} \varphi(n) \ , \quad \text{for } n \geq 3 \ , \tag{6.27}$$

where $\sum_{a|n}$ denotes summation over all divisors, a, of n. The first formula, (6.26), is well-known (see for instance [Lothaire (1983); Hao and Zheng (1998)] for a derivation). To arrive at (6.27) we count the number of distinct (up to cyclic permutation) $\mathcal{S}[l,m,n]$ for an arbitrary fixed n: When $l = 1$ there is one distinct rotational symbol sequence, namely LR^{n-1}. Similarly when $l = n-1$ there is only $L^{n-1}R$. There are $(n-3)$ values of l left to consider. For each of these, there are $\varphi(n)$ possible values for m. By Lemma 6.8f, the $\frac{\varphi(n)}{2}$ values of m that are less than $\frac{n}{2}$ yield distinct symbol sequences and by Lemma 6.8d, the remaining values of m only produce cyclic permutations. Thus we have (6.27).

The first few values of $\mathcal{N}_n$ and $\mathcal{N}_n^{\rm rot}$ are:

n	1	2	3	4	5	6	7	8	9	10	11	12
$\mathcal{N}_n$	2	1	2	3	6	9	18	30	56	99	186	335
$\mathcal{N}_n^{\rm rot}$	0	1	2	3	6	5	14	12	20	16	42	20

All primitive symbol sequences of length $n < 6$ are rotational symbol sequences except when $n = 1$ because L and R are not considered to be rotational. There are four distinct primitive symbol sequences of length six that are not rotational. These are: $LRLRRR$, $LLRLRR$, $LLRRLR$ and $LLLRLR$. Roughly, as n increases the number of distinct primitive symbol sequences of length n that are not rotational also increases. To quantify this statement, $\mathcal{N}_n$ (6.26), grows like e^n, whereas $\mathcal{N}_n^{\rm rot}$ (6.27), grows like n^2 (since $\varphi(n)$ grows linearly ($\varphi(p) = p - 1$ for any prime p)). Thus for large n the majority of primitive symbol sequences are not rotational.

6.6 Shrinking Points

Roughly speaking, as in [Yang and Hao (1987)], this book refers to points where lens-chain-shaped resonance tongues have zero width, shrinking points. The aim of this section is to provide a rigorous foundation for the unfolding of shrinking points described in Sec. 6.7. Two classes of shrinking points are defined: terminating and non-terminating. This categorization provides a distinction between shrinking points that lie at the end of a lens-chain (terminating) and those that lie in the middle (non-terminating), see Chapter 7. In this section it is assumed that associated symbol sequences are rotational; this assumption is crucial to the analysis. The main results are Lemma 6.9, which states that at a shrinking point there exists an invariant, nonplanar (though planar in special cases) polygon, and Corollary 6.2, which shows that shrinking points are a hub for the singularity of important matrices.

Near a shrinking point, a lens-shaped resonance tongue corresponding to the existence of an admissible, stable n-cycle, $\{x_i\}$, has two boundaries. These correspond to nonsmooth folds of $\{x_i\}$ with an unstable orbit of the same period. At each boundary, one point on the orbit lies on the switching manifold. The first question to address is the following: which points lie on the switching manifold at the two resonance tongue boundaries?

Suppose $\{x_i\}$ has an associated symbol sequence that is rotational, $\mathcal{S}[l, m, n]$. We may picture the points, x_i, lying on a topological circle, as in Fig. 6.7. The switching manifold intersects the circle at two points. If this structure is maintained as parameters vary, then it seems reasonable that only points that lie adjacent to an intersection can collide with the switching manifold. When $l \neq 1, n-1$, there are four such points: x_0, x_{-d}, $x_{(l-1)d}$ and x_{ld}. (For example, if $[l, m, n] = [3, 2, 7]$, as in Fig. 6.7, then $d = 4$ and the four adjacent points are x_0, x_3, x_1 and x_5.)

Suppose w.l.o.g. that x_0 lies on the switching manifold at one resonance tongue boundary. In the interior of the tongue the symbol sequence of the corresponding unstable periodic solution then differs from $\mathcal{S}$ in the 0th element, i.e., is $\mathcal{S}^{\overline{0}}$. If we assume that $\{x_i\}$ collides and annihilates with the same unstable periodic solution on the second boundary, it must be $x_{(l-1)d}$ that lies on the switching manifold there because the only index $i \neq 0$ for which $\mathcal{S}^{\overline{id}}$ is a cyclic permutation of $\mathcal{S}^{\overline{0}}$, is $i = l-1$. Consequently, in view of Corollary 6.1, since $\{x_i\}$ is not always well-defined, non-terminating shrinking points will be defined by the singularity of $P_{\mathcal{S}}$ and $P_{\mathcal{S}^{((l-1)d)}}$ (Definition 6.2 below). In Sec. 6.7 it will be shown that resonance tongue boundaries at which the remaining two points, x_{-d} and x_{ld}, lie on the switching manifold, form a second lens-shaped resonance tongue emanating from the shrinking point.

As we will see, the cases $l = 1$ and $l = n - 1$ correspond to terminating shrinking points. It suffices to consider only one of these cases, we choose $l = n - 1$, because they are interchangeable via swapping L and R. For the remainder of this chapter it will be assumed that $l \neq 1$ which greatly simplifies the analysis. A consequence of this is that it is always reasonable to assume that the unstable periodic solution described above always exists (though it may not be admissible). That is $(I - M_{\mathcal{S}^{\overline{0}}})$ is nonsingular and therefore the n-cycle solution system, (6.9), of $\mathcal{S}^{\overline{0}}$ has the unique solution

$$p = p_0 = \mu(I - M_{\mathcal{S}^{\overline{0}}})^{-1} P_{\mathcal{S}^{\overline{0}}} b \,. \tag{6.28}$$

The $i^{\rm th}$ iterate of p via the symbol sequence $\mathcal{S}^{\overline{0}}$ will be denoted by p_i and t_i will denote its first component. The orbit, $\{p_i\}$, will play a pivotal role in the analysis.

Definition 6.2. Consider the map (6.2) with $N \geq 2$ and suppose that $\mu \neq 0$ and $\varrho^{\mathsf{T}} b \neq 0$. Let $\mathcal{S} = \mathcal{S}[l, m, n]$ be a rotational symbol sequence with $1 < l < n - 1$. Suppose

$$P_{\mathcal{S}} \text{ and } P_{\mathcal{S}^{((l-1)d)}} \text{ are singular.} \qquad \text{(the singularity condition)}$$

Let $\check{\mathcal{S}} = \mathcal{S}^{\overline{0}}$ and $\hat{\mathcal{S}} = \mathcal{S}^{\overline{ld}}$ and assume $(I - M_{\check{\mathcal{S}}})$ and $(I - M_{\hat{\mathcal{S}}})$ are nonsingular. Suppose the orbit, $\{p_i\}$, of (6.28), is admissible. Then (6.2) is said to be at a *non-terminating shrinking point.*

Definition 6.2 essentially characterizes non-terminating shrinking points as the codimension-two phenomenon at which the border-collision matrices $P_{\mathcal{S}}$ and $P_{\mathcal{S}^{((l-1)d)}}$ are simultaneously singular. The following definition for terminating shrinking points is quite different because the defining characteristic of these points is the codimension-two requirement that one fixed point (here $x^{*(L)}$ because $l = n - 1$ by assumption) is admissible and has a pair of associated multipliers on the unit circle with a particular rational angular frequency.

Definition 6.3. Consider the map (6.2) with $N \geq 2$, suppose $(I - A_L)$ is nonsingular and $s^{*(L)} = \frac{\mu \varrho^{\mathsf{T}} b}{\det(I - A_L)} < 0$ (i.e. the fixed point $x^{*(L)}$ is admissible, see (1.25)). Let $\mathcal{S} = \mathcal{S}[l, m, n]$ be a rotational symbol sequence with $l = n - 1$ and $n \geq 3$. Let $\check{\mathcal{S}} = \mathcal{S}^{\overline{0}}$ and suppose $(I - M_{\check{\mathcal{S}}})$ is nonsingular. Suppose

$$\mathrm{e}^{\pm \frac{2\pi i m}{n}} \text{ are multipliers of } A_L \,. \qquad \text{(the singularity condition)}$$

Then we say (6.2) is at a *terminating shrinking point.*

In two dimensions, terminating shrinking points are center bifurcations for rational rotation numbers, m/n, studied in [Sushko and Gardini (2008, 2006)]. These authors show there exists an invariant polygon that has one side on the switching manifold, within which all points other than the fixed point belong to periodic orbits with rotation number, m/n. As shown below, in higher dimensions this behavior occurs on the center manifold of $x^{*(L)}$ corresponding to the multipliers $\mathrm{e}^{\pm\frac{2\pi im}{n}}$, call it E_c. It will shown that E_c is two-dimensional and must intersect the switching manifold.

The following lemma concerns the orbit, $\{p_i\}$, of (6.28). The reader should take care to notice that singularity of the matrix $P_{\mathcal{S}^{((l-1)d)}}$, by Corollary 6.1, implies that p_{ld} (not $p_{(l-1)d}$) lies on the switching manifold because

$$P_{\check{\mathcal{S}}^{(ld)}} = P_{\mathcal{S}^{\overline{0}(ld)}} = P_{\mathcal{S}^{((l-1)d)\overline{0}}} = P_{\mathcal{S}^{((l-1)d)}} \;, \tag{6.29}$$

where the second equality is Lemma 6.8e and (6.13) is used for the last equality.

Lemma 6.9. *Suppose (6.2) is at a shrinking point. Then,*

(a) $t = t_{ld} = 0$;
(b) if the shrinking point is non-terminating, then $t_d, t_{(l-1)d} < 0$ *and* $t_{(l+1)d}, t_{-d} > 0$ *as in Fig. 6.8; if the shrinking point is terminating, then* $t_{id} < 0$ *for all* $i \neq 0, -1$;
(c) $\{p_i\}$ *has period* n;
(d) the iterates of p *are the vertices of an invariant, nonplanar* n*-gon,* $\mathcal{P}$*, that is comprised of uncountably many* $\mathcal{S}$*-cycles and the restriction of (6.2) to* $\mathcal{P}$ *is homeomorphic to rigid rotation with rotation number,* m/n.

Proofs of Lemma 6.9 for non-terminating and terminating shrinking points are given in Appendix A. If the shrinking point is terminating, then the polygon $\mathcal{P}$ is planar. In general, when $N > 2$, $\mathcal{P}$ is nonplanar as in Fig. 6.9. In any case $\mathcal{P}$ has two vertices on the switching manifold.

$\mathcal{P}$ is comprised of many $\mathcal{S}$-cycles, in other words the n-cycle solution system of $\mathcal{S}$ has more than one distinct solution. Therefore at a shrinking point the matrix $(I - M_{\mathcal{S}})$ is singular. Furthermore, for each i, by Lemma 6.7, $(I - M_{\mathcal{S}^{(i)}})$ is singular. However, there exist solutions to the n-cycle solution system of $\mathcal{S}^{(i)}$ (such as p_i), so to avoid a contradiction with Lemma 6.4, $P_{\mathcal{S}^{(i)}}$ must be singular. Consequently we have the following:

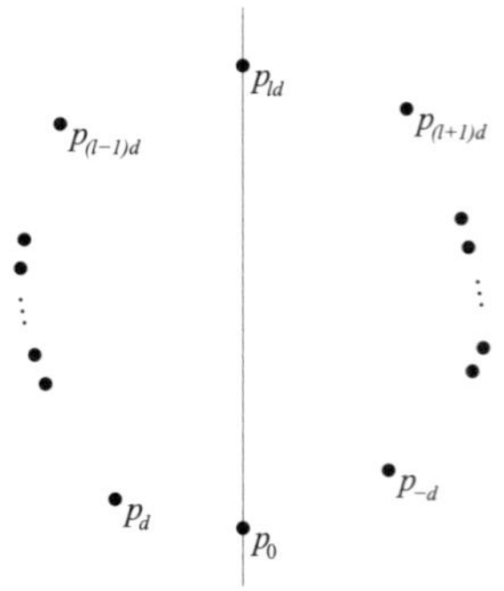

Fig. 6.8: A schematic illustrating iterates of p, (6.28), on or near the switching manifold at a non-terminating shrinking point.

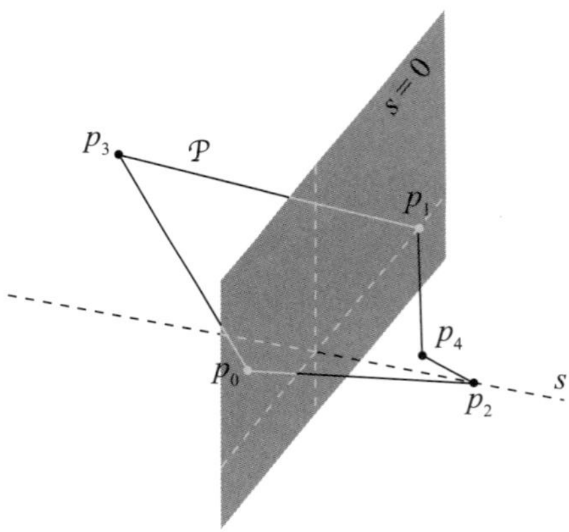

Fig. 6.9: The invariant, nonplanar polygon, $\mathcal{P}$, at a non-terminating shrinking point corresponding to the rotational symbol sequence $\mathcal{S}[2,2,5] =$ LRRLR for the map (6.2) when A_L and A_R are companion matrices (1.27), with first columns $(0, 1, \frac{28}{87})^{\mathsf{T}}$ and $(-\frac{23}{14}, 0, \frac{3}{2})^{\mathsf{T}}$, respectively. Here $b = e_1$ and $p_0 = (0, -1, \frac{3}{2})^{\mathsf{T}}$. The switching manifold, $s = 0$, is shaded gray.

Corollary 6.2. *Suppose (6.2) is at a shrinking point. Then,*

(a) $(I - M_{\mathcal{S}})$ *is singular,*
(b) $P_{\mathcal{S}^{(i)}}$ *is singular, for all* i.

The singularity of a matrix is a codimension-one phenomenon. Thus by Corollary 6.2, it is expected there will be curves in two-dimensional parameter space passing through shrinking points along which $(I - M_{\mathcal{S}})$

and each $P_{\mathcal{S}^{(i)}}$ is singular. Some of these curves form resonance tongue boundaries, and it is these that form the focus of the next section.

6.7 Unfolding Shrinking Points

This section presents an unfolding of the dynamics near terminating and non-terminating shrinking points. The first step is to perform a coordinate transformation such that, locally, in two dimensions, two tongue boundaries lie on the positive coordinate axes.

Recall (6.2) varies with $\eta \in \mathbb{R}^{M-1}$. Here assume $M \geq 3$ and (6.2) is C^{k+1} ($k \geq 2$). Then A_L and A_R are C^k functions of η. Suppose (6.2) is at a shrinking point when $\eta = 0$, for some fixed $\mu \neq 0$. Then, $\det(P_{\mathcal{S}}(\eta))$ and $\det(P_{\mathcal{S}^{((l-1)d)}}(\eta))$ are C^k scalar functions that are zero when $\eta = 0$. If the 2×2 Jacobian $\frac{\partial(\det(P_{\mathcal{S}}),\det(P_{\mathcal{S}^{((l-1)d)}}))}{\partial(\eta_i,\eta_j)}$ is nonsingular for some $i^{\rm th}$ and $j^{\rm th}$ components of η, we may utilize the implicit function theorem to obtain a coordinate transformation of η, such that $\det(P_{\mathcal{S}}) = 0$ when $\eta_i = 0$ and $\det(P_{\mathcal{S}^{((l-1)d)}}) = 0$ when $\eta_j = 0$. The remaining $(M-3)$ components of η are not important for the present analysis so will be ignored for the remainder of this section. As in [Simpson and Meiss (2009a)], the scalar parameters $\eta = \eta_i$ and $\nu = \eta_j$ will be used (take care to notice the new definition for η).

Let $\check{x}(\eta,\nu)$ denote the unique solution to the n-cycle solution system, (6.9), of $\check{\mathcal{S}} = \mathcal{S}^{\overline{0}}$ for small values of η and ν. Let $p = \check{x}(0,0)$ (to coincide with (6.28)). If the shrinking point is non-terminating, let $\hat{x}(\eta,\nu)$ denote the unique solution to the n-cycle solution system of $\hat{\mathcal{S}} = \mathcal{S}^{\overline{ld}}$ for small values of η and ν. Let $\check{s}_i$ and $\hat{s}_i$ denote the first components of $\check{x}_i$ and $\hat{x}_i$, respectively. From the above assumptions on η and ν, by Corollary 6.1 we have $\check{s}(0,\nu) = 0$ and $\check{s}_{ld}(\eta,0) = 0$, (see (6.29)). If $\frac{\partial}{\partial\eta}\check{s}(0,0)$ and $\frac{\partial}{\partial\nu}\check{s}_{ld}(0,0)$ are both nonzero, η and ν may be redefined via a nonlinear scaling so that

$$\check{s}(\eta,\nu) = \eta(1+O(1)) \,, \tag{6.30}$$

$$\check{s}_{ld}(\eta,\nu) = \nu(1+O(1)) \,. \tag{6.31}$$

Theorem 6.1. *Suppose the map (6.2) is a C^{k+1} ($k \geq 2$) function of parameters η and ν, has a shrinking point when $\eta = \nu = 0$ and (6.30) and (6.31) are satisfied. Write*

$$\det(I - M_{\mathcal{S}}(\eta,\nu)) = k_1\eta + k_2\nu + O(2) \,, \tag{6.32}$$

and assume $k_1, k_2 \neq 0$. As usual assume $N \geq 2$. When $(I - M_{\mathcal{S}}(\eta,\nu))$ is nonsingular, let $x_0(\eta,\nu)$ be given by (6.12) and x_{i+1} be given by (6.7).

Let $\Psi_1 = \{(\eta, \nu) \mid \eta, \nu \geq 0\}$. *Then for small* η *and* ν, *the* $\mathcal{S}$ *and* $\check{\mathcal{S}}$*-cycles are admissible in* Ψ_1 *and collide in nonsmooth folds at the boundaries,* $\eta = 0$ *and* $\nu = 0$ *and* $s_0(0, \nu) = 0$ *whenever* $\nu \neq 0$ *and* $s_{(l-1)d}(\eta, 0) = 0$ *whenever* $\eta \neq 0$.

If the shrinking point is non-terminating, there exist C^k *functions* $g_1, g_2 : \mathbb{R} \to \mathbb{R}$ *such that for small* η *and* ν, *the* $\mathcal{S}$ *and* $\hat{\mathcal{S}}$*-cycles are admissible in* $\Psi_2 = \{(\eta, \nu) \mid \eta \leq g_1(\nu), \nu \leq g_2(\eta)\}$ *and collide in nonsmooth folds at the boundaries,* $\eta = g_1(\nu)$ *and* $\nu = g_2(\eta)$ *and* $s_{ld}(g_1(\nu), \nu) = 0$ *whenever* $\nu \neq 0$ *and* $s_{-d}(\eta, g_2(\eta)) = 0$ *whenever* $\eta \neq 0$. *Furthermore, the lowest-order, nonzero terms of* g_1 *and* g_2 *are* $g_1''(0), g_2''(0) < 0$.

A proof is given in Appendix A. Near non-terminating shrinking points, Theorem 6.1 predicts the bifurcation set sketched in Fig. 6.10. The four resonance tongue boundaries that emanate from the shrinking point correspond to the collision of the points of the $\mathcal{S}$-cycle: x_0, x_{-d}, $x_{(l-1)d}$, x_{ld}, with the switching manifold. As mentioned in Sec. 6.6 these are the four points adjacent to the switching manifold in the interior of the tongue.

By applying the implicit function theorem to (6.32) there exists a C^k function

$$h(\eta) = -\frac{k_1}{k_2}\eta + O(\eta^2) , \tag{6.33}$$

such that $\det(I - M_{\mathcal{S}}(\eta, h(\eta))) = 0$. But by Theorem 6.1, $\mathcal{S}$-cycles exist throughout the first quadrant of parameter space therefore the curve $\nu =$

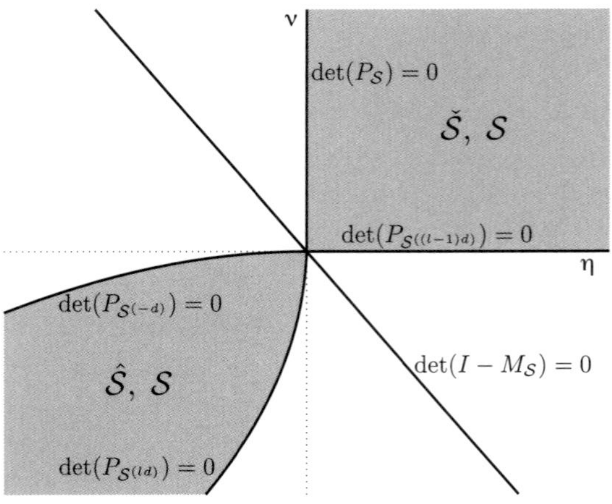

Fig. 6.10: A schematic of the generic bifurcation set in a neighborhood of a non-terminating shrinking point.

$h(\eta)$ must not enter this quadrant. Hence, since $h'(0) = -\frac{k_1}{k_2}$,

$$k_1 k_2 < 0 \ , \tag{6.34}$$

as in Fig. 6.10.

By Corollary 6.2, the border-collision matrix (6.11) of every cyclic permutation of $\mathcal{S}$ is singular at a shrinking point. Theorem 6.1 describes curves along which some of these matrices remain singular. All other such curves are now described. If p_0 and p_{ld} are the only points of the n-cycle, $\{p_i\}$, that lie on the switching manifold (as is generically the case), then for each $i \neq 0, l-1, l, -1$ if the shrinking point is non-terminating, and for each $i \neq 0, -2, -1$ if the shrinking point is terminating, there exists a C^k function

$$q_i(\eta) = -\frac{k_1 t_{(i+1)d}}{k_2 t_{id}} \eta + O(\eta^2) \ , \tag{6.35}$$

that satisfies $\det(P_{\mathcal{S}^{(id)}}(\eta, q_i(\eta))) = 0$. In each case $t_{(i+1)d}$ and t_{id} have the same sign, thus by (6.34) we have $q_i'(0) < 0$. Although the border-collision matrix, $P_{\mathcal{S}^{(id)}}$, is singular along $\nu = q_i(\eta)$, these curves are not seen in resonance tongue diagrams because the related periodic solutions are virtual.

So far curves passing through the shrinking point along which $P_{\mathcal{S}^{(id)}}$ is singular have been described for each i, except for $i = -1$ when the shrinking point is terminating. Here $\mathcal{S}^{(-d)} = RL^{n-1}$, hence $P_{\mathcal{S}^{(-d)}} = I + A_L + \cdots + A_L^{n-1} = (I - A_L)^{-1}(I - A_L^n)$. The matrix A_L^n has a multiplier 1 with an algebraic multiplicity of at least two, therefore $\det(P_{\mathcal{S}^{(-d)}}(\eta, \nu)) = O(2)$. Thus in this case, for small η and ν, generically $P_{\mathcal{S}^{(-d)}}$ is singular only when $\eta = \nu = 0$.

In addition, for terminating shrinking points there generically exists a curve in parameter space passing through the shrinking point along which the matrix, A_L, has a pair of complex multipliers on the unit circle. As one moves along the curve, the angular frequency of multipliers changes. Whenever the angular frequency is rational there may be additional terminating shrinking points. Such curves correspond to $s_R = 1$ in diagrams shown in Sec. 7.4 and are also described in [Sushko and Gardini (2008, 2006); Zhusubaliyev and Mosekilde (2008b)].

The border-collision bifurcations occurring on resonance tongue boundaries are described by piecewise-linear maps of the form (6.14). Theorem 6.1 states that these bifurcations are nonsmooth folds, thus we may use Feigin's results [di Bernardo *et al.* (1999)] (see Sec. 1.6) to determine the relative stability of periodic solutions near shrinking points. Let $a_i[\mathcal{S}]$ denote the number of real multipliers of $M_{\mathcal{S}}$ that are greater than one in the interior of Ψ_i. Then $a_1[\mathcal{S}] + a_1[\check{\mathcal{S}}]$ and $a_2[\mathcal{S}] + a_2[\hat{\mathcal{S}}]$ are odd. Also, $\det(I - M_{\mathcal{S}}(\eta, \nu)) = O(1)$, (6.32), thus $|a_1[\mathcal{S}] - a_2[\mathcal{S}]| = 1$. Consequently, if $\mathcal{S}$-cycles are stable in the interior of Ψ_1, then $\check{\mathcal{S}}$-cycles are unstable and $\mathcal{S}$-cycles are unstable in the interior of Ψ_2.

Finally, notice that at shrinking point an orbit intersects a switching manifold at two distinct points. For this reason shrinking points do not occur for the one-dimensional map (1.39) because here the switching manifold is a single point. The one-dimensional, piecewise-linear, circle map studied in [Yang and Hao (1987)] exhibits lens-chain structures because it has two switching manifolds.

Furthermore, shrinking points typically do not occur for a map that has nonlinear terms. In this case a lens-chain geometry only appears in the limit of parameter values towards the border-collision bifurcation of the fixed point (i.e. $\mu \to 0$ for (1.14)). It remains to be determined exactly how shrinking points break apart as μ is varied from zero.

Chapter 7

Neimark-Sacker Bifurcations in Planar, Piecewise-Smooth, Continuous Maps

As a fixed point of a piecewise-smooth, continuous map crosses a switching manifold under parameter variation, its associated multipliers may change discontinuously. This chapter studies border-collision bifurcations that result when the map is two-dimensional and the multipliers are complex and jump from inside to outside the unit circle at the bifurcation. This basic behavior has recently been demonstrated in power converters [Zhusubaliyev *et al.* (2008, 2006); Dai *et al.* (2007)]. It will be shown that these border-collision bifurcations exhibit diverse dynamical behavior, some akin to that of a Neimark-Sacker bifurcation in a smooth map, some like that of other border-collision bifurcations, and some phenomena that is possibly original.

Recall, in Sec. 1.2 it was argued that structurally stable dynamical behavior local to border-collision bifurcations is independent of nonlinear terms of the piecewise-smooth, continuous map (1.14). Furthermore, as detailed in Sec. 1.4, when (1.14) is observable, (1.11) may be transformed to (1.34). Since this chapter assumes (1.14) is two-dimensional and has associated complex-valued multipliers, by Corollary 1.1, the map is indeed observable. In this chapter the constant vector b in (1.14), is chosen to be e_1 in order to match [Simpson and Meiss (2008a)], instead of e_N as given in the canonical form and used elsewhere in this book. By Theorem 1.1, this difference is immaterial, the only effect is to alter the precise x and y values of orbits.

This chapter studies the map $(x', y') = F_\mu(x, y; \tau_L, \tau_R, \delta_L, \delta_R)$, where F_μ is piecewise-linear, defined by

$$\begin{bmatrix} x' \\ y' \end{bmatrix} = \begin{cases} \mu \begin{bmatrix} 1 \\ 0 \end{bmatrix} + A_L \begin{bmatrix} x \\ y \end{bmatrix}, \; x \le 0 \\ \mu \begin{bmatrix} 1 \\ 0 \end{bmatrix} + A_R \begin{bmatrix} x \\ y \end{bmatrix}, \; x \ge 0 \end{cases}, \tag{7.1}$$

$$\text{where} \quad A_L = \begin{bmatrix} \tau_L & 1 \\ -\delta_L & 0 \end{bmatrix}, \quad A_R = \begin{bmatrix} \tau_R & 1 \\ -\delta_R & 0 \end{bmatrix}. \tag{7.2}$$

As discussed in Sec. 1.2, as with other piecewise-linear systems (7.5) has the scaling symmetry

$$F_{\lambda\mu}(\lambda x, \lambda y; \tau_L, \tau_R, \delta_L, \delta_R) = \lambda F_\mu(x, y; \tau_L, \tau_R, \delta_L, \delta_R), \ \forall \lambda > 0. \tag{7.3}$$

Consequently if $\mathcal{I}$ is an invariant set under F_μ, then $\lambda\mathcal{I}$ is an invariant set for $F_{\lambda\mu}$, $(\lambda > 0)$. Therefore every bounded invariant set collapses onto the origin as $\mu \to 0$, and it is sufficient to consider only $\mu \in \{1, 0, -1\}$.

The parameters τ_i and δ_i are the trace and determinant of A_i (7.2), respectively, for $i = L, R$. By Lemma 1.1, (7.1) is a homeomorphism of $\mathbb{R}^2$ if and only if $\delta_L\delta_R > 0$. If $\delta_L, \delta_R > 0$, the inverse, F_μ^{-1}, is related to F_μ simply by

$$F_\mu^{-1}(x, y; \tau_L, \tau_R, \delta_L, \delta_R) = F_{-\mu}(y, x; \frac{\tau_R}{\delta_R}, \frac{\tau_L}{\delta_L}, \frac{1}{\delta_R}, \frac{1}{\delta_L}) . \tag{7.4}$$

Thus the dynamical properties of F for $\mu > 0$ are the same as those of F^{-1} for $\mu < 0$ for a different combination of parameter values.

The remainder of this chapter is organized as follows. First Sec. 7.1 defines a new map (7.5), which is the same as (7.1) except an alternative parameter set that is better suited for complex multipliers is used in place of $\{\tau_L, \tau_R, \delta_L, \delta_R\}$. Rotation numbers are then defined for (7.5). Section 7.2 describes basic Neimark-Sacker-like behavior of (7.5) and illustrates the effect of adding nonlinear terms. Section 7.3 rigorously proves the existence or absence of attracting sets for limiting values of the parameters. In some cases it is shown that the attracting set persists for nearby parameter values.

Particularly for the case of complex multipliers, it is of interest to compute resonance tongues. As detailed in Chapter 6, resonance tongues of piecewise-smooth, continuous maps commonly exhibit a lens-chain structure. Section 7.4 provides a detailed discussion of resonance tongues and related dynamics for the map (7.5). Curves of shrinking points (see Sec. 6.6 and Sec. 6.7), solutions with irrational rotation numbers and non-rotational periodic solutions (see Sec. 6.4) are then discussed in Sec. 7.5. Finally Sec. 7.6 looks at the case that the fixed point on the switching manifold is a saddle and presents a brief investigation into coexisting attractors.

Throughout this chapter the variable z will be used to denote a point $(x, y)^{\mathsf{T}} \in \mathbb{R}^2$. Also, "right half-plane" will be abbreviated to RHP; similarly "left half-plane" will be abbreviated to LHP.

7.1 A Two-Dimensional Map

This chapter studies the map (7.1) when A_L and A_R, (7.2), have multipliers $r_L e^{\pm 2\pi i \omega_L}$ and $\frac{1}{s_R} e^{\pm 2\pi i \omega_R}$ respectively, where $r_L, s_R \in (0,1)$ and without loss of generality $\omega_L, \omega_R \in (0, \frac{1}{2})$. Note that this corresponds to the case that $0 < \delta_L = r_L^2 < 1$ and $1 < \delta_R = 1/s_R^2$, so that the map (7.1) is an orientation preserving homeomorphism. With these new parameters, the map becomes

$$\begin{bmatrix} x' \\ y' \end{bmatrix} = \begin{cases} \mu \begin{bmatrix} 1 \\ 0 \end{bmatrix} + \begin{bmatrix} 2r_L \cos(2\pi\omega_L) & 1 \\ -r_L^2 & 0 \end{bmatrix} \begin{bmatrix} x \\ y \end{bmatrix}, & x \leq 0 \\ \mu \begin{bmatrix} 1 \\ 0 \end{bmatrix} + \begin{bmatrix} \frac{2}{s_R} \cos(2\pi\omega_R) & 1 \\ -\frac{1}{s_R^2} & 0 \end{bmatrix} \begin{bmatrix} x \\ y \end{bmatrix}, & x \geq 0 \end{cases}. \tag{7.5}$$

In this chapter, the map (7.5) will be denoted $(x', y') = f_\mu(x, y; r_L, s_R, \omega_L, \omega_R)$.

The existence of fixed points and 2-cycles may be determined by Feigin's results [di Bernardo *et al.* (1999)] (see Sec. 1.6). Both A_L and A_R have no real-valued multipliers, hence using the notation defined in Sec. 1.6, $\sigma_1^+ = \sigma_2^+ = \sigma_1^- = \sigma_2^- = 0$. Thus (7.5) has a unique fixed point for all $\mu \in \mathbb{R}$ and the map has no 2-cycles for any values of the parameters. Explicitly, the fixed point is

$$z^*(\mu) = \begin{bmatrix} x^*(\mu) \\ y^*(\mu) \end{bmatrix} = \begin{cases} \dfrac{\mu}{r_L^2 - 2r_L \cos(2\pi\omega_L) + 1} \begin{bmatrix} 1 \\ -r_L^2 \end{bmatrix}, & \mu \leq 0 \\ \dfrac{\mu}{s_R^2 - 2s_R \cos(2\pi\omega_R) + 1} \begin{bmatrix} s_R^2 \\ -1 \end{bmatrix}, & \mu \geq 0 \end{cases}. \tag{7.6}$$

The fixed point moves from the left half-plane when $\mu < 0$ where it is a stable focus, to the origin at $\mu = 0$, and then to the right half-plane when $\mu > 0$ where it is an unstable focus.

Since (7.5) is a homeomorphism its inverse is given by the symmetry (7.4). Explicitly the inverse is

$$f_\mu^{-1}(x, y) = \begin{cases} \mu \begin{bmatrix} 0 \\ -1 \end{bmatrix} + \begin{bmatrix} 0 & -s_R^2 \\ 1 & 2s_R \cos(2\pi\omega_R) \end{bmatrix} \begin{bmatrix} x \\ y \end{bmatrix}, & y \leq 0 \\ \mu \begin{bmatrix} 0 \\ -1 \end{bmatrix} + \begin{bmatrix} 0 & -\frac{1}{r_L^2} \\ 1 & \frac{2}{r_L} \cos(2\pi\omega_L) \end{bmatrix} \begin{bmatrix} x \\ y \end{bmatrix}, & y \geq 0 \end{cases}. \tag{7.7}$$

Rotation numbers

The rotation number (or winding number) of an orbit of a map is a characterization of the average increase in angle per iteration. It is most easily defined for maps on circles or annuli. The rotation number for a map on $\mathbb{R}^2$ can be defined relative to a fixed point, z^*, if one exists, because removing z^* from the plane leaves an annulus [Le Calvez (2001); Franks (1990); Schwartzman (1957)]. An alternative, intrinsic definition is called the "self-rotation number" [Dullin *et al.* (2000); Peckham (1990)]. The first definition is dependent upon the choice of z^*, but since (7.5) has a unique fixed point for any combination of parameter values, it is natural to define the rotation number about this point.

Since orbits of (7.5) rotate in a clockwise direction about z^*, (7.6), the angle $\phi : \mathbb{R}^2 \setminus \{z^*\} \to (-\pi, \pi]$ will be defined as

$$\begin{aligned}\phi(z) &= \text{polar angle of } (x - x^*) - \mathrm{i}(y - y^*) \\ &= \mathrm{atan2}(y^* - y, x - x^*) \ ,\end{aligned}$$

where atan2 is the two argument arctangent. To compute the rotation number one simply averages the changes in ϕ over iterations of the map. Let $\Delta\phi : \mathbb{R}^2 \setminus \{z^*\} \to [0, 2\pi)$, be

$$\Delta\phi(z) = \phi(f(z)) - \phi(z) \bmod 2\pi \ .$$

Then the *rotation number* of an orbit is

$$\rho(z) = \lim_{n\to\infty} \frac{1}{2\pi n} \sum_{i=0}^{n-1} \Delta\phi(z_i) \ , \tag{7.8}$$

if this limit exists. The following lemma tells us that the limit always exists and gives upper and lower bounds.

Lemma 7.1. *Suppose* $0 < r_L, s_R < 1$, $0 < \omega_L, \omega_R < \frac{1}{2}$ *and* $\mu \in \mathbb{R}$, *as usual. Then for any* $z \in \mathbb{R}^2 \setminus \{z^*\}$ *(where* z^* *is given by (7.6)) the limit (7.8) exists and*

$$0 < \rho(z) < \frac{1}{2} \ . \tag{7.9}$$

A proof is given in Appendix A. Clearly if z belongs to an n-cycle, the rotation number is rational: $\rho = m/n$. If (7.5) has an invariant circle, then since f_μ is a homeomorphism, so is its restriction to the circle. Since any invariant circle bounds an invariant disk, the Brouwer fixed point theorem implies the disk must contain a fixed point, which must be z^*, since the fixed point of (7.5) is unique. This is consistent with Lemma 7.1 which implies that the rotation number of the circle is nonzero. The usual definition for the rotation number on a circle [Kuznetsov (2004); Cornfeld *et al.*

(1982)] coincides with (7.8). Furthermore, all points on the invariant circle have the same rotation number [Kuznetsov (2004); Cornfeld *et al.* (1982)]. Similar behavior has been observed for a piecewise-linear, continuous, area-preserving map [Lagarias and Rains (2005a)].

7.2 Basic Dynamics

Recall that a Neimark-Sacker bifurcation in a sufficiently smooth map generically occurs when an attracting fixed point has a complex pair of multipliers, say $\lambda(\mu) = r(\mu)e^{2\pi i\omega(\mu)}$ and $\bar{\lambda}(\mu)$, that cross the unit circle, at say $\mu = 0$ [Kuznetsov (2004)]. If the map satisfies a non-degeneracy assumption and $\lambda^n(0) \neq 1$ for $n = 1, 2, 3$, or 4, then an invariant circle is created or destroyed as μ crosses zero. The circle emerges from the fixed point with a size $O(|\mu|^{\frac{1}{2}})$. The criticality of the bifurcation is determined by quadratic and cubic terms in the normal form; when it is *supercritical* the invariant circle is stable and exists when $|\lambda(\mu)| > 1$ and when it is *subcritical*, the opposite is true. For small μ, dynamical behavior on the circle is determined by the rotation number $\omega(\mu)$. If $\omega(\mu) = \frac{m}{n}$ with $\gcd(m, n) = 1$ and $n > 4$, the motion is called *weakly resonant* or *mode-locked* and there generically exist two or more n-cycles on the invariant circle. When $n \leq 4$, the dynamics is *strongly resonant* and an invariant circle need not exist [Kuznetsov (2004); Arnold (1988)]. When $\omega(\mu)$ is irrational, all orbits on the circle are dense and quasiperiodic.

The behavior of the piecewise-smooth map (7.5) can be much more complicated when μ crosses zero. However for some parameter values, its behavior is similar to the classical Neimark-Sacker bifurcation; two example bifurcation diagrams for (7.5) are shown in Fig. 7.1. For panel (a) of the figure, when $\mu < 0$ the stable fixed point is a global attractor, see Fig. 7.2(a). When $\mu > 0$ the fixed point is unstable and is encircled by a stable invariant circle with rotation number $\rho \approx 0.1601$ whose basin of attraction appears to be the entire phase space except for the fixed point, Fig. 7.2(b). Consequently, this bifurcation is analogous to a supercritical Neimark-Sacker bifurcation. An important difference is that, unlike in smooth systems, the size of the invariant circle grows linearly with respect to μ. This is a simple consequence of the scaling symmetry (7.3) and is common in other bifurcations in piecewise-smooth systems.

By contrast, the bifurcation shown in Fig. 7.1(b) is analogous to a subcritical Neimark-Sacker bifurcation. When $\mu < 0$, there is an unstable

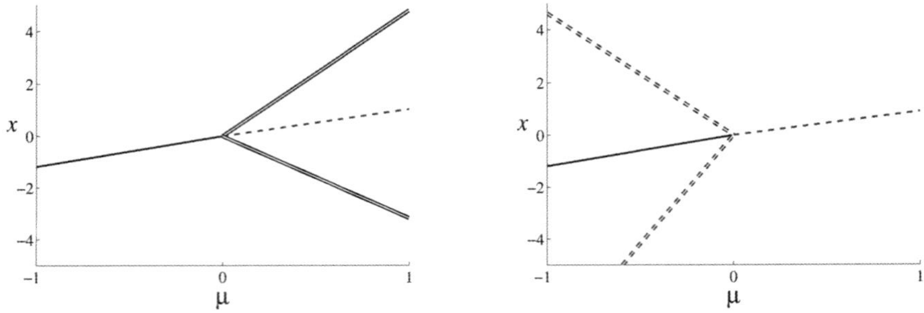

(a) $s_R = 0.95$ (b) $s_R = 0.85$

Fig. 7.1: Bifurcation diagrams of (7.5) when $\omega_L = \omega_R = 0.16$, $r_L = 0.9$, for two different values of s_R. Solid [dotted] lines denote stable [unstable] solutions. In each panel one fixed point exists for each value of μ and the maximum and minimum x values of an invariant circle that is created at $\mu = 0$ are shown as double lines. Phase portraits corresponding to panel (a) are shown in Fig. 7.2.

invariant circle whose radius shrinks linearly to zero as $\mu \to 0^-$. Points inside the circle are attracted to the stable fixed point, whereas the forward orbits of points outside the circle are unbounded. When $\mu > 0$ the orbit of every initial condition, except for the unstable fixed point, is unbounded.

A fundamental question regarding the map (7.5) is: what governs the criticality of the Neimark-Sacker-like bifurcation? As will be shown, the

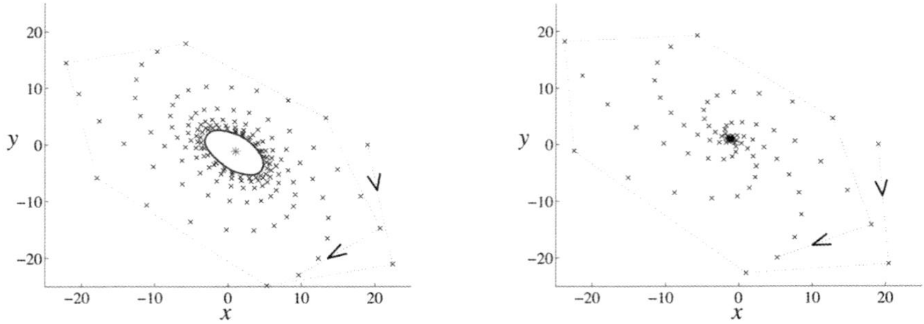

(a) $\mu = -1$ (b) $\mu = 1$

Fig. 7.2: Phase portraits of (7.5) with the same parameter values as Fig. 7.1, panel (a). In both panels one orbit is shown and the first few iterates are connected to illustrate the direction of motion. The fixed point is indicated by an asterisk.

answer is not simple. For instance there may exist no invariant circles for any value of μ, or both supercritical and subcritical behavior may be observed together, see Sec. 7.6. Furthermore, the invariant circle is not, geometrically speaking, an ellipse. It will be seen below that the attracting set may appear fractal. For piecewise-linear, continuous, area-preserving maps, invariant circles may be the piecewise union of arcs of conic sections [Lagarias and Rains (2005b)].

Numerical simulations suggest (7.5) does not exhibit strong resonance. Significantly different dynamical phenomena has not been observed when $\omega_L = \omega_R = 1/3$ or $1/4$.

The addition of nonlinear terms to (7.5) does not affect structurally stable dynamics for sufficiently small values of μ. For instance, bifurcation diagrams are shown in Fig. 7.3 for the map

$$\begin{bmatrix} x' \\ y' \end{bmatrix} = f_\mu(x, y) - \begin{bmatrix} x^2 \\ 0 \end{bmatrix}. \tag{7.10}$$

In panel (a), the parameter values of Fig. 7.1, panel (a) are used. For small $\mu > 0$, an attracting invariant circle grows in amplitude approximately linearly with respect to μ. For larger values of μ the invariant circle undergoes mode-locking.

In panel (b), parameters are chosen such that the map, (7.5), has a stable 3-cycle for $\mu > 0$. For small $\mu > 0$, the nonlinear map exhibits the

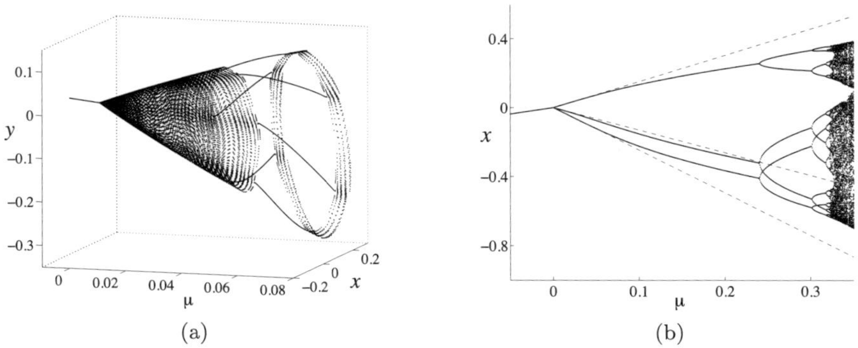

Fig. 7.3: Bifurcation diagrams of the nonlinear map, (7.10), computed numerically by plotting forward orbits after transients have decayed. In panel (a) the parameter values are the same as for Fig. 7.1(a); in panel (b), $\omega_L = \omega_R = 0.4$, $r_L = 0.2$ and $s_R = 0.7$. In panel (b) the corresponding bifurcation diagram of (7.5) is superimposed - shown as dashed lines.

same periodic solution which initially grows at the same linear rate as for the piecewise-linear map. However, as μ is increased the 3-cycle undergoes a period-doubling cascade. The cascade is comprised of usual smooth period-doubling bifurcations where the attracting set is bounded away from the switching manifold.

7.3 Limiting Parameter Values

This section provides theorems and proofs regarding the existence or absence of an attractor for some limiting values of the parameters. When either r_L or ω_L is sufficiently small it is proved below that there exists an asymptotically stable invariant set (Theorems 7.1 and 7.2). In both cases the invariant set is first shown to exist in the limit that the parameter is zero, then, as in [Nusse and Yorke (1992)], persistence of the invariant set for nearby parameter values is proved. When $\omega_R = 0$ almost all forward orbits are unbounded (Theorem 7.3). Throughout this section $f = f_1$ denotes the map (7.5), when $\mu = 1$.

Theorem 7.1. *Consider (7.5) with* $\mu = 1$ *and assume* $0 < s_R < 1$, $0 < \omega_L, \omega_R < 1/2$ *as usual. Then there is an* $\varepsilon > 0$ *such that whenever* $0 \leq r_L < \varepsilon$, *the map has an asymptotically stable invariant set.*

Proof. First suppose $r_L = 0$. Since $\mu = 1$, the unique fixed point of (1.14) lies in the RHP. The fixed point is an unstable focus, thus any point in the RHP other than the fixed point maps into the LHP in finitely many steps. Moreover, if $z = (x, y)^{\mathsf{T}}$ is the first point in the LHP, then it is easy to see that $y < 0$. The image of this point is $f(z) = (y + 1, 0)^{\mathsf{T}}$, with an x component less than one. If this image is still in the LHP then the second iterate $f^2(z) = (1, 0)^{\mathsf{T}} \in$ RHP. Consequently, any point in the LHP maps into the RHP in at most two steps.

Thus the forward orbit of every point other than the fixed point intersects the segment $\mathcal{I} = \{(x, 0)^{\mathsf{T}} \mid x \in [0, 1]\}$. Since $\mathcal{I}$ is compact, there is an $N \in \mathbb{N}$ such every point in $\mathcal{I}$ maps back into $\mathcal{I}$ in at most N steps. Hence

$$\Omega = \bigcup_{i=0}^{N} f^i(\mathcal{I}) ,$$

is a compact, bounded, forward invariant set of f. Note that $z^* \notin \Omega$. Let Σ be any compact neighborhood of Ω, $\Omega \subset \text{int}(\Sigma)$, that does not contain the fixed point. Since Σ is compact, every point in Σ maps to $\mathcal{I}$, thus there is an $\hat{N} \in \mathbb{N}$ such that $f^{\hat{N}}(\Sigma) \subset \Omega$, i.e., Σ is a trapping set for $f^{\hat{N}}$.

When r_L is small, a similar construction must hold: since (7.5) depends continuously on r_L, there is an $\varepsilon > 0$ such that whenever $0 \le r_L < \varepsilon$, Σ is still a trapping set for $f^{\hat{N}}$. Let

$$\Lambda = \bigcap_{i=0}^{\infty} f^i(\Sigma) . \tag{7.11}$$

Then Λ is an attracting set for (7.5); it is necessarily asymptotically stable and invariant. □

The attracting set (7.11) may not be an "attractor" as it need not be minimal or chain-transitive [Perko (2001); Robinson (1999)]. Indeed, it will be seen in Sec. 7.6 that f can have multiple attractors.

Theorem 7.2. *Consider (7.5) with $\mu = 1$ and $0 < r_L, s_R < 1$, $0 < \omega_R < 1/2$ as usual. Then there is an $\varepsilon > 0$ such that whenever $0 \le \omega_L < \varepsilon$, the map has an asymptotically stable invariant set.*

Proof. When $\omega_L = 0$, the left half of (7.5) becomes

$$z' = A_L z + \begin{bmatrix} 1 \\ 0 \end{bmatrix} , \quad A_L = \begin{bmatrix} 2r_L & 1 \\ -r_L^2 & 0 \end{bmatrix} .$$

This linear map has a unique (virtual) fixed point $z_L^* = \frac{1}{(1-r_L)^2}(1, -r_L^2)^{\mathsf{T}}$ in the RHP, see Fig. 7.4(a). A_L has a repeated eigenvalue r_L, and a single eigenvector $(1, -r_L)^{\mathsf{T}}$. The one-dimensional invariant manifold E^{r_L}, of the virtual fixed point intersects the y-axis at $\frac{r_L}{1-r_L} > 0$. Observe E^{r_L} separates orbits of the left half-map (i.e., if z is below [above] E^{r_L}, then so is z'). Let $\Xi = \{(x, y)^{\mathsf{T}} \mid x, y \ge 0, y \le r_L(\frac{1}{1-r_L} - x)\}$, see Fig. 7.4(a). Now consider the forward orbit of any $z \in \Xi$ under the full map (7.5). Since the map in the RHP corresponds to an unstable focus, z maps to a point $\hat{z}$ in the LHP in finitely many steps, and since for any $x > 0$, $y' = -\frac{1}{s_R^2}x < 0$, $\hat{z}$ lies below the x-axis. Every point in the LHP below E^{r_L} maps into the RHP below E^{r_L} in finitely many steps. Since for any $x < 0$, $y' = -r_L^2 x > 0$, the first iterate in the RHP will lie above the x-axis, thus in Ξ. Since Ξ is compact, there is an $N \in \mathbb{N}$, such that for all $z \in \Xi$, $f^n(z) \in \Xi$ for some $n \le N$. Hence,

$$\Omega = \bigcup_{i=0}^{N} f^i(\Xi) ,$$

is a compact, bounded, forward invariant set for f.

Following the argument in Theorem 7.1, this implies that for $0 \le \omega_L < \epsilon$ there exists an attracting set Λ (7.11) for f, where Σ is a neighborhood of Ω. □

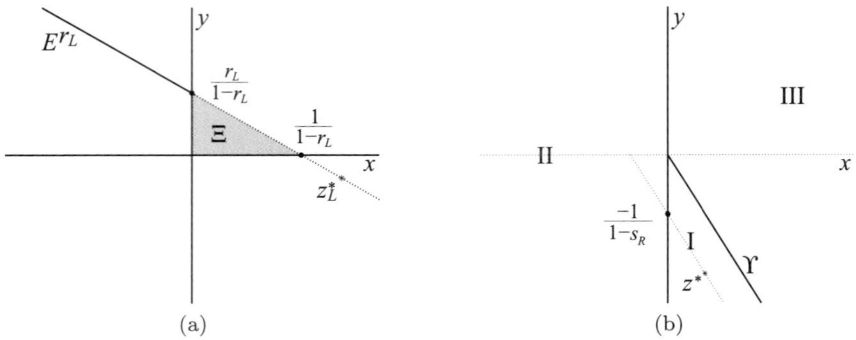

Fig. 7.4: Partitions of phase space when $\mu = 1$, applied in Theorems 7.2 and 7.3. In panel (a) $\omega_L = 0$; in panel (b) $\omega_R = 0$.

Theorem 7.3. *Consider (7.5) with* $\mu = 1$ *and the limiting case* $\omega_R = 0$. *Assume the remainder of the parameter values satisfy* $0 < r_L, s_R < 1$, $0 < \omega_L < 1/2$ *as usual. Then the forward orbit of any point other than the fixed point is unbounded.*

Proof. When $\omega_R = 0$ the right half of (7.5) becomes

$$z' = A_R z + \begin{bmatrix} 1 \\ 0 \end{bmatrix}, \quad A_R = \begin{bmatrix} \frac{2}{s_R} & 1 \\ -\frac{1}{s_R^2} & 0 \end{bmatrix}.$$

This linear map has a unique fixed point $z^* = -\frac{1}{(1-s_R)^2}(s_R^2, -1)^{\mathsf{T}}$ in the RHP. A_R has a repeated eigenvalue $\frac{1}{s_R}$, and a single eigenvector $(s_R, -1)^{\mathsf{T}}$ whose associated one-dimensional invariant manifold Σ, intersects the y-axis at $-\frac{1}{1-s_R} < 0$. Let Υ be the line parallel to Σ which passes through the origin. Υ partitions phase space into three regions as shown in Fig. 7.4(b). Region III, the set $\{(x, y)^{\mathsf{T}} \mid x \ge 0,\ x > -s_R y\}$, is forward invariant for (7.5) because for any z in region III, $z' = (\frac{2}{s_R}x + y + 1, -\frac{1}{s_R^2}x)^{\mathsf{T}}$, that is,

$$x' = \frac{2}{s_R}x + y + 1 > \frac{2}{s_R}x - \frac{1}{s_R}x + 1 \quad \Rightarrow \quad x' > \frac{1}{s_R}x > 0\,.$$

Using $x = -s_R^2 y'$, we see $x' > -s_R y'$, hence z' is in region III. Since the map in the LHP corresponds to a virtual stable focus, and since for any

$x < 0$, $y' = -r_L^2 x > 0$, every point in region II is mapped into region III in finitely many iterations. Thus the forward orbit of any point either starts in the RHP and remains there, or enters region III via the LHP in finitely many steps and remains in region III. In any case the tail of the forward orbit is contained in the RHP and since dynamics in the RHP are governed by an unstable linear map, the forward orbit of every point except z^* is unbounded. □

The limiting case, $s_R = 1$ has been described previously, [Sushko and Gardini (2008, 2006); Puu *et al.* (2006); Gallegati *et al.* (2003)]. It will be discussed further in the next section.

7.4 Resonance Tongues

Numerical computation of resonance tongues

Figs. 7.5, 7.6, and 7.7 show numerically computed regions of existence of stable periodic orbits with periods shown in the bar at the bottom of Fig. 7.5. As is commonly observed in piecewise-smooth systems, these tongues have the form of a chain of lens-shaped regions (or sausages) [Zhusubaliyev and Mosekilde (2003); Puu and Sushko (2006)]. The rotation number of each lens-chain is fixed and, as will be discussed below, within a given chain the symbol sequence of the associated stable periodic solution changes from lens to lens. The tongues emanate from $s_R = 1$ and their sizes are ordered by the Farey sequence as indicated along the top of Fig. 7.5. In some places tongues overlap corresponding to the coexistence of multiple stable periodic solutions.

The figures are computed on a grid of 1024 frequency values and 128 values of s_R. For each set of parameters, the admissibility conditions, see (6.8), and the stability conditions (that the multipliers of the stability matrix (6.10) lie inside the unit circle) were checked for periodic solutions up to period 30 with symbol sequences that are rotational (see Sec. 6.4). (Some non-rotational symbol sequences were also checked near codimension-two, nonsmooth period-doubling bifurcations, see Sec. 7.5.) If no stable n-cycle was found, N iterates were computed along the forward orbit of the point $z = (M, 0)^{\mathsf{T}}$ for N up to 10^4. If $\max_{i=0}^{n} |x^{(i)}|$ appeared to grow steadily as $n \to N$, this orbit is declared to be unbounded, and the corresponding point is shaded white. Otherwise the point is shaded black and presumably corresponds to bounded motion that has a period larger than 30 or

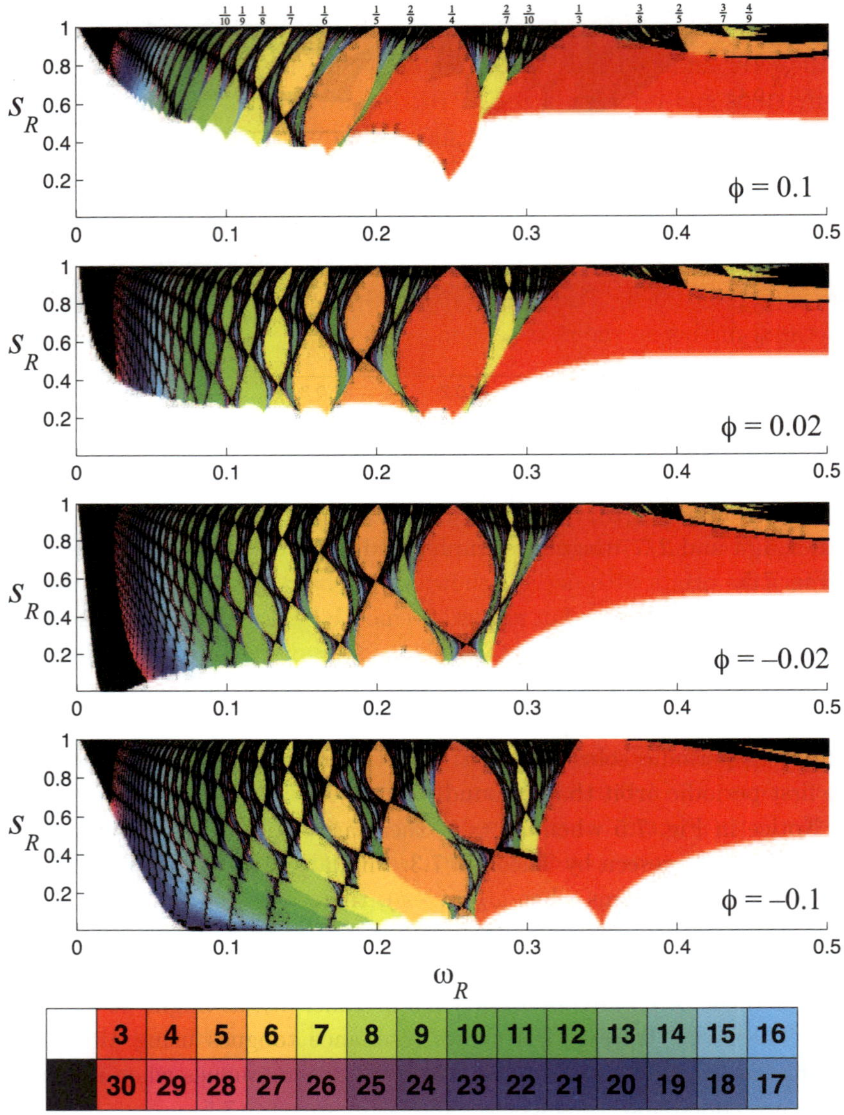

Fig. 7.5: Resonance tongues of (7.5) when $\mu = 1$, $r_L = 0.2$, $\omega_L = \omega_R + \phi$ for four different values of ϕ. Each region corresponds to the existence of a stable periodic orbit with period shown in the bar at the bottom. White regions correspond to unbounded orbits, and black to orbits with period larger than 30.

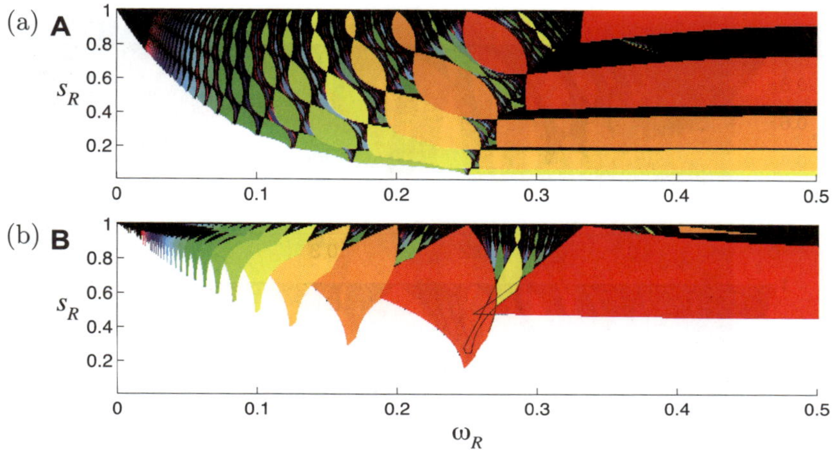

Fig. 7.6: Resonance tongues of (7.5) when $\mu = 1$. In panel (a), $r_L = 0.3$, $\omega_L = 0.09$; in panel (b), $r_L = 0.16$, $\omega_L = 0.38$. In panel (b), the overlapping of 1/3, 1/4 and 2/7 tongues is emphasized. The color scheme is the same as Fig. 7.5.

is non-periodic. Since the orbits for small ω_R appear to range over a large domain, the large value $M = 10^{12}$ was used. In some cases, multiple attracting periodic solutions exist, and the algorithm used simply chose the first periodic orbit that it found.

Notice in Fig. 7.5 when $\omega_R = 0$, there are no stable solutions for any value of s_R as foreseen by Theorem 7.3. Similarly when $\omega_L = 0$ (as seen in the lower two diagrams when $\omega_R = -\phi$), there is a stable solution for all values of s_R in accordance with Theorem 7.2. Notice in Fig. 7.7 the diagram corresponding to the smaller value of r_L (panel (b)) has stable orbits over a larger range of parameter space than in panel (a), as might be expected from Theorem 7.1. Fig. B.4 shows resonance tongues using alternative parameters on the axes. Fig. B.5 shows resonance tongues when nonlinear terms are added and for a smooth approximation to (7.5).

Dynamics near $s_R = 1$

First the dynamics for the limiting case, $s_R = 1$, is described. This corresponds to the top edge of Figs. 7.5, 7.6, and 7.7. When $\mu = 1$, the fixed point, z^*, (7.6), lies in the RHP and is a center. Points sufficiently near z^*

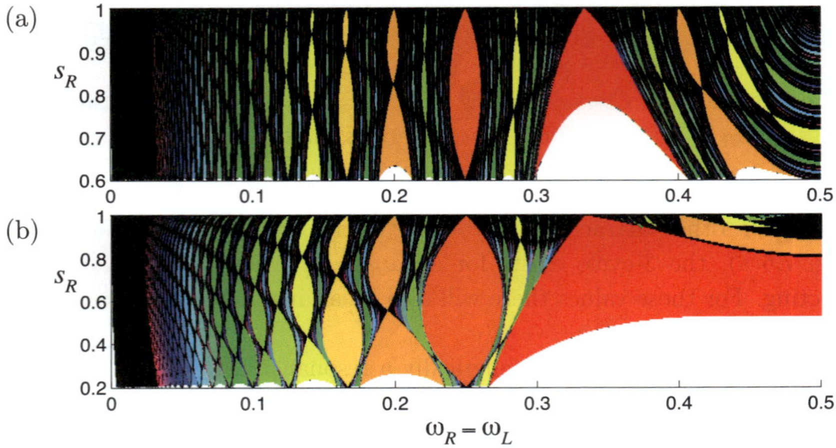

Fig. 7.7: Resonance tongues of (7.5) for equal polar angles, i.e. $\omega_L = \omega_R$, when $\mu = 1$. In panel (a), $r_L = 0.6$; in panel (b), $r_L = 0.2$. The color scheme is the same as Fig. 7.5.

rotate around it with rotation number ω_R, on invariant ellipses contained in the RHP. The largest of this family of nested ellipses has one point on the switching manifold. The boundary of the region in phase space within which this rigid rotation occurs has a geometry dependent upon the rationality of ω_R [Sushko and Gardini (2008)]. When ω_R is irrational the boundary is the largest ellipse. When $\omega_R = m/n$ is rational, the boundary is an invariant n-sided polygon, $\mathcal{P}$. In this case there are points in the region bounded by the largest ellipse and $\mathcal{P}$. These points simply belong to m/n-cycles contained in the RHP.

Whenever $\omega_R = m/n$ is rational for $n \geq 3$, the point $(\omega_R, s_R) = (\frac{m}{n}, 1)$ is a terminating shrinking point, see Sec. 6.6. The associated rotational symbol sequence is $\mathcal{S}[1, m, n]$, (6.18). With $l = 1$, the required non-degeneracy condition for Definition 6.3 is that the matrix $(I - M_{\hat{\mathcal{S}}})$ is nonsingular, where $\hat{\mathcal{S}} = \mathcal{S}[2, m, n]$. When this condition is satisfied, the polygon $\mathcal{P}$ has one side on the switching manifold and is otherwise contained in the RHP. If we let $z_0 = \mu(I - M_{\hat{\mathcal{S}}})^{-1} P_{\hat{\mathcal{S}}} b$ (compare with (6.28)), then the iterates, z_i, are the vertices of $\mathcal{P}$. The vertices z_0 and z_d lie on the switching manifold (where d is the multiplicative inverse of m modulo n). For example when $m/n = 3/7$, $d = 5$ since $5 \times 3 \bmod 7 = 15 \bmod 7 = 1$, see Fig. 7.8(a),

Numerically it has been observed that the boundary, whether a polygon or an ellipse, attracts nearby outside points prompting the following conjecture.

Conjecture 7.1. *Consider (7.5) with* $\mu = 1$ *and assume* $0 < r_L < 1$, $0 < \omega_L, \omega_R < 1/2$ *as usual. Then there is an* $\varepsilon > 0$ *such that whenever* $1 - \varepsilon < s_R \leq 1$, *the map has an asymptotically stable invariant set.*

Notice other attractors may exist when $s_R = 1$. For example in Fig. 7.6(a), the 3-cycle exists for all $\omega_R > \frac{1}{3}$ when $s_R = 1$ and is attracting. For these values there will also be an invariant polygon or ellipse with rotation number ω_R.

Theorem 6.1 states that generically a resonance tongue emanates from any terminating shrinking point. This is seen clearly in Figs. 7.5, 7.6, and 7.7: many tongues emanate from the line $s_R = 1$. Each lens-shaped region has two boundaries that are defined by the codimension-one conditions $x_0 = 0$ and $x_d = 0$. Within the region there exist both stable and saddle orbits with rotation number m/n. These orbits collide and annihilate on the boundaries in a nonsmooth fold. The stable orbits have the symbol sequence $\mathcal{S}[2, m, n]$, whereas the saddle orbits have the symbol sequence $\mathcal{S}[1, m, n]$. An example of a phase portrait showing these two orbits

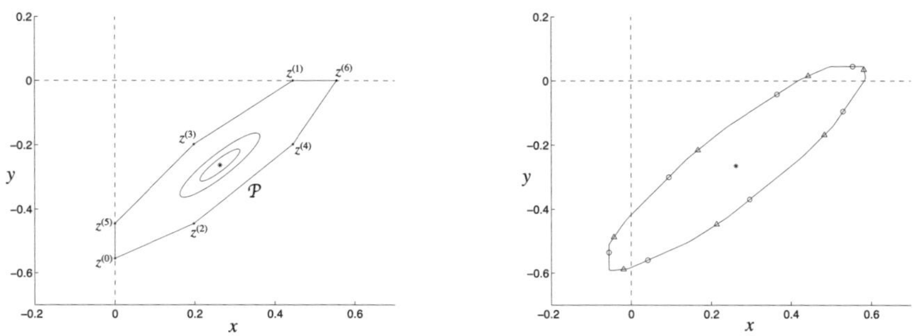

(a) (b)

Fig. 7.8: Phase portraits of (7.5) when $\mu = 1$, $\omega_L = \omega_R = 3/7$, $r_L = 0.9$ and $s_R = 1$ in panel (a) and $s_R = 0.995$ in panel (b). The fixed point z^*, is indicated by an asterisk. Panel (a) shows the invariant heptagon $\mathcal{P}$ and two of the uncountably many invariant ellipses near the center z^*, each consisting of infinitely many 3/7-cycles. In panel (b) the right half-map is no longer area preserving; $\mathcal{P}$ appears to become an invariant circle attracting from both sides. The stable [saddle] 3/7-cycle is indicated by triangles [circles].

and the invariant circle formed from the unstable manifolds of the saddle is shown in Fig. 7.8(b).

Resonance tongue boundaries

As detailed in Sec. 6.3, codimension-one, resonance tongue boundaries correspond to either the collision of one or more points of the stable periodic solution with the switching manifold, or a loss of stability. As an example, Fig. 7.9 illustrates the lens-chain for a 1/5-cycle, and also shows phase portraits of the stable cycle in the two lenses and on their boundaries. In the upper lens, two points of the stable 1/5-cycle lie in the LHP; in the lower lens three points lie in the LHP. The left and right boundaries correspond to nonsmooth folds along which the stable cycle collides and annihilates with a saddle cycle. Along the bottom boundary a multiplier associated with the stable cycle passes through 1.

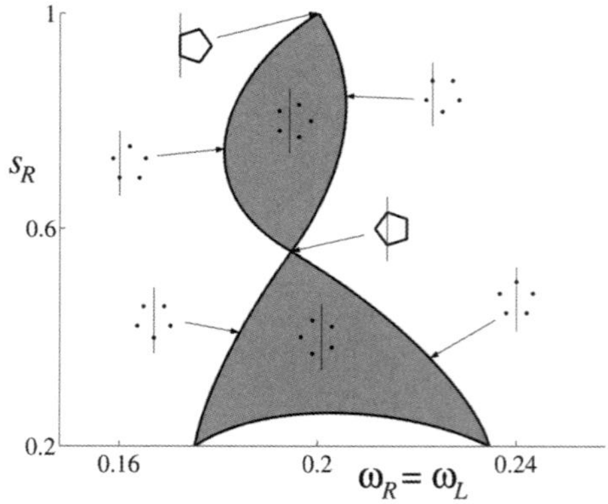

Fig. 7.9: The 1/5-resonance tongue of (7.5) when $\mu = 1$, $r_L = 0.2$, as seen in Fig. 7.7(b). Schematic phase portraits of the stable 1/5-cycle in relation to the switching manifold are shown.

As this lower bound is approached, the amplitude of the stable 1/5-cycle becomes unbounded. Of course, this is a somewhat artificial consequence of the lack of nonlinear terms in (7.5), see Sec. 6.3. A saddle 1/5-cycle exists on both sides of the lower boundary.

If Conjecture 7.1 holds, the bounding invariant polygon $\mathcal{P}$ for $s_R = 1$ is an attracting invariant set, and so it persists as such a set, typically as an invariant circle for $s_R < 1$. There are many ways in which an invariant circle of a planar, piecewise-linear, continuous map can break-up [Sushko and Gardini (2006); Zhusubaliyev *et al.* (2008, 2007)]. For instance, the stable and unstable manifolds of the saddle cycle may transversely intersect, replacing the invariant circle with a homoclinic tangle [Sushko and Gardini (2008); Zhusubaliyev *et al.* (2006, 2007)]. The invariant circle consisting of a stable cycle, a saddle cycle and the unstable invariant manifolds of the saddle may disappear upon collision and annihilation of the two cycles in a border-collision bifurcation. Also a loss of stability of the stable cycle may lead to the destruction of the invariant circle [Sushko and Gardini (2006); Maistrenko *et al.* (1998)]. These mechanisms are similar to those that occur in smooth maps [Aronson *et al.* (1982); Agliari *et al.* (2005)]. Note that standard results concerning normal hyperbolicity [Robinson (1999); Hirsch *et al.* (1977); Fenichel (1971)] do not immediately apply because the map is not a diffeomorphism.

Periodic solutions in resonance tongues that emanate from $(\omega_R, s_R) = (\frac{m}{n}, 1)$, have symbol sequences that are rotational. When the m/n-cycle has l points in the LHP, its symbol sequence is $\mathcal{S}[l, m, n]$, (6.18). Stable m/n-cycles associated with the top-most lens in a tongue are comprised of two points in the LHP and $(n-2)$ points in the RHP. Numerically it has been observed that other lenses correspond to stable m/n-cycles with one more point in the LHP than the connected lens above. The saddle n-cycle has one fewer point in the LHP and one more point in the RHP than its stable counterpart.

Fig. 7.7 shows resonance tongues when $\omega_L = \omega_R$ for two different fixed values of r_L. Lens-chains emanate from $s_R = 1$ and may extend to $s_R = r_L$. If n is odd, the bottom-most lens corresponds to stable n-cycles with $\frac{n+1}{2}$ points in the LHP and the boundary of this lens intersects $s_R = r_L$ at one point. Alternatively if n is even, the bottom-most lens corresponds to stable n-cycles with $\frac{n}{2}$ points in the LHP and there are possibly two intersection points of the boundary of this lens with $s_R = r_L$. The boundary curve connecting these two points bends above $s_R = r_L$ as in Fig. 7.9 and corresponds to the associated stability matrix, (6.10), having a multiplier 1. No stable solutions have been discovered when $s_R < r_L$ prompting the following conjecture.

Conjecture 7.2. *Suppose* $\mu = 1$, $0 < \omega_L = \omega_R < 1/2$ *and* $0 < s_R < r_L < 1$. *Then (7.5) has no stable solutions.*

Fig. 7.10 shows two typical phase portraits corresponding to points in parameter space just beyond resonance tongue boundaries that correspond to stability loss of a periodic solution via an associated multiplier passing through the value -1. The nonsmooth period-doubling bifurcations that occur at these boundaries were described from a general viewpoint in Sec. 6.3. For the map (7.5), in the case that the period-doubled cycle is unstable and exists for $\lambda_{PD} < -1$ (where λ_{PD} is the critical multiplier) it has been observed that it is typically embedded in a complicated attractor. In some cases, this attractor coincides with the period-doubled solution when it is created and grows in size as $\lambda_{PD} + 1$ decreases, see Fig. 7.10(a). Alternatively the attractor can be large and contain the undoubled solution when it is created, see Fig. 7.10(b). In both cases the Lyapunov exponent for the attractor appears to be positive ($\gamma_{\rm Lyap} \approx 0.0747$ in panel (a), $\gamma_{\rm Lyap} \approx 0.0129$ in panel (a)) suggesting the attractor is chaotic. As the parameters are varied further the multiple-piece attractor may undergo merging [Maistrenko *et al.* (1998)], see also [Avrutin *et al.* (2008); Avrutin and Schanz (2008)].

Just as the fixed point can undergo a Neimark-Sacker bifurcation, so can a periodic solution. When this bifurcation does not coincide with a border-collision, the stability matrix $M_{\mathcal{S}}$ will have a pair of complex eigenvalues on the unit circle. For a two-dimensional map such as (7.5), this implies $\det(M_{\mathcal{S}}) = 1$. Consequently $r_L^l = s_R^{n-l}$, where l is the number of L's in $\mathcal{S}$. Fig. 7.11 provides an example with $\mathcal{S} = LLR$. Here, when

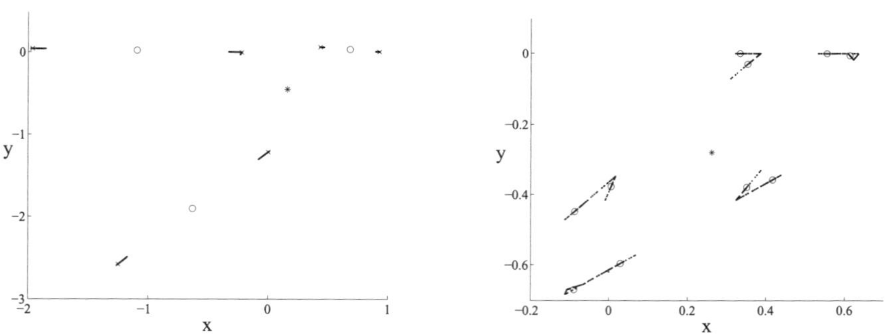

(a) (b)

Fig. 7.10: Phase portraits of (7.5) when $\lambda_{PD} \approx -1.01$, $\mu = 1$. In panel (a), $n = 3$, in panel (b), $n = 10$. The fixed point z^* is indicated by an asterisk. A saddle n-cycle is indicated by circles. In panel (a), a saddle $2n$-cycle is indicated by crosses. In both panels a complicated attractor is indicated by dots. The parameter values are $(r_L, s_R, \omega_L, \omega_R) = (0.18202, 0.6, 0.09, 0.38)$ in panel (a) and $(r_L, s_R, \omega_L, \omega_R) = (0.0193, 0.964, 0.38, 0.41)$ in panel (b).

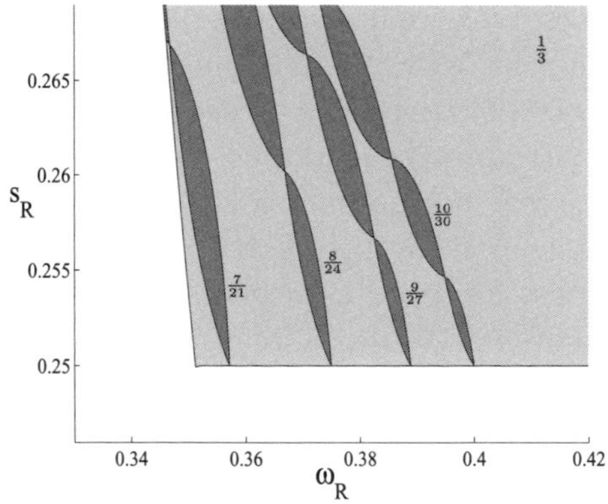

Fig. 7.11: The large region is a resonance tongue for 3-cycles of (7.5) when $r_L = 0.5$, $\omega_L = 0.25$ and $\mu = 1$. When $s_R = 0.25$ the 3-cycles lose stability via associated complex multipliers crossing the unit circle at $e^{2\pi i(\frac{1}{2}-\omega_R)}$. Admissibility regions emanate from points on this boundary for which $(\frac{1}{2} - \omega_R)$ is rational. Four such regions are shown. In each case the corresponding periodic solutions are unstable.

$s_R = 0.25$ and $\omega_R \in (\hat{\omega}, \frac{1}{2})$ where $\hat{\omega} = \frac{1}{2\pi}\cos^{-1}(\frac{19}{32}) \approx 0.3512$, $M_{\mathcal{S}}$ has multipliers $e^{2\pi i(\frac{1}{2}-\omega_R)}$ and the $\mathcal{S}$-cycle is admissible. Whenever $(\frac{1}{2} - \omega_R) = p/q$ is an irreducible fraction there will exist uncountably many qn-cycles with rotation numbers qm/qn at the bifurcation. Generically one of these cycles has two points on the switching manifold. In a similar manner as for terminating shrinking points described in Sec. 6.6, regions of admissibility of the qn-cycles emanate from this codimension-two point in parameter space, Fig. 7.11. However no example has been found with (7.5) for which the qn-cycle is stable.

7.5 Complex Phenomena Relating to Resonance Tongues

Shrinking point curves

By Lemma 6.9, at a shrinking point there exists an invariant polygon with two vertices on the switching manifold. Now denote these two vertices by $w_1 = (0, y_1)^{\mathsf{T}}$ and $w_2 = (0, y_2)^{\mathsf{T}}$ and assume without loss of generality, $y_1 < y_2$. Since these points are on a periodic orbit then $f^{\kappa}(w_1) = w_2$ for some κ. For example, in Fig. 7.12, $\kappa = 5$. The polygon persists as an attracting invariant set as parameters are continuously varied, though it will no longer necessarily contain only periodic orbits. When the polygon persists as an invariant circle the points of intersection of the invariant circle with the switching manifold, w_1 and w_2, also vary continuously. Moreover it is a codimension-one phenomenon for w_1 to map into w_2 in κ iterations. Thus there exists a curve in two-dimensional parameter space along which $f^{\kappa}(w_1) = w_2$. Such a curve will be referred to as a *shrinking point curve.* These curves also exist for one-dimensional, piecewise-linear circle maps, and in [Yang and Hao (1987)], the authors were able to obtain analytical expressions.

Shrinking point curves can be computed numerically by first finding an approximate invariant circle by the algorithm described in [Edoh and Lorenz (2003)], and then estimating the points w_1 and w_2 by interpolation. Via an iterative scheme, the parameters are chosen to minimize $|f^{\kappa}(w_1) - w_2|$ for some fixed κ; an example is shown in Fig. 7.13.

For instance when $\kappa = 5$, there is a curve that extends from the top of the 1/4-resonance tongue to the bottom of the 3/10-resonance tongue, see Fig. 7.13. This curve intersects the 2/7-resonance tongue at which point

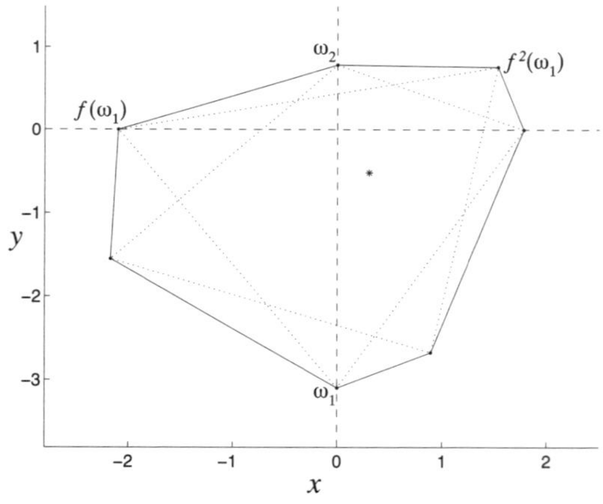

Fig. 7.12: A phase portrait of (7.5) for parameter values ($r_L = 0.6$, $s_R \approx 0.7583$, $\omega_L = \omega_R \approx 0.2841$) corresponding to a non-terminating shrinking point on the 2/7 resonance tongue. The invariant circle is a heptagon that consists entirely of 2/7-cycles. As when $s_R = 1$, the vertices map to one another as do the sides. The fixed point z^* is indicated by an asterisk. The dotted lines connect the vertices with their images.

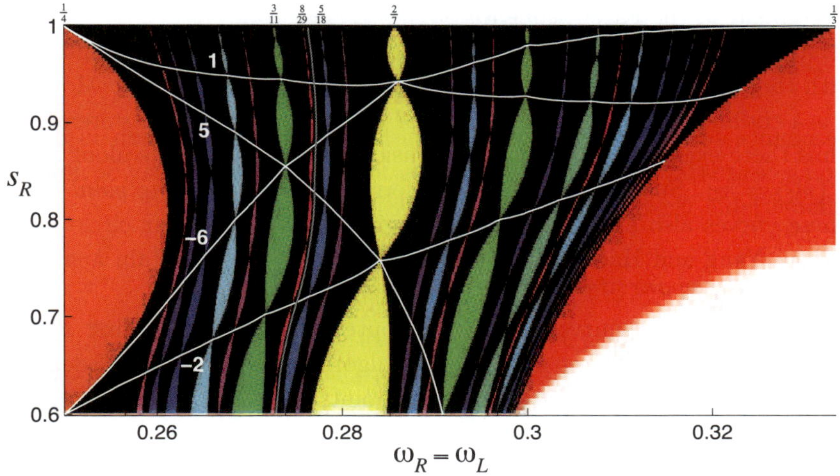

Fig. 7.13: A magnified view of Fig. 7.7(a) with additional curves. The white curves are shrinking point curves for $\kappa = -6$, -2, 1, and 5. The gray curve corresponds to stable solutions for which the rotation number is $\frac{1}{2+\gamma} \approx 0.2764$ where γ is the golden ratio.

the associated 2/7-polygon appears as in Fig. 7.12. As seen in this figure, w_1 maps into w_2 upon five iterations of (1.14). At this point w_2 maps into w_1 in two iterations, hence the shrinking point curve for which $\kappa = -2$ also crosses this point. Similarly the curves for $\kappa = 1$ and $\kappa = -6$ intersect at the upper non-terminating shrinking point of the 2/7-resonance tongue.

At shrinking points, shrinking point curves appear non-differentiable. This suggests that any shrinking point curve is non-differentiable at a dense set of points on the curve.

Irrational rotation numbers

In addition to invariant circles with rational rotation numbers, irrational circles may be approximated by following a sequence of rational lens-chains whose rotation numbers limit on a given irrational number. For example, in Fig. 7.13, the gray one-dimensional curve corresponds to the existence of an invariant circle whose rotation number is computed to be $\frac{1}{2+\gamma}$, where $\gamma = \frac{\sqrt{5}+1}{2}$ is the golden ratio, within an error of 10^{-10}. Note that this curve is sandwiched between resonance tongues whose rotation numbers correspond to the Farey sequence

$$\frac{1}{3}, \frac{1}{4}, \frac{2}{7}, \frac{3}{11}, \frac{5}{18}, \frac{8}{29} \to \frac{1}{2+\gamma} \approx 0.2764 .$$

Resonance tongues associated with the first six rotation numbers in this sequence are shown in Fig. 7.13. Curves defined in this way may be nowhere differentiable [Sushko and Gardini (2008)].

Lyapunov exponents

To distinguish between regular and chaotic orbits, Lyapunov exponents were numerically computed for initial conditions on a 512×512 grid in parameter space, see Fig. 7.14. To create panel (b), for each choice of parameter values, a randomly chosen initial condition was iterated for 10^4 steps after removing transients. When there are multiple attractors, the value computed for the Lyapunov exponent is dependent upon which basin of attraction the initial random point is located. For example, the black and white area centered at $(\omega_R, s_R) = (0.294, 0.43)$ corresponds to the coexistence of a stable 1/4-cycle and a chaotic attractor born through the loss of stability of the 2/9-cycle via a nonsmooth period-doubling bifurcation,

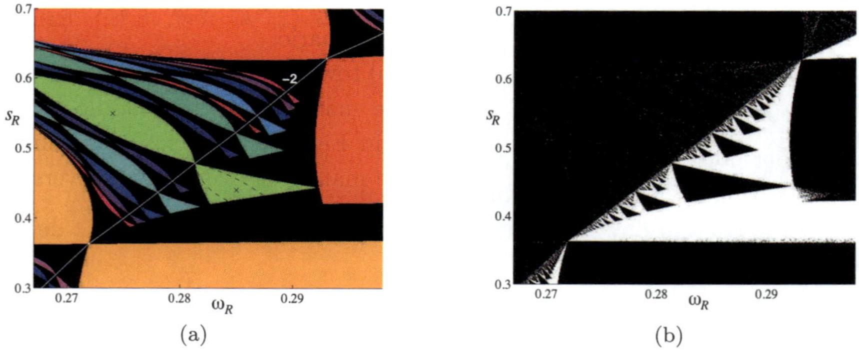

Fig. 7.14: Panel (a) shows a magnified view of Fig. 7.6(a), ($r_L = 0.3$, $\omega_L = 0.09$ and $\mu = 1$). The white curve is the shrinking point curve with $\kappa = -2$; the dashed black curves correspond to first homoclinic tangencies. Phase portraits for parameter values indicated by crosses are shown in Fig. 7.15. Panel (b) shows numerically calculated Lyapunov exponents over the same parameter range. Black [white] areas correspond to negative [positive] Lyapunov exponents. Gray areas correspond to numerically computed Lyapunov exponents with a magnitude less than 0.0005.

see Sec. 6.3.

If there exists an attracting invariant circle with an irrational rotation number, the map restricted to the circle is semi-conjugate to rigid rotation [Cornfeld *et al.* (1982)]. Thus in this case, orbits on the circle are quasiperiodic and have zero Lyapunov exponent. For this reason the upper-left half of Fig. 7.14(b) contains many data points for which the numerically computed Lyapunov exponent has a value within 0.0005 of zero. These grey curves also seem to fall between the lens-chains corresponding to rational rotation numbers, which is consistent with their rotation number being irrational.

The line that runs diagonally through Fig. 7.14 is the $\kappa = -2$ shrinking point curve. To the left of this curve, there appear to be no chaotic solutions, or at least they are very much less common. Fig. 7.15 shows stable and unstable invariant manifolds of the saddle 2/9-cycle on either side of the shrinking point curve. In panel (a) the associated eigenvalues of both the stable and saddle 2/9-cycles are positive and the unstable manifolds form an attracting invariant circle. In panel (b) the associated eigenvalues of the stable 2/9-cycle are negative and the unstable manifolds spiral into the stable 2/9-cycle. Similar behavior is described in [Aronson *et al.* (1982)] (in particular Fig. 8.2 of this reference). A similar situation is observed near the shrinking point curve for all lens-chains shown in Fig. 7.14(a), thus Fig. 7.15 is believed to illustrate a typical scenario.

Variation of the parameters of panel (b) towards a nonsmooth fold leads to a collision of the stable and unstable manifolds of the saddle cycle. Beyond curves of first tangency (shown in Fig. 7.14) the invariant circle no longer exists. The stable and unstable manifolds intersect transversely forming a homoclinic tangle.

Non-rotational periodic solutions

The rotation number of an n-cycle of the map (7.5) is a fraction m/n, where m is the number of times the orbit revolves around the fixed point in n steps. It is possible that m and n contain a common factor, i.e. the rotation number is a reducible fraction. In this case the periodic solution will usually have a symbol sequence that is non-rotational (an exception is the 2/6-cycle shown in Fig. 7.17(a) for which $\mathcal{S} = LLRRRR$). As shown below resonance tongues associated with non-rotational periodic solutions may not exhibit the familiar lens-chain structure.

An example is shown in Fig. 7.16. Here there are three regions corre-

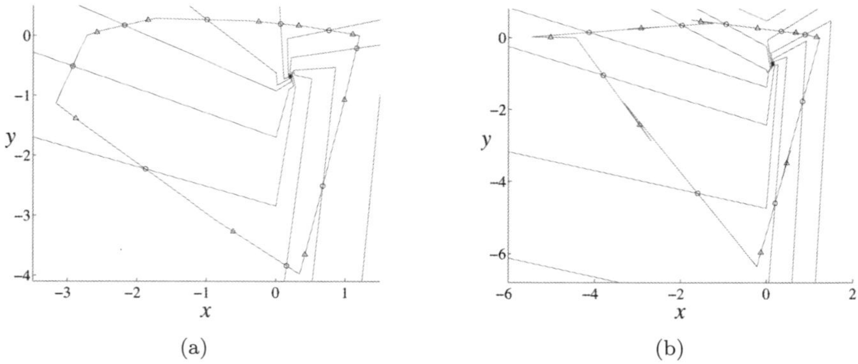

Fig. 7.15: Stable and unstable invariant manifolds of a saddle 2/9-cycle of (7.5) when $r_L = 0.3$, $\omega_L = 0.09$ and $\mu = 1$. In panel (a), $\omega_R = 0.274$ and $s_R = 0.55$. In panel (b), $\omega_R = 0.285$ and $s_R = 0.44$. Triangles [circles] denote stable [saddle] 2/9-cycles. In each panel stable invariant manifolds of the saddle cycle limit on the unstable fixed point z^* in the RHP.

sponding to stable solutions with three different symbol sequences. The region (abe) corresponds to parameters for which there exists a stable 3/9-cycle with symbol sequence LRLLRRRRL. The right-hand boundary of this region, (a-e), corresponds to a loss of stability via a nonsmooth period-doubling bifurcation. To the right of this boundary the 3/9-cycle exists but

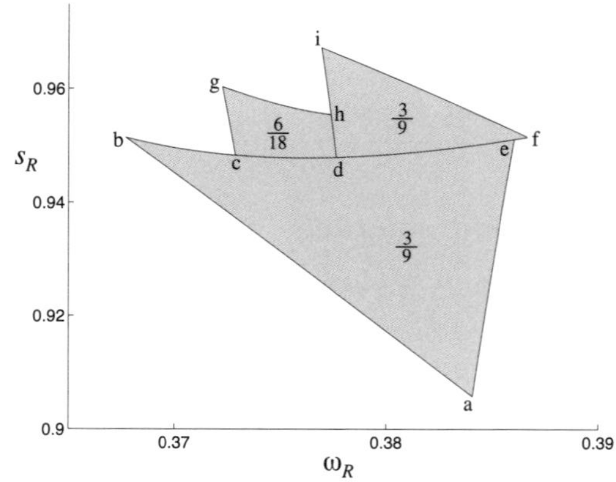

Fig. 7.16: Resonance tongues of (7.5) when $\mu = 1$, $r_L = 0.475$, $\omega_L = 0.09$. The bottom and upper right regions correspond to a stable 3/9-cycle. The upper left region corresponds to a stable 6/18-cycle.

is not attracting. Throughout this region there also exists a saddle 3/9-cycle (with symbol sequence LRLLRLRRL) that collides and annihilates with the stable cycle in a usual nonsmooth fold on the boundary (a-b).

The boundary (b-e) also corresponds to a nonsmooth fold; however, here the 3/9-cycle persists above the boundary. Above this boundary the 3/9-cycle has symbol sequence RRLLRRRRL and is stable to the right of the boundary (d-i), which corresponds to a nonsmooth period-doubling bifurcation. Along (b-d) a 6/18-cycle is created that exists above the boundary and is stable to the right of (c-g), which also corresponds to a nonsmooth period-doubling bifurcation. This period-doubled cycle has a symbol sequence that is the concatenation of the symbol sequences of the two stable 3/9-cycles. Finally, the boundaries (g-h) and (f-i) correspond to border-collision bifurcations beyond which saddles of the same rotation number exist.

Of all the border-collision boundaries so far discussed, the boundary (d-e) is the only one for which stable n-cycles exist on both sides of the bifurcation. Also of all the nonsmooth period-doubling bifurcations so far discussed, the boundary (d-h) is the only one along which the period-doubled solution is stable. Scenarios similar to Fig. 7.16 have been found for different rotation numbers. In summary, non-rotational periodic solutions appear to exhibit a wider variety of codimension-one bifurcations than rotational periodic solutions. In particular the familiar lens-chain structure may not arise.

Codimension-two, nonsmooth period-doubling bifurcations

As discussed in Sec. 6.3, at a generic, nonsmooth period-doubling bifurcation, a single point, $q_{\delta_{\max}}$, of the $2n$-cycle with symbol sequence $(\mathcal{S}^2)^{\overline{0}}$, lies on the switching manifold. Let us now consider the codimension-two situation that two points of this cycle, say $q^{(i)}_{\delta_{\max}}$ and $q^{(j)}_{\delta_{\max}}$, lie on the switching manifold when $M_{\mathcal{S}}$ has a multiplier $\lambda_{PD} = -1$.

In this case there will be two curves that cross at the codimension-two point; these correspond to the the vanishing of the x-components of $q^{(i)}_{\delta_{\max}}$ and $q^{(j)}_{\delta_{\max}}$ individually. These two curves divide the neighborhood of the codimension-two point into four quadrants. In one quadrant there will be no admissible $2n$-cycle, and in two quadrants exactly one of the two new orbits will be admissible; one with symbol sequence $(\mathcal{S}^2)^{\overline{i}}$ and the other with symbol sequence $(\mathcal{S}^2)^{\overline{j}}$. In the final quadrant, a new $2n$-cycle will be admissible, corresponding to the symbol sequence $(\mathcal{S}^2)^{\overline{ij}}$.

Fig. 7.17 shows two examples where the doubly-flipped $2n$-cycles are stable. In these cases the primary orbits, 1/3 and 2/5 respectively, are seen to lose stability with $\lambda_{PD} = -1$, and there is a narrow tongue corresponding to stable orbits with rotation numbers 2/6 and 4/10 respectively, that emanates from a codimension-two point on the period-doubling curve. Interestingly, there also exist additional resonance tongues in the neighborhood of this codimension-two point. These appear in sequences with rotation numbers $2k/6k$ for k up to 7 in panel (a) and rotation numbers $4k/10k$ for k up to 3 in panel (b). Resonance tongues corresponding to larger values of k were not found. In panel (a) the doubled orbits have symbol sequences $(LLR)^{2k-1}RRR$; note that only the first three tongues in the sequence appear to emanate from the codimension-two point. In panel (b) the doubled orbits have symbol sequences $(LRLRR)^{2k-1}RRRRR$, and only the first two tongues emanate from the codimension-two point.

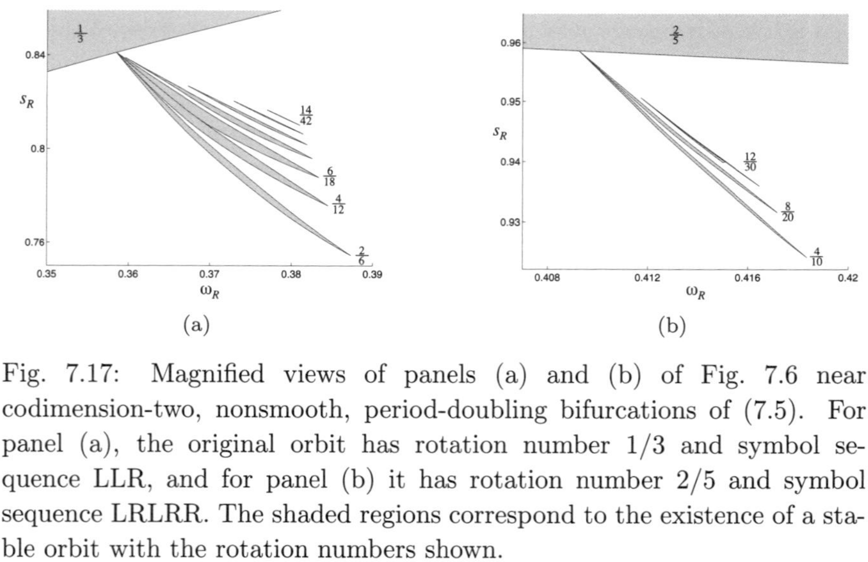

Fig. 7.17: Magnified views of panels (a) and (b) of Fig. 7.6 near codimension-two, nonsmooth, period-doubling bifurcations of (7.5). For panel (a), the original orbit has rotation number 1/3 and symbol sequence LLR, and for panel (b) it has rotation number 2/5 and symbol sequence LRLRR. The shaded regions correspond to the existence of a stable orbit with the rotation numbers shown.

7.6 More Complex Phenomena

Saddle fixed points for $\mu = 0$

The white regions in Fig. 7.7 correspond to parameter values where no attractor exists when $\mu = 1$. This section begins with a description of the corresponding border-collision bifurcation that occurs at $\mu = 0$. The most significant feature is that no invariant circle is created thus the bifurcation bears little relation to a Neimark-Sacker bifurcation in a smooth system.

As an example, Fig. 7.18 shows phase portraits of (7.5) at parameter values corresponding to a point in a white region of Fig. 7.7 for $\mu = -1, 0, 1$. When $\mu = -1$, the fixed point z^* (7.6) of (7.5) lies in the left half-plane and is stable. However its basin of attraction is not the entire plane: there is a saddle 3-cycle with symbol sequence LLR. The one-dimensional, piecewise-linear stable manifolds of this orbit form the boundary of the basin of attraction for the fixed point. Orbits of points in the interior of the complement of this region are unbounded. When $\mu = 1$, the opposite situation occurs: the fixed point lies in the right half-plane and is a repellor. The unstable manifolds of a saddle 3-cycle with symbol sequence RRL form the boundary for the basin of repulsion of the fixed point.

The LLR-cycle that exists when $\mu < 0$ is destroyed at $\mu = 0$; however, its unstable manifolds persist as an invariant set of piecewise-linear curves contained in the basin of repulsion; see panel (c) of Fig. 7.18. These manifolds extend to infinity and connect to the fixed point. Similarly the stable manifolds of the RRL-cycle become an invariant set of piecewise-linear curves in the basin of attraction of the fixed point when $\mu = 1$.

When $\mu = 0$, the two 3-cycles collide with the fixed point at the origin. The stable manifolds of the RRL-cycle and the unstable manifolds of the LLR-cycle become stable and unstable manifolds of the origin, which is now a saddle with six hyperbolic sectors.

The above dynamical behavior is believed to be generic. When $\omega_L = \omega_R$

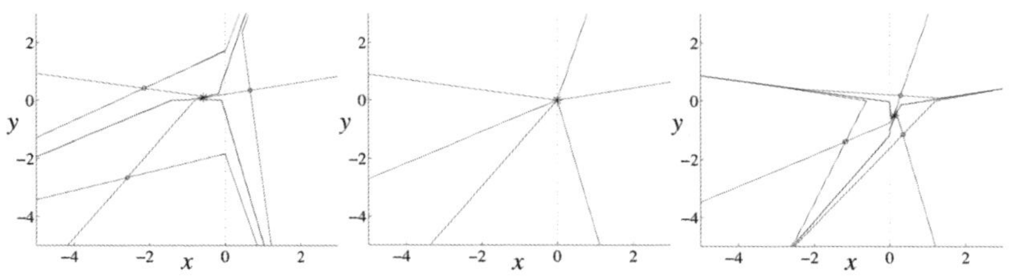

(a) (b) (c)

Fig. 7.18: Stable and unstable manifolds of a saddle 3-cycle (indicated by circles) for (7.5) when $(r_L, s_R, \omega_L, \omega_R) = (0.4, 0.5, 0.38, 0.38)$ for $\mu = -1, 0, 1$. The fixed point, z^*, is indicated with an asterisk and is stable in panel (a), a saddle in panel (b), and unstable in panel (c).

and the origin is of saddle-type for $\mu = 0$, it is expected that n stable invariant rays and n unstable invariant rays emanate from the origin as observed above for $n = 3$. In general when $\omega_L \neq \omega_R$ more complications may arise. A more thorough understanding of saddle fixed points on a switching manifold of (7.5) may follow by considering attractors at infinity [Gardini (2008)].

Multiple attractors

The overlapping of resonance tongues corresponds to the coexistence of multiple stable periodic solutions. An overlap is shown, for example in Fig. 7.6(b) near $\omega_R = 0.27$ and $s_R = 0.5$. Multiple attractors in piecewise-smooth maps have been described previously, see for instance [Zhusubaliyev *et al.* (2006); Sushko and Gardini (2008); Nusse *et al.* (1994); Maistrenko *et al.* (1998)]. As $\mu \to 0$ attractors contract to the origin, thus if small noise is added to the system there is an inherent unpredictability in the time evolution of solutions near the bifurcation point [Dutta *et al.* (1999); Kapitaniak and Maistrenko (1998)].

Two examples of coexisting attractors are shown in Fig. 7.19. For the parameters of panel (a), there exists a stable 3-cycle and a stable 4-cycle. For this case there is also a saddle 3-cycle and a saddle 4-cycle, and the stable manifold of the latter saddle forms the boundary between the basins of attraction of the two stable orbits.

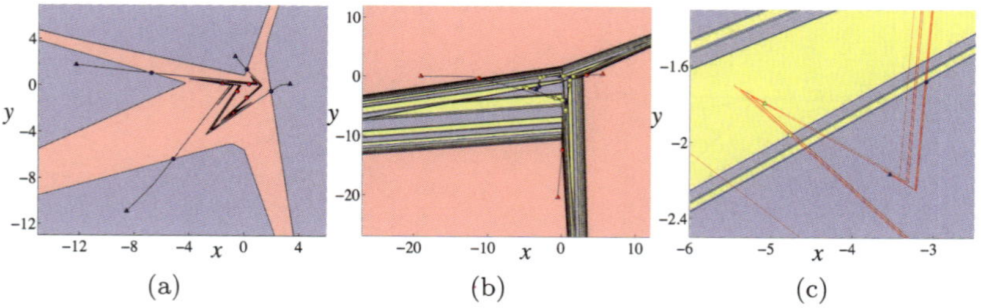

Fig. 7.19: Attractors and their basins of attraction of (7.5) when $\mu = 1$ and $(r_L, s_R, \omega_L, \omega_R) = (0.45, 0.55, 0.2, 0.4)$ in panel (a) and $(0.16, 0.52, 0.38, 0.265)$ in panel (b). Stable [saddle] periodic orbits are indicated by triangles [circles]. and shaded red for period 3, blue for period 4 and yellow for period 7. Panel (c) is a magnification of panel (b); here stable [unstable] invariant manifolds are colored blue [red].

The coexistence of three attractors is illustrated in Fig. 7.19(b); these correspond to the mutual intersection of the 1/3, 1/4 and 2/7 resonance tongues as seen in Fig. 7.6(b). The one-dimensional stable manifold of the saddle 3-cycle forms the boundary of the basin of attraction of the stable 3-cycle. However the boundary between the remaining two basins is more complicated. The stable manifolds of the saddle 4-cycle and the saddle 7-cycle transversely intersect the unstable manifolds of all three saddles forming a collection of homoclinic and heteroclinic orbits. The stable manifolds of the saddle 4-cycle and the saddle 7-cycle appear to accumulate and form a fractal boundary between the two basins. Note that six coexisting attractors have been found for the map (7.5) (specifically when $(\mu, r_L, s_R, \omega_L, \omega_R) = (1, 0.68, 0.8, 0.38, 0.27)$ there exist attractors with symbol sequences $LRLL(RLL)^k$ for $k = 7, \ldots, 12$); it is hypothesized that arbitrarily many distinct stable solutions can coexist.

Further complications

This chapter has mostly studied the case $\mu > 0$ for the map (7.5). However, using the symmetry property (7.4), each stable [unstable] solution for $\mu > 0$ corresponds to a unstable [stable] solution of the same period and rotation number for $\mu < 0$ if the R and L parameters are exchanged. For some choices of parameter values, non-trivial periodic orbits can be observed for both signs of μ. For example in panel (a) of Fig. 7.20, we see an unstable 4-cycle for $\mu < 0$ and a stable 3-cycle for $\mu > 0$. Alternatively a stable n-cycle can coexist with the stable fixed point. For example in panel (b)

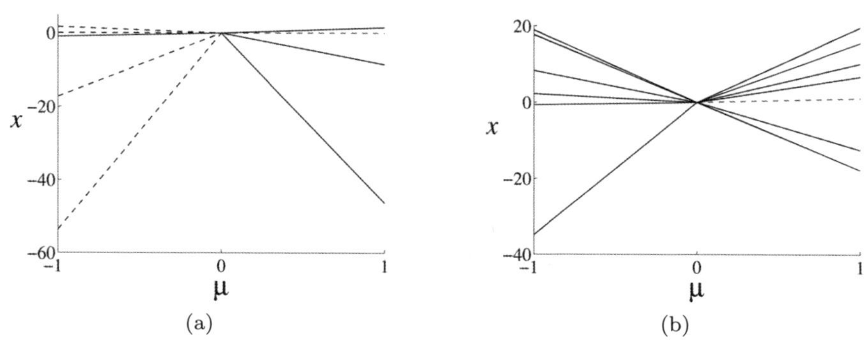

Fig. 7.20: Bifurcation diagrams of (7.5) when $(r_L, s_R, \omega_L, \omega_R) = (0.25, 0.18, 0.25, 0.38)$ in panel (a) and $(0.2, 0.7, 0.49, 0.12)$ in panel (b). Solid [dashed] lines denote stable [unstable] solutions.

of Fig. 7.20 a stable 5-cycle coexists with the stable fixed point for $\mu < 0$, and a stable 6-cycle is created for $\mu > 0$.

Multiple attractors may coexist. Similarly stable and unstable solutions with different periods may coexist, and various combinations of these phenomena may occur. In short, the bifurcation at $\mu = 0$ may be extremely complex.

Appendix A

Selected Proofs

Proof. [Lemma 1.3] For clarity here we write $A_L = A_L(0)$. Firstly suppose (1.10) [(1.14)] is observable. Let $y = \mathcal{O}^{-1}e_1$ and let $S = \left[\begin{array}{c|c|c|c} A_L^{N-1}y & \cdots & A_L y & y \end{array}\right]$. Notice, $\forall i \neq 1$, the ith column of $A_L S$ is, $A_L(A_L^{N-i}y) = A_L^{N-i+1}y$ equals the $(i-1)^{\text{th}}$ column of S, thus

$$A_L S = SC \; , \tag{A.1}$$

for some companion matrix C. To show S is nonsingular we show that its columns are linearly independent. We search for scalars α_i which solve

$$\alpha_1 y + \alpha_2 A_L y + \cdots + \alpha_N A_L^{N-1} y = 0 \; . \tag{A.2}$$

Multiplying (A.2) on the left by e_1^T and using $\mathcal{O}y = e_1$ we obtain $\alpha_N = 0$. Successive left multiplications of (A.2) by $e_1^\mathsf{T} A_L, e_1^\mathsf{T} A_L^2, \ldots, e_1^\mathsf{T} A_L^{N-1}$ gives $\alpha_{N-1} = \alpha_{N-2} = \cdots = \alpha_1 = 0$ verifying independence. Thus A_L is similar to C and $C = C_L$. Let

$$T = S^{-1} \; .$$

Then $TA_LT^{-1} = C_L$ as required. Notice the ith component of $e_1^\mathsf{T} S$ is $[e_1^\mathsf{T} S]_i = e_1^\mathsf{T}(A_L^{N-i}y) = (e_1^\mathsf{T} A_L^{N-i})y = [\mathcal{O}y]_i = [e_1^\mathsf{T}]_i$, thus $e_1^\mathsf{T} S = e_1^\mathsf{T}$. Multiplying this last equation on the right by T produces (1.30). Conversely suppose (1.30) and (1.31) are satisfied for some nonsingular T. Then $\mathcal{O}T^{-1} = \begin{bmatrix} e_1^\mathsf{T} A_L^{N-1} T^{-1} \\ \vdots \\ e_1^\mathsf{T} A_L T^{-1} \\ e_1^\mathsf{T} T^{-1} \end{bmatrix} = \begin{bmatrix} e_1^\mathsf{T} T^{-1}(TA_LT^{-1})^{N-1} \\ \vdots \\ e_1^\mathsf{T} T^{-1}(TA_LT^{-1}) \\ e_1^\mathsf{T} T^{-1} \end{bmatrix} = \begin{bmatrix} e_1^\mathsf{T} C_L^{N-1} \\ \vdots \\ e_1^\mathsf{T} C_L \\ e_1^\mathsf{T} \end{bmatrix}$. It is easily established this last matrix has the form $\begin{bmatrix} \theta_{N-1} & \theta_{N-2} & \cdots & \theta_1 & 1 \\ \theta_{N-2} & & & \cdot^{\cdot^{\cdot}} & 1 \\ \vdots & & \cdot^{\cdot^{\cdot}} & \cdot^{\cdot^{\cdot}} & \\ \theta_1 & 1 & & & \\ 1 & & & & \end{bmatrix}$

and is therefore nonsingular. Hence $\mathcal{O}$ is nonsingular. □

Proof. [Theorem 1.1] For clarity we prove the result for just the continuous-time case. The discrete-time case may be proved similarly. Let T be the similarity transformation matrix for $A_L(0)$ as described in Lemma 1.3. Let $d \in \mathbb{R}^N$ with

$$e_N^{\mathsf{T}} T d \neq 0 \,. \tag{A.3}$$

Let $\alpha \in \mathbb{R}$, $\psi \in \mathbb{R}^N$ with $\psi_1 = 0$ and let $\phi = (\alpha, \psi_2, \ldots, \psi_N)^{\mathsf{T}}$. Define a C^k function $G : \mathbb{R}^N \times \mathbb{R} \to \mathbb{R}^N$ by

$$G(\phi, \mu) = f^{(L)}(\psi; \mu) - \alpha d \,.$$

Then $G(0,0) = 0$, $\frac{\partial G}{\partial \mu}(0,0) = b(0)$ and $D_\phi G(0,0) = \hat{A}$, where $\hat{A}$ is the $N \times N$ matrix identical to $A_L(0)$ except in the first column, which is $-d$. We now show $\hat{A}$ is nonsingular. Notice $\hat{C} = T\hat{A}T^{-1}$ is a companion matrix and from our particular formulation of companion matrices, (1.27), it follows $\det(\hat{A}) = \det(\hat{C}) = (-1)^{N+1} e_N^{\mathsf{T}} \hat{C} e_1$. By judiciously using the relation $e_1^{\mathsf{T}} T = e_1^{\mathsf{T}} T^{-1} = e_1^{\mathsf{T}}$, we obtain

$$\hat{C} = T(A_L(0) - A_L(0) e_1 e_1^{\mathsf{T}} - d e_1^{\mathsf{T}}) T^{-1} = C_L(0)(I - T e_1 e_1^{\mathsf{T}}) - T d e_1^{\mathsf{T}} \,,$$

and therefore

$$\det(\hat{A}) = \det(C_L(0))(1 - e_1^{\mathsf{T}} T e_1) - e_N^{\mathsf{T}} T d = -e_N^{\mathsf{T}} T d \neq 0 \,. \tag{A.4}$$

Using (A.4), by the implicit function theorem, for small μ

$$G(\phi(\mu), \mu) \equiv 0 \,,$$

for a C^k function ϕ. Notice then

$$\alpha'(0) = e_1^{\mathsf{T}} \phi'(0) = -e_1^{\mathsf{T}} \hat{A}^{-1} b(0) = -\frac{\varrho_{\mathrm{ode}}^{\mathsf{T}}(0) b(0)}{\det(\hat{A})} \neq 0 \,,$$

by the earlier assumption, (1.24).

It remains to compute $\hat{T}$. Let

$$\tilde{A}_L(\mu) = D_x f^{(L)}(\psi(\mu); \mu) \,.$$

Notice $\tilde{A}_L(\mu)$ is C^{k-1} and $\tilde{A}_L(0) = A_L(0)$ therefore we may apply Lemma 1.3 to $\tilde{A}_L(\mu)$ for small μ. There exists an $N \times N$ matrix, $\hat{T}(\mu)$, C^{k-1} dependent on μ, with first row e_1^{T} and such that $\tilde{C}_L(\mu) = \hat{T}(\mu)\tilde{A}_L(\mu)\hat{T}^{-1}(\mu)$ is a companion matrix. Then by applying (1.32) and (1.33) we obtain

$$\begin{aligned}\hat{f}^{(L)}(\hat{x};\hat{\mu}) &= \hat{T}(\mu)f^{(L)}(x;\mu) \\ &= \hat{\mu}q + \tilde{C}_L(\hat{\mu})\hat{x} + O(|\hat{x}|^2) + o(k) ,\end{aligned}$$

where $q = \hat{T}(\mu)d(\mu)$, by (A.3) must satisfy

$$e_N^{\mathsf{T}}q \neq 0 .$$

Our choice of $q = e_N$ satisfies this condition. □

Proof. [Theorem 2.1]

Step 1: *Define* $\mathcal{P}_L$ *and* $\mathcal{P}_R$.

Denote the left and right half-flows of (1.10) with (1.34) by $\varphi_t^L(x, y; \mu)$ and $\varphi_t^R(x, y; \mu)$, respectively, for $(x, y; \mu)$ in some neighborhood of the origin. For sufficiently small $\delta_j > 0$ and $\delta_{\mu_j} > 0$, we define the first return map

$$\mathcal{P}_R : [0, \delta_R] \times [-\delta_{\mu_R}, 0] \to \mathbb{R}^- ,$$

by $\mathcal{P}_R(y_0; \mu) = y_1$ where y_1 is the first intersection of $\varphi_t^R(0, y_0; \mu)$ with the negative y-axis or origin for $t \geq 0$, and

$$\mathcal{P}_L : [-\delta_L, 0] \times [-\delta_{\mu_L}, 0] \to \mathbb{R}^+ ,$$

by $\mathcal{P}_L(y_1; \mu) = y_2$ where y_2 is the first intersection of $\varphi_t^L(0, y_1; \mu)$ with the positive y-axis or origin for $t > 0$. This is illustrated in Fig. 2.1(a).

Step 2: *Show* $\mathcal{P}_R$ *and* $\mathcal{P}_L$ *are well-defined when* $\mu = 0$.

When $\mu = 0$ the origin is a hyperbolic equilibrium of both left and right half-systems; consequently, $\mathcal{P}_R(0; 0) = \mathcal{P}_L(0; 0) = 0$. Since when $\mu = 0$ the matrices C_j (of (1.34)) have complex-valued eigenvalues, the trajectories of the linearized systems spiral about the origin repeatedly intersecting the y-axis. The spiral behavior is clockwise because in both linear systems $\dot{x}(0, y; 0) = y$. The same can be said of trajectories of the nonlinear half-systems sufficiently close to the origin because, by Hartman's theorem [Hartman (1964)], there exist neighborhoods about the origin within which the half-systems are C^1 conjugate to their linearizations. Thus when $\mu = 0$, $\mathcal{P}_R$ and $\mathcal{P}_L$ are well-defined.

Step 3: *Show* $\mathcal{P}_R$ *is well-defined for small* $\mu < 0$.

When $\mu < 0$, the origin is no longer an equilibrium of (1.10). The required properties of $\mathcal{P}_R$ can be obtained by considering the local behavior of the trajectory that passes through the origin. This can be deduced by computing approximations to the right half-flow on the y-axis. Since

$$\dot{x}(0, y; \mu) = y + o(y) ,$$

there is a neighborhood of the origin such that the positive y-axis flows into the right half-plane and the negative y-axis flows into the left half-plane. Moreover, since

$$\dot{y}(0, y; \mu) = \mu + o(y) ,$$

for small y and small $\mu < 0$, we have $\dot{y} < 0$. Thus the trajectory that passes through the origin, does so tangent to the y-axis and without entering the right half-plane. Since the flow is continuous, there exists a non-empty interval $[0, \delta_R]$ on the positive y-axis that maps into the negative y-axis. Thus the map, $\mathcal{P}_R$, is well-defined for some $\delta_{\mu_R} > 0$. Furthermore $\mathcal{P}_R(0; \mu) = 0$ (since the definition of $\mathcal{P}_R$ allows for an intersection at $t = 0$). Finally $\mathcal{P}_R$ is C^k since the right half-flow is C^k.

Step 4: *Show $\mathcal{P}_L$ is well-defined for small $\mu < 0$.*
When $\mu < 0$, the initial velocity of the trajectory $\varphi_t^L(0, 0; \mu)$ is directly downwards. The equilibrium of the left half-flow lies in the left half-plane when μ is small and negative. For small μ, the equilibrium lies close to the origin so that we may use the linearization of the flow at the equilibrium to approximate $\varphi_t^L(0, 0; \mu)$. Recall that the eigenvalues of $C_L(0)$ are $\nu_L \pm i\omega_L$, where $\nu_L, \omega_L > 0$, thus the equilibrium of the left half-system is a repelling focus at $\mu = 0$, and by continuity remains a repelling focus when μ is small enough. Consequently the flow $\varphi_t^L(0, 0; \mu)$ initially spirals clockwise around the equilibrium solution within the left half-plane. Since the equilibrium is repelling, before $\varphi_t^L(0, 0; \mu)$ has completed $360°$ about the equilibrium it will intersect the y-axis at some point $\mathcal{P}(0; \mu) = \hat{y}(\mu) > 0$, see Fig. 2.1. Since the flow is continuous, $\mathcal{P}_L$ must map a non-empty interval $[-\delta_L, 0]$ to points on the positive y-axis above $\hat{y}(\mu)$. Thus the map, $\mathcal{P}_L$, is well-defined for some $\delta_{\mu_L} > 0$. Also $\mathcal{P}_L$ is C^k since the left half-flow is C^k.

Step 5: *Define $\mathcal{P}$ and compute its derivatives at $(0_+; 0_-)$.*
The results above imply that there are $\delta, \delta_\mu > 0$ such that the Poincaré map, $\mathcal{P} : [0, \delta] \times [-\delta_\mu, 0] \to \mathbb{R}^+$ defined by (2.11) is C^k. This map is sketched in Fig. A.1. Note that since $\mathcal{P}_R(0; \mu) = 0$, its left-sided derivative with respect to μ as $\mu \to 0_-$ is $D_\mu P_R(0; 0_+) = 0$. Moreover, since $\mathcal{P}_L(0; \mu) = \hat{y}(\mu) = O(\mu)$ is positive for $\mu < 0$, we have

$$\alpha \equiv D_\mu \mathcal{P}_L(0; 0_-) = \lim_{\mu \to 0^-} \frac{\hat{y}(\mu)}{\mu} < 0 .$$

Thus the left-sided derivative of $\mathcal{P}$ with respect to μ is also negative:

$$D_\mu \mathcal{P}(0; 0_-) = D_\mu \mathcal{P}_L(0; 0_-) = \alpha < 0 \, .$$

To compute the value of $D_y \mathcal{P}(0_+; 0)$, consider the linear system

$$\dot{z} = \begin{bmatrix} a & 1 \\ b & 0 \end{bmatrix} z \, ,$$

with eigenvalues $\nu \pm i\omega$. For points on the y-axis, this system has the flow

$$\varphi_t(0, y) = \frac{e^{\nu t}}{\omega} \begin{bmatrix} \sin(\omega t) \\ (-a + \nu)\sin(\omega t) + \omega\cos(\omega t) \end{bmatrix} y \, .$$

Consequently it maps the positive y-axis onto the negative y-axis in time $\tau = \frac{\pi}{\omega}$ so that $\varphi_\tau(0, y) = (0, -e^{\nu\tau} y)$. Moreover, this system approximates the right half-flow in the right half-plane near the origin. Therefore as $y_0 \to 0_+$, $\mathcal{P}_R(y_0; 0) \to -e^{\frac{\nu_R \pi}{\omega_R}} y_0$. By a similar argument, as $y_1 \to 0_-$, $\mathcal{P}_L(y_1; 0) \to -e^{\frac{\nu_L \pi}{\omega_L}} y_1$. Therefore as $y_0 \to 0_+$,

$$\mathcal{P}(y_0; 0) \to \left(-e^{\frac{\nu_L \pi}{\omega_L}}\right)\left(-e^{\frac{\nu_R \pi}{\omega_R}}\right) y_0 = e^{\Lambda\pi} y_0 \, .$$

Thus

$$D_y \mathcal{P}(0_+; 0) = e^{\Lambda\pi} \, . \qquad \text{(A.5)}$$

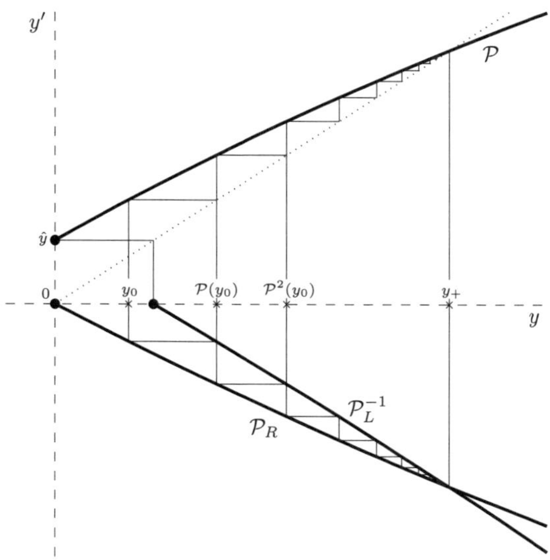

Fig. A.1: A sketch of a cobweb diagram for the Poincaré map, $\mathcal{P}$ (2.11). $\mathcal{P}$ has the stable fixed point y_+. Also shown are $\mathcal{P}_R$ and $\mathcal{P}_L^{-1}$.

Step 6: *Show the periodic orbit exists when* $\Lambda < 0$.
The function $V : [0,\delta] \times [-\delta_\mu, 0] \to \mathbb{R}$ defined by $V(y_0,\mu) = \mathcal{P}(y_0;\mu) - y_0$ is C^k. Zeros of V correspond to fixed points of the Poincaré map, $\mathcal{P}$, and thus periodic orbits in the system (1.10). In order to apply the implicit function theorem to V at the origin we must first smoothly extend its definition to a neighborhood of the origin. Let $\tilde{V} : [-\delta,\delta] \times [-\delta_\mu, \delta_\mu] \to \mathbb{R}$ be any C^k function with $\tilde{V}(y_0,\mu) = V(y_0,\mu)$ whenever $y_0 \geq 0$, $\mu \leq 0$. Note that

(i) $\tilde{V}(0,0) = V(0,0) = \mathcal{P}(0;0) = 0$,
(ii) $D_y\tilde{V}(0,0) = D_y\mathcal{P}(0_+;0) - 1 = e^{\Lambda\pi} - 1$,
(iii) $D_\mu\tilde{V}(0,0) = D_\mu\mathcal{P}(0;0_-) = \alpha < 0$.

Therefore, if $\Lambda \neq 0$, then the implicit function theorem implies that there is a C^k function y_+ satisfying $\tilde{V}(y_+(\mu),\mu) = 0$ for all μ in some neighborhood of $\mu = 0$ such that $y_+(0) = 0$ and

$$D_\mu y_+(0) = -\frac{D_\mu\tilde{V}(0,0)}{D_y\tilde{V}(0,0)} = \frac{\alpha}{1 - e^{\Lambda\pi}}\,. \tag{A.6}$$

Consequently, when $\Lambda < 0$, we have $D_\mu y_+(0) < 0$. Thus for small $\mu < 0$, $y_+(\mu) > 0$ so that $\tilde{V}(y_+(\mu),\mu) = V(y_+(\mu),\mu) = 0$ implying that $y_+(\mu)$ is a periodic point. In other words, the graph of $\mathcal{P}(y;\mu)$ necessarily intersects the diagonal as sketched in Fig. A.1, and when $-\delta_\mu < \mu < 0$, (1.10) has a periodic orbit that intersects the positive y-axis at $y_+(\mu)$.

The periodic orbit intersects the negative y-axis, at, say, $y_-(\mu) = \mathcal{P}_R(y_+(\mu);\mu)$. The function $y_-(\mu)$ is also C^k and vanishes at $\mu = 0$. Moreover,

$$\begin{aligned} D_\mu y_-(0_-) &= D_y\mathcal{P}_R(0_+;0)D_\mu y_+(0) + D_\mu\mathcal{P}_R(0;0_-) \\ &= \frac{\alpha}{e^{\frac{\nu_L\pi}{\omega_L}} - e^{\frac{-\nu_R\pi}{\omega_R}}} > 0\,. \end{aligned}$$

The resulting bifurcation diagram is sketched in Fig. 2.5. The radius of the periodic orbits, by any sensible definition, grows linearly with respect to $|\mu|$, to first order.

Step 7: *Show the periodic orbit is stable.*
The stability of periodic orbits can be deduced by calculating the value of $D_y\mathcal{P}(y_+(\mu);\mu)$. When $\Lambda < 0$, (A.5) implies that $0 < D_y\mathcal{P}(0_+;0) < 1$. Since this function is C^{k-1} and $k \geq 1$, we have $0 < D_y\mathcal{P}(y_+(\mu);\mu) < 1$, for all sufficiently small $\mu < 0$. Thus the periodic orbit is stable, as we should expect since the equilibrium is repelling.

Step 8: *Show there are no periodic orbits for* $\Lambda > 0$.
If $\Lambda > 0$, (A.6) implies that $y_+'(0) > 0$, so that the locus of zeros, $y_+(\mu)$,

fails to enter the quadrant $y_0 > 0, \mu < 0$, near $(0,0)$. Since, by the implicit function theorem, $y_+(\mu)$ is the unique solution that emerges from the origin, it follows that $\tilde{V}(y_0,\mu) \neq 0$, for all sufficiently small $y_0 > 0$, $\mu < 0$. Hence in this case, there are no periodic orbits. □

Proof. [Theorem 3.1] We first show $v(0,0)$ is not orthogonal to e_1. Suppose for a contradiction, $v_1(0,0) = 0$. Let B_{ij} denote the $(N-1) \times (N-1)$ matrix formed by removing the ith row and j^{th} column from $A_L(0,0)$. Let $\tilde{v} = (v_2(0,0) \; \ldots \; v_N(0,0))^{\mathsf{T}} \in \mathbb{R}^{N-1}$. Then $\tilde{v} \neq 0$ and $B_{i1}\tilde{v} = 0$ for each i. Therefore each $\det(B_{i1}) = 0$, i.e., each element in the first column of the cofactor matrix of $A_L(0,0)$ is zero. By definition the adjugate matrix is the transpose of the cofactor matrix, thus $\varrho^{\mathsf{T}}(0,0) = 0$ which is a contradiction by assumption (ii). Therefore we may indeed choose the length of $v(0,0)$ such that $e_1^{\mathsf{T}}(0,0)v(0,0) = 1$.

By putting $A = A_L(0,0)$ into (C.1), multiplying on the left by e_1^{T} and using (1.22) we obtain

$$\varrho^{\mathsf{T}}(0,0)A_L(0,0) = 0 \,.$$

That is, the vector $\varrho^{\mathsf{T}}(0,0)$ is the left eigenvector of $A_L(0,0)$ for the eigenvalue $\lambda(0,0) = 0$. Therefore $p_0 \neq 0$.

Let $v^{(1)}, \ldots, v^{(N)}$ be N generalized eigenvectors of $A_L(0,0)$ that form a basis of $\mathbb{R}^N$ with $v^{(1)} = v(0,0)$. Let $V = \left[v^{(1)} \; \cdots \; v^{(N)}\right]$. We introduce the linear change of coordinates

$$\hat{x} = V^{-1}x \,. \tag{A.7}$$

Let

$$F = \begin{bmatrix} \dot{\hat{x}} \\ \dot{\mu} \\ \dot{\eta} \end{bmatrix} = \begin{bmatrix} V^{-1}f^L(V\hat{x};\mu,\eta) \\ 0 \\ 0 \end{bmatrix} , \tag{A.8}$$

denote the $(N+2)$-dimensional, C^k,extended left half-flow in the basis of generalized eigenvectors. The Jacobian

$$DF(0;0,0) = \left[\begin{array}{c|c|c} V^{-1}A_L(0,0)V & V^{-1}b(0,0) & 0 \\ \hline 0 & 0 & 0 \end{array}\right] , \tag{A.9}$$

has a three-dimensional nullspace, E^c. There exists $\varphi \in \mathbb{R}^N$ such that the three $(N+2)$-dimensional vectors

$$\{n_1, n_2, n_3\} = \{e_1, \begin{bmatrix} \varphi \\ 1 \\ 0 \end{bmatrix}, e_{N+2}\} \,,$$

span E^c. Note, the vector, n_2, is a generalized eigenvector, because, if not, then $V^{-1}A_L(0,0)V\varphi + V^{-1}b = 0$ which implies $\varrho^{\mathsf{T}}(0,0)A_L(0,0)V\varphi = -\varrho^{\mathsf{T}}(0,0)b(0,0)$ which is a contradiction because $\varrho^{\mathsf{T}}(0,0)A_L(0,0) = 0$ and $\varrho^{\mathsf{T}}(0,0)b(0,0) \neq 0$. Furthermore, $DF(0;0,0)n_2 = \frac{b_0}{p_0}e_1$. Any point on E^c may be written in terms of $\hat{x}_1$, μ and η as the following linear combination

$$\begin{bmatrix} \hat{x} \\ \mu \\ \eta \end{bmatrix} = (\hat{x}_1 - \mu e_1^{\mathsf{T}}\varphi)n_1 + \mu n_2 + \eta n_3 \ ,$$

thus, in particular, $\hat{x} = \hat{x}_1 e_1 + \mu\zeta$, where $\zeta \in \mathbb{R}^N$ is equal to φ except that its first element is zero. The local center manifold, W^c, is tangent to E^c, thus on W^c,

$$\hat{x} = H(\hat{x}_1;\mu,\eta) = \hat{x}_1 e_1 + \mu\zeta + O(2) \ .$$

Notice $\varrho^{\mathsf{T}}(0,0)V = p_0 e_1^{\mathsf{T}}$, where we have used (3.1), thus

$$\hat{x}_1 = e_1^{\mathsf{T}}\hat{x} = \frac{1}{p_0}\varrho^{\mathsf{T}}(0,0)V\hat{x} = \frac{1}{p_0}\varrho^{\mathsf{T}}(0,0)x \ .$$

Restricted to W^c the dynamics (A.8) become the C^{k-1} system

$$\begin{aligned} \dot{\hat{x}}_1 = {} & \frac{\mu}{p_0}\varrho^{\mathsf{T}}(0,0)b(\mu,\eta) + \frac{1}{p_0}\varrho^{\mathsf{T}}(0,0)A_L(\mu,\eta)VH(\hat{x}_1;\mu,\eta) \\ & + \frac{1}{p_0}\varrho^{\mathsf{T}}(0,0)g^L(VH(\hat{x}_1;\mu,\eta);\mu,\eta) \ , \end{aligned} \tag{A.10}$$

where $g^L(x;\mu,\eta)$ denotes all terms of f^L that are nonlinear in x. By expanding each term in (A.10) to second order we obtain

$$\dot{\hat{x}}_1(\hat{x}_1;\mu,\eta) = c_1\mu + c_2\hat{x}_1^2 + c_3\hat{x}_1\eta + c_4\hat{x}_1\mu + c_5\mu^2 + c_6\mu\eta + O(3) \ , \tag{A.11}$$

where, in particular

$$\begin{aligned} c_1 &= \frac{b_0}{p_0} \ , \\ c_2 &= \frac{a_0}{2p_0} \ , \\ c_3 &= \frac{1}{p_0}\varrho^{\mathsf{T}}(0,0)\frac{\partial A_L}{\partial\eta}(0,0)v(0,0) = \frac{\partial\lambda}{\partial\eta}(0,0) = 1 \ . \end{aligned}$$

Let $\hat{x}_1^*$ be an equilibrium of (A.11). Since $c_1 \neq 0$ by hypothesis, by the implicit function theorem there exists a unique C^{k-1} function, $\mu_{\rm eq}(\hat{x}_1^*,\eta)$ such that $\dot{\hat{x}}_1(\hat{x}_1^*;\mu_{\rm eq}(\hat{x}_1^*,\eta),\eta) = 0$ and

$$\mu_{\rm eq}(\hat{x}_1^*,\eta) = -\frac{c_2}{c_1}{\hat{x}_1^*}^2 - \frac{c_3}{c_1}\hat{x}_1^*\eta + O(3) \ .$$

Near $(\mu, \eta) = (0, 0)$, the linearization about the equilibrium $\hat{x}_1^*$, has an associated eigenvalue of 0 exactly when

$$0 = \frac{\partial \dot{\hat{x}}_1}{\partial \hat{x}_1}(\hat{x}_1^*; \mu_{\text{eq}}(\hat{x}_1^*, \eta), \eta) = 2c_2\hat{x}_1^* + c_3\eta + O(2) . \tag{A.12}$$

Since $a_0 \neq 0$ by hypothesis, the implicit function theorem again implies there exists a unique C^{k-2} function, $\hat{h} : \mathbb{R} \to \mathbb{R}$ such that, (A.12) is satisfied when $\hat{x}_1^* = \hat{h}(\eta)$ and

$$\hat{h}(\eta) = -\frac{c_3}{2c_2}\eta + O(\eta^2) . \tag{A.13}$$

The function

$$h(\eta) = \mu_{\text{eq}}(\hat{h}(\eta), \eta) = \frac{c_3^2}{4c_1c_2}\eta^2 + O(\eta^3) , \tag{A.14}$$

is C^{k-2} and agrees with (3.2). We now show saddle-node bifurcations occur for the left half-flow on the curve $\mu = h(\eta)$ when η is small by verifying the three conditions of the saddle-node bifurcation theorem, see for instance [Guckenheimer and Holmes (1986)]:

(i) by construction, $D_x f^{(L)}(x^*; h(\eta), \eta)$ has a zero eigenvalue of algebraic multiplicity 1, and there are no other eigenvalues with zero real part when η is sufficiently small,
(ii) $w(\mu, \eta)\frac{\partial f^L}{\partial \mu}(x^*; h(\eta), \eta) = b_0 + O(\eta) \neq 0$ where $w(\mu, \eta)$ is the left eigenvector of $A_L(\mu, \eta)$ for $\lambda(\mu, \eta)$,
(iii) $w(\mu, \eta)(D_x^2 f^L(x^*; h(\eta), \eta)(v(\mu, \eta), v(\mu, \eta))) = a_0 + O(\eta) \neq 0$.

Finally, notice on W^c when $\mu = h(\eta)$ we have

$$x_1 = e_1^{\mathsf{T}} V H(\hat{x}_1; \mu, \eta) = e_1^{\mathsf{T}} V e_1 \hat{x}_1 + e_1^{\mathsf{T}} V \zeta h(\eta) + O(2) = \hat{x}_1 + O(2) ,$$

and by (A.13)

$$x_1^* = -\frac{c_3}{2c_2}\eta + O(\eta^2) = -\frac{p_0}{a_0}\eta + O(\eta^2) .$$

The equilibrium at the saddle-node bifurcation is admissible when $x_1^* < 0$, i.e. when $\text{sgn}(\eta) = \text{sgn}(a_0 p_0)$. □

Proof. [Theorem 3.2] For ease of notation we expand the vector field in the left half-plane to second order in x and y

$$\begin{aligned} f^{(L)}(x, y; \mu, \eta) &= \eta x + y + a_1 x^2 + a_2 xy + a_3 y^2 + O(|x, y|^3) , \\ g^{(L)}(x, y; \mu, \eta) &= \mu - \delta^{(L)} x + b_1 x^2 + b_2 xy + b_3 y^2 + O(|x, y|^3) , \end{aligned} \tag{A.15}$$

where coefficients vary with μ and η.

Step 1: *Compute equilibria, eigenvalues and h_1.*
By the implicit function theorem, there exists a unique C^k function, $\begin{bmatrix} x^{*(L)} \\ y^{*(L)} \end{bmatrix} : \mathbb{R}^2 \to \mathbb{R}^2$, such that $\begin{bmatrix} f^{(L)}(x^{*(L)}(\mu,\eta), y^{*(L)}(\mu,\eta); \mu,\eta) \\ g^{(L)}(x^{*(L)}(\mu,\eta), y^{*(L)}(\mu,\eta); \mu,\eta) \end{bmatrix} = 0$, for small μ, η. Via substitution of a series expansion of $x^{*(L)}$ and $y^{*(L)}$ into (A.15) it is readily determined that

$$\begin{aligned} x^{*(L)}(\mu,\eta) &= \frac{1}{\omega^2}\mu + \left(\frac{b_1}{\omega^6} + \frac{\delta_\mu^{(L)}(0,0)}{\omega^4}\right)\mu^2 - \frac{\delta_\eta^{(L)}(0,0)}{\omega^4}\mu\eta + O(3) \ , \\ y^{*(L)}(\mu,\eta) &= -\frac{a_1}{\omega^4}\mu^2 - \frac{1}{\omega^2}\mu\eta + O(3) \ , \end{aligned} \tag{A.16}$$

where μ and η subscripts denote derivatives. Near $(x, y; \mu, \eta) = (0, 0; 0, 0)$, $(x^{*(L)}, y^{*(L)})^{\mathsf{T}}$ is the only equilibrium of the left half-flow and by (A.16) it is an equilibrium of the full flow (i.e. admissible), exactly when $\mu \leq 0$. If $\delta_R \neq 0$, the corresponding equilibrium in the right half-plane may be determined similarly.

Let $J^{(L)}$ denote the Jacobian of (3.7) when $x < 0$. For small μ and η the matrix

$$J^{(L)}(x^{*(L)}(\mu,\eta), y^{*(L)}(\mu,\eta); \mu,\eta) = \begin{bmatrix} j_1 \ j_2 \\ j_3 \ j_4 \end{bmatrix} \ , \tag{A.17}$$

has the complex conjugate eigenvalue pair

$$\lambda_\pm = \nu \pm \mathrm{i}\xi \ ,$$

where ν and ξ are C^{k-1} functions of μ and η. To first order we have

$$\nu(\mu,\eta) = \frac{2a_1 + b_2}{2\omega^2}\mu + \frac{1}{2}\eta + O(2) \ . \tag{A.18}$$

By the implicit function theorem, there exists a unique C^{k-1} function $h_1 : \mathbb{R} \to \mathbb{R}$, such that $\nu(\mu, h_1(\mu)) = 0$ for small μ. Furthermore

$$h_1(\mu) = -\frac{2a_1 + b_2}{\omega^2}\mu + O(\mu^2) \ ,$$

thereby confirming (3.13) of the theorem. Let

$$\eta_1 = \eta - h_1(\mu) \tag{A.19}$$

represent the deviation from the Hopf bifurcation curve, then

$$\nu(\mu,\eta_1) = \eta_1\left(\frac{1}{2} + O(1)\right) \ . \tag{A.20}$$

Step 2: *Compute a_0 and introduce polar coordinates.*
Let $v_{\lambda_\pm}$ denote the complex-valued eigenvector associated with $\lambda_\pm$ for (A.17). In a standard manner, we construct a matrix P using the real and imaginary parts of $v_{\lambda_\pm}$

$$P = \begin{bmatrix} 1 & 0 \\ \frac{1}{2j_2}(-j_1 + j_4) & -\frac{1}{2j_2}\sqrt{-(j_1 - j_4)^2 - 4j_2 j_3} \end{bmatrix} ,$$

which is well-defined (because $j_2 \neq 0$ for sufficiently small μ, η) and non-singular. Let

$$\begin{bmatrix} u \\ v \end{bmatrix} = P^{-1}\left(\begin{bmatrix} x \\ y \end{bmatrix} - \begin{bmatrix} x^{*(L)}(\mu,\eta) \\ y^{*(L)}(\mu,\eta) \end{bmatrix}\right) .$$

Then the left half-system in (u, v) coordinates becomes

$$\begin{bmatrix} \dot{u} \\ \dot{v} \end{bmatrix} = \begin{bmatrix} \check{f}(u, v; \mu, \eta) \\ \check{g}(u, v; \mu, \eta) \end{bmatrix} = D^{(L)} \begin{bmatrix} u \\ v \end{bmatrix} + O(2) ,$$

where

$$D^{(L)} = P^{-1} J^{(L)} P = \begin{bmatrix} \nu & -\xi \\ \xi & \nu \end{bmatrix} .$$

Letting

$$z = u + \mathrm{i}v , \tag{A.21}$$

then

$$\dot{z} = \lambda_+ z + O(2) . \tag{A.22}$$

Following standard proofs of the Hopf bifurcation theorem [Guckenheimer and Holmes (1986); Kuznetsov (2004); Glendinning (1999)], there exists a two-variable polynomial ζ, comprised of only quadratic and cubic terms such that the near identity transformation

$$w = z + \zeta(z, \bar{z}) , \tag{A.23}$$

removes all quadratic terms and all but one cubic term from the left half-system (A.22)

$$\dot{w} = \lambda_+ w + A w^2 \bar{w} + O(4) , \tag{A.24}$$

where $A \in \mathbb{C}$ and

$$\mathcal{R}e(A(\mu,\eta)) = \left[\frac{1}{16}(\check{f}_{uuu} + \check{g}_{uuv} + \check{f}_{uvv} + \check{g}_{vvv}) + \frac{1}{16\xi}\left(\check{f}_{uv}(\check{f}_{uu} + \check{f}_{vv}) - \check{g}_{uv}(\check{g}_{uu} + \check{g}_{vv}) - \check{f}_{uu}\check{g}_{uu} + \check{f}_{vv}\check{g}_{vv}\right)\right]\Bigg|_{(u,v)=(0,0)} . \tag{A.25}$$

Notice $\mathcal{R}e(A(\mu,\eta))$ is a C^{k-3} function of μ and η. Let

$$a_0 = \mathcal{R}e(A(0,0)) \, .$$

Using (A.25) and since $P(0,0) = \begin{bmatrix} 1 & 0 \\ 0 & -\omega \end{bmatrix}$, it follows that a_0 appears as given in the statement of the theorem, (3.10).

To prove that generic Hopf bifurcations occur along $\mu = h_1(\eta)$ it remains to verify the non-degeneracy conditions of the Hopf bifurcation theorem [Guckenheimer and Holmes (1986); Kuznetsov (2004); Glendinning (1999)].

(i) By construction, $J^{(L)}(x^{*(L)}(\mu,h_1(\mu)), y^{*(L)}(\mu,h_1(\mu)); \mu, h_1(\mu))$ has purely imaginary eigenvalues, $\pm \mathrm{i}\xi = \pm \mathrm{i}(\omega + O(\mu)) \neq 0$.
(ii) $\frac{\partial \nu}{\partial \eta}(\mu, h_1(\mu)) = \frac{1}{2} + O(\mu) \neq 0$.
(iii) By assumption, $\mathcal{R}e(A(\mu,h_1(\mu))) = a_0 + O(\mu) \neq 0$.

Therefore if $a_0 < 0$, the curve $\eta = h_1(\mu)$ corresponds to supercritical Hopf bifurcations and stable periodic orbits exist for small $\eta > h_1(\mu)$. Conversely, if $a_0 > 0$, the curve $\eta = h_1(\mu)$ corresponds to subcritical Hopf bifurcations and unstable periodic orbits exist for small $\eta < h_1(\mu)$. Consequently we have proved (1) of the theorem.

We now introduce polar coordinates. Let

$$w = r\mathrm{e}^{\mathrm{i}\theta} \, . \tag{A.26}$$

In polar coordinates the left half-system is

$$\begin{aligned} \dot{r} &= \nu r + \mathcal{R}e(A) r^3 + O(r^4) \, , \\ \dot{\theta} &= \xi + \mathcal{I}m(A) r^2 + O(r^3) \, . \end{aligned} \tag{A.27}$$

Denote the components of the C^k flow by $R(r_0,\theta_0,t;\mu,\eta_1)$ and $\Theta(r_0,\theta_0,t;\mu,\eta_1)$ respectively. Expressions for R and Θ may be derived by expanding each as a series in r_0 and computing coefficients by solving initial value problems. After much algebraic manipulation we obtain

$$R(r_0,\theta_0,t;\mu,\eta_1) = \mathrm{e}^{\nu t} r_0 + \mathcal{R}e(A) \frac{\mathrm{e}^{3\nu t} - \mathrm{e}^{\nu t}}{2\nu} r_0^3 + O(r_0^4) \, , \tag{A.28}$$

$$\Theta(r_0,\theta_0,t;\mu,\eta_1) = \theta_0 + \xi t + \mathcal{I}m(A) \frac{\mathrm{e}^{2\nu t} - 1}{2\nu} r_0^2 + O(r_0^3) \, . \tag{A.29}$$

Step 3: *Define the Poincaré section,* Π.
By using (A.15) it is straightforward to show

$$\lim_{\mu \to 0} \left(\frac{y^{*(L)}}{x^{*(L)}} \right) = -\eta \, .$$

Thus the quotient

$$S = \frac{y^{*(L)}}{x^{*(L)}} ,$$

is a well-defined C^k function for small μ, η. For our analysis we only need the Taylor series of S to first order, which from (A.16) is found to be

$$S(\mu, \eta) = -\frac{a_1}{\omega^2}\mu - \eta + O(2) .$$

For $\mu < 0$, $y = Sx$ is the line intersecting $(x^{*(L)}, y^{*(L)})^{\mathsf{T}}$ and the origin. Let

$$\Pi = \left\{(x, y) \mid y = Sx, x \geq x^{*(L)}\right\} , \tag{A.30}$$

as in Fig. 3.8. In the polar coordinates centered at the equilibrium, Π is described by a C^k function

$$\theta_\Pi(r; \mu, \eta_1) = \check{\theta} + \rho_1 r + O(r^2) , \tag{A.31}$$

where $\check{\theta}$ and ρ_1 are coefficients dependent on μ and η_1. The coefficient $\check{\theta}$ describes the angle of Π at the equilibrium in (u, v) coordinates; ρ_1 arises from the nonlinear coordinate change, (A.23). Since explicit forms for these coefficients will not be required, we do not derive them.

Step 4: *Derive* $\mathcal{P}_{\text{lhf}}$ *and compute* h_2.

We now wish to determine the left half Poincaré map, $\mathcal{P}_{\text{lhf}}$. To do this we compute the trajectory of a point, $(r_0, \theta_\Pi(r_0; \mu, \eta_1))$ on Π, and find its next intersection, $(r_1, \theta_\Pi(r_1; \mu, \eta_1))$. The transition time is

$$T_4(r_0; \mu, \eta_1) = \frac{2\pi}{\omega} + O(1) . \tag{A.32}$$

Substituting (A.20) and (A.32) into (A.28) produces

$$\begin{aligned} r_1(r_0; \mu, \eta_1) &= R(r_0, \theta_\Pi(r_0; \mu, \eta_1), T_4(r_0; \mu, \eta_1); \mu, \eta_1) \\ &= \left(1 + \eta_1(\frac{\pi}{\omega} + O(|\mu, \eta_1|^1))\right) r_0 + \frac{2\pi a_0}{\omega} r_0^3 + O(4) . \end{aligned} \tag{A.33}$$

Let s denote the distance from $(x^{*(L)}, y^{*(L)})$ in (u, v), (A.21), coordinates. That is

$$s = |z| = (u^2 + v^2)^{\frac{1}{2}} .$$

By (A.23) and (A.26)

$$s(r) = r + \sigma_2 r^2 + O(r^3) , \tag{A.34}$$

for some coefficient σ_2, determined by ζ, (A.23). By combining (A.33) and (A.34) we are able to obtain s_1 as a function of s_0, where s_0 is a point on Π and s_1 is the next point on Π

$$s_1(s_0;\mu,\eta_1) = \left(1+\eta_1(\frac{\pi}{\omega}+O(|\mu,\eta_1|^1))\right)s_0 + \frac{\pi\sigma_2}{\omega}\eta_1 s_0^2 + \frac{2\pi a_0}{\omega}s_0^3 + O(4)\,. \tag{A.35}$$

Let $(\varepsilon, S\varepsilon)^{\mathsf{T}}$ be a point on Π in (x,y) coordinates. Let

$$\breve{\varepsilon} = \varepsilon - x^{*(L)}\,. \tag{A.36}$$

Since $\breve{\varepsilon}$ is a scalar multiple of s and when $\mu = \eta_1 = 0$ we have $\breve{\varepsilon} = s$, it follows that to third order the map between $\breve{\varepsilon}_0$ and $\breve{\varepsilon}_1$ is the same as (A.35), i.e.

$$\breve{\varepsilon}_1(\breve{\varepsilon}_0;\mu,\eta_1) = \left(1+\eta_1(\frac{\pi}{\omega}+O(|\mu,\eta_1|^1))\right)\breve{\varepsilon}_0 + \frac{\pi\sigma_2}{\omega}\eta_1\breve{\varepsilon}_0^2 + \frac{2\pi a_0}{\omega}\breve{\varepsilon}_0^3 + O(4)\,. \tag{A.37}$$

The system (3.7) has a periodic orbit that grazes the y-axis when $\breve{\varepsilon}_0 = \breve{\varepsilon}_1 = -x^{*(L)} \neq 0$. By substituting this into (A.37) and dividing through by $-x^{*(L)}$ we obtain

$$1 = 1+\eta_1(\frac{\pi}{\omega}+O(|\mu,\eta_1|^1)) - \frac{\pi\sigma_2}{\omega}\eta_1 x^{*(L)}(\mu,\eta_1) + \frac{2\pi a_0}{\omega}x^{*(L)^2}(\mu,\eta_1) + O(3)\,. \tag{A.38}$$

Using (A.16) and the implicit function theorem we find (A.38) is satisfied when $\eta_1 = \hat{h}_2(\mu)$ for a C^k function

$$\hat{h}_2(\mu) = -\frac{2a_0}{\omega^4}\mu^2 + O(\mu^3)\,.$$

Let

$$\begin{aligned} h_2 &= h_1 + \hat{h}_2\,, \\ \eta_2 &= \eta_1 - \hat{h}_2(\mu)\,. \end{aligned} \tag{A.39}$$

We have therefore derived (3.14) and proved (2) of the theorem. We now write $\mathcal{P}_{\mathrm{lhf}}$ in terms of ε, μ and η_2. Combining (A.36), (A.37) and (A.39) produces

$$\begin{aligned} \mathcal{P}_{\mathrm{lhf}}(\varepsilon_0;\mu,\eta_2) = \mu\eta_2(-\frac{\pi}{\omega^3}+O(|\mu,\eta_2|^1)) + \Big(1 + \frac{\pi}{\omega}\eta_2 + q_1\mu^2 + q_2\mu\eta_2 \\ + q_3\eta_2^2 + O(|\mu,\eta_2|^3)\Big)\varepsilon_0 + O(\varepsilon_0^2)\,, \end{aligned} \tag{A.40}$$

where

$$q_1 = \frac{4\pi a_0}{\omega^5}\,, \tag{A.41}$$

and explicit forms for q_2 and q_3 will not be required subsequently.

Step 5: *Derive the discontinuity map, $\mathcal{P}_{\rm dm}$.*
In order to avoid singularities, here we assume $\mu < 0$ and introduce the spatial scaling

$$\begin{bmatrix} \hat{x} \\ \hat{y} \end{bmatrix} = -\frac{1}{\mu} \begin{bmatrix} x \\ y \end{bmatrix} . \tag{A.42}$$

In the left half-plane

$$\begin{bmatrix} \dot{\hat{x}} \\ \dot{\hat{y}} \end{bmatrix} = \begin{bmatrix} \hat{f}^{(L)}(\hat{x},\hat{y};\mu,\eta) \\ \hat{g}^{(L)}(\hat{x},\hat{y};\mu,\eta) \end{bmatrix} = -\frac{1}{\mu} \begin{bmatrix} f^{(L)}(\mu\hat{x},\mu\hat{y};\mu,\eta) \\ g^{(L)}(\mu\hat{x},\mu\hat{y};\mu,\eta) \end{bmatrix} . \tag{A.43}$$

Let

$$X^{(L)}(\hat{y}_0,t;\mu,\eta) = A_1 t + A_2 t^2 + A_3 t^3 + O(t^4) , \tag{A.44}$$
$$Y^{(L)}(\hat{y}_0,t;\mu,\eta) = \hat{y}_0 + B_1 t + B_2 t^2 + B_3 t^3 + O(t^4) , \tag{A.45}$$

denote the components of the C^k left half-flow for an initial condition $(0,\hat{y}_0)$, on the switching manifold. The parameter dependent coefficients are obtained by solving (A.43) using (A.15)

$$\begin{aligned} A_1 &= \hat{y}_0 - a_3\mu\hat{y}_0^2 + O(\hat{y}_0^3) , \\ A_2 &= -\frac{1}{2} + (\frac{1}{2}\eta + a_3\mu)\hat{y}_0 + O(\hat{y}_0^2) , \\ A_3 &= -\frac{1}{3}a_3\mu - \frac{1}{6}\eta + O(\hat{y}_0) , \\ B_1 &= -1 - b_3\mu\hat{y}_0^2 + O(\hat{y}_0^3) , \\ B_2 &= (-\frac{1}{2}\delta^{(L)} + b_3\mu)\hat{y}_0 + O(\hat{y}_0^2) , \\ B_3 &= -\frac{1}{3}b_3\mu + \frac{1}{6}\delta^{(L)} + O(\hat{y}_0) . \end{aligned} \tag{A.46}$$

We now derive $\mathcal{P}_1$, $\mathcal{P}_2$ and $\mathcal{P}_3$ in scaled coordinates, (A.42), beginning with $\mathcal{P}_3$. The point $p_4 = \mathcal{P}_3(p_3)$ (see Fig. 3.8) lies on Π, therefore to compute the corresponding transition time, T_3, we solve $Y^{(L)} = SX^{(L)}$ for t. The function

$$\begin{aligned} G_1(\hat{y}_0,t;\mu,\eta) &= Y^{(L)}(\hat{y}_0,t;\mu,\eta) - SX^{(L)}(\hat{y}_0,t;\mu,\eta) \\ &= \hat{y}_0 - t - S\hat{y}_0 t + \frac{1}{2}St^2 + O(3) , \end{aligned}$$

is C^k and by the implicit function theorem there exists a unique C^k function $T_3 : \mathbb{R}^3 \to \mathbb{R}$ such that $G_1(\hat{y}_0, T_3(\hat{y}_0;\mu,\eta);\mu,\eta) = 0$. Furthermore

$$T_3(\hat{y}_0;\mu,\eta) = \hat{y}_0 - \frac{1}{2}S\hat{y}_0^2 + O(\hat{y}_0^3) . \tag{A.47}$$

Combining (A.44) and (A.47) yields

$$\mathcal{P}_3(\hat{y}_0;\mu,\eta) = \frac{1}{2}\hat{y}_0^2 + \frac{1}{3}(\eta - a_3\mu)\hat{y}_0^3 + O(\hat{y}_0^4) . \tag{A.48}$$

Notice (see Fig. 3.8) that $\mathcal{P}_1^{-1}$ is the same as $\mathcal{P}_3$ except the y-component of the initial condition has opposite sign. That is, $\hat{y}_0 = \mathcal{P}_1(\hat{\varepsilon}_0;\mu,\eta)$, whenever $\hat{\varepsilon}_0 = \mathcal{P}_3(-\hat{y}_0;\mu,\eta)$. By inverting (A.48) we obtain

$$\mathcal{P}_1(\hat{\varepsilon}_0;\mu,\eta) = \sqrt{2}\hat{\varepsilon}_0^{\frac{1}{2}} - \frac{2}{3}(\eta - a_3\mu)\hat{\varepsilon}_0 + O(\hat{\varepsilon}_0^{\frac{3}{2}}) . \tag{A.49}$$

In a similar manner as for $\mathcal{P}_3$ we are able to use a series expansion of the right half-flow to determine $\mathcal{P}_2$. We obtain

$$\mathcal{P}_2(\hat{y}_0;\mu,\eta) = -\hat{y}_0 - \frac{2}{3}(\tau^{(R)} - a_3\mu)\hat{y}_0^2 + O(\hat{y}_0^3) . \tag{A.50}$$

The discontinuity map is $\mathcal{P}_{\rm dm} = \mathcal{P}_3 \circ \mathcal{P}_2 \circ \mathcal{P}_1$. Composition of (A.48)-(A.50) produces

$$\hat{\varepsilon}_4 = \mathcal{P}_{\rm dm}(\hat{\varepsilon}_1;\mu,\eta) = \hat{\varepsilon}_1 + \frac{4\sqrt{2}}{3}(\tau^{(R)} - \eta)\hat{\varepsilon}_1^{\frac{3}{2}} + O(\hat{\varepsilon}_1^2) . \tag{A.51}$$

Step 6: *Obtain the full Poincaré map, $\mathcal{P}$ and compute h_3.*
The full Poincaré map is $\mathcal{P} = \mathcal{P}_{\rm lhf} \circ \mathcal{P}_{\rm dm}$. In scaled coordinates (A.40) becomes

$$\begin{aligned}\hat{\varepsilon}_6 = \eta_2\Big(\frac{\pi}{\omega^3} + O(|\mu,\eta_2|^1)\Big) + \Big(1 + \frac{\pi}{\omega}\eta_2 + q_1\mu^2 + q_2\mu\eta_2 \\ + q_3\eta_2^2 + O(|\mu,\eta_2|^3)\Big)\hat{\varepsilon}_4 + O(\hat{\varepsilon}_4^2) .\end{aligned} \tag{A.52}$$

Composing (A.51) and (A.52) produces

$$\hat{\varepsilon}_6 = \mathcal{P}(\hat{\varepsilon}_1;\mu,\eta_2) = \Omega_0 + \Omega_1\hat{\varepsilon}_1 + \Omega_2\hat{\varepsilon}_1^{\frac{3}{2}} + O(\hat{\varepsilon}_1^2) , \tag{A.53}$$

where

$$\begin{aligned}\Omega_0(\mu,\eta_2) &= \frac{\pi}{\omega^3}\eta_2 + O(2) , \\ \Omega_1(\mu,\eta_2) &= 1 + \frac{\pi}{\omega}\eta_2 + q_1\mu^2 + q_2\mu\eta_2 + q_3\eta_2^2 + O(3) , \\ \Omega_2(\mu,\eta_2) &= \frac{4\sqrt{2}}{3}\tau_R + O(1) .\end{aligned}$$

To remove fractional powers we introduce $\chi = \sqrt{\hat{\varepsilon}}$. The function

$$G_2(\chi;\mu,\eta_2) = \mathcal{P}(\chi^2;\mu,\eta_2) - \chi^2 ,$$

is C^{k-1}, and by the implicit function theorem, there exists a unique C^{k-1} function, $\mathcal{F}$, such that $G_2(\chi;\mu,\mathcal{F}(\chi;\mu)) = 0$. Via a series expansion it is straightforward to obtain

$$\mathcal{F}(\chi;\mu) = \left(-\frac{\omega^3 q_1}{\pi}\mu^2 + O(\mu^3)\right)\chi^2 + \left(-\frac{4\sqrt{2}\omega^3\tau_R}{3\pi} + O(\mu)\right)\chi^3 + O(\chi^4)\,. \tag{A.54}$$

The fixed point (A.54), has an associated multiplier of one when the C^{k-2} function

$$\begin{aligned} G_3(\chi;\mu) &= \frac{\partial \mathcal{P}}{\partial \varepsilon}(\chi^2;\mu,\mathcal{F}(\chi;\mu)) - 1 \\ &= (q_1\mu^2 + O(\mu^3)) + (2\sqrt{2}\tau_R + O(\mu))\chi + O(\chi^2)\,, \end{aligned}$$

is zero. By the implicit function theorem, there exists a unique C^{k-2} function, $\hat{h}_3$, such that $G_3(\hat{h}_3(\mu);\mu) = 0$. Furthermore, using (A.41),

$$\hat{h}_3(\mu) = -\frac{\sqrt{2}\pi a_0}{\omega^5\tau_R}\mu^2 + O(\mu^3)\,.$$

Notice this fixed point is valid when $\hat{\varepsilon}_1^{\frac{1}{2}} = \chi = \hat{h}_3(\mu) \geq 0$ which is true when a_0 and τ_R have opposite signs. Finally let

$$\begin{aligned} h_3(\mu) &= h_2(\mu) + \mathcal{F}(\hat{h}_3(\mu);\mu) \\ &= h_2(\mu) - \frac{8\pi^2 a_0^3}{3\omega^{12}\tau_R^2}\mu^6 + o(\mu^6)\,. \end{aligned}$$

Then h_3 is the C^{k-2} function (3.15) and we have verified (3) and (4) of the theorem. □

Proof. [Theorem 3.3] Since the real and imaginary parts of the eigenvectors $z_\pm$, $u^{(1)}$ and $u^{(2)}$, are linearly independent, there exists $j \neq 1$ such that

$$U = \begin{bmatrix} u_1^{(1)} & u_1^{(2)} \\ u_j^{(1)} & u_j^{(2)} \end{bmatrix}\,,$$

is nonsingular. For the remainder of this proof we will set $j = 2$, w.l.o.g. It is easily verified that

$$A_L\left[u^{(1)}\ u^{(2)}\right] = \left[u^{(1)}\ u^{(2)}\right] D\,, \tag{A.55}$$

where

$$D = \begin{bmatrix} \nu & \omega \\ -\omega & \nu \end{bmatrix}.$$

Define two new vectors by

$$\left[v^{(1)}\ v^{(2)}\right] = \left[u^{(1)}\ u^{(2)}\right] U^{-1} , \tag{A.56}$$

let

$$V = \left[v^{(1)}\ v^{(2)}\ e_3\ \cdots\ e_N\right] ,$$

and introduce the new coordinate system

$$\hat{x} = V^{-1}x . \tag{A.57}$$

(The inclusion of the matrix, U^{-1}, in (A.56) allows for simplification below, in particular, $e_i^{\mathsf{T}}V = e_i^{\mathsf{T}}V^{-1} = e_i^{\mathsf{T}}$ for $i = 1, 2$.) As in the given proof of Theorem 3.1, the $(N+2)$-dimensional, C^k, extended left half-flow in the new coordinates is given by (A.8). Here the Jacobian, $DF(0;0,0)$, (A.9), has a four-dimensional, linear, center manifold, E^c, spanned by

$$\{n_1, n_2, n_3, n_4\} = \left\{e_1, e_2, \begin{bmatrix} -(A_L(0,0)V)^{-1}b \\ 1 \\ 0 \end{bmatrix}, e_{N+2}\right\} .$$

Notice W^c is not tangent to the switching manifold by condition (iv) of the theorem. On the local center manifold

$$\hat{x} = H(\hat{x}_1, \hat{x}_2; \mu, \eta) = \hat{x}_1 e_1 + \hat{x}_2 e_2 + \zeta\mu + O(2) ,$$

where $\zeta \in \mathbb{R}^N$ is equal to $-(A_L(0,0)V)^{-1}b$ except that its first two elements are zero. The dynamics on W^c are described by

$$\begin{bmatrix} \dot{\hat{x}}_1 \\ \dot{\hat{x}}_2 \end{bmatrix} = \begin{bmatrix} e_1^{\mathsf{T}} \\ e_2^{\mathsf{T}} \end{bmatrix} \Big(V^{-1}\mu b + V^{-1}A_L V H(\hat{x}_1, \hat{x}_2; \mu, \eta) \\ + V^{-1}g^{(L)}(VH(\hat{x}_1, \hat{x}_2; \mu, \eta); \mu, \eta) \Big) ,$$

where $g^{(L)}$ represents all terms of $f^{(L)}$ that are nonlinear in x. By using (A.55) and (A.56) we obtain

$$\begin{bmatrix} \dot{\hat{x}}_1 \\ \dot{\hat{x}}_2 \end{bmatrix} = \mu\hat{b}(\mu, \eta) + \hat{A}_L(\mu, \eta) \begin{bmatrix} \hat{x}_1 \\ \hat{x}_2 \end{bmatrix} + O(2) , \tag{A.58}$$

where, at $\mu = \eta = 0$,

$$\hat{A}_L = \begin{bmatrix} e_1^{\mathsf{T}} \\ e_2^{\mathsf{T}} \end{bmatrix} V^{-1}A_L V \left[e_1\ e_2\right] = UDU^{-1} , \tag{A.59}$$

and

$$\hat{b} = \begin{bmatrix} e_1^{\mathsf{T}} \\ e_2^{\mathsf{T}} \end{bmatrix} V^{-1} b + \begin{bmatrix} e_1^{\mathsf{T}} \\ e_2^{\mathsf{T}} \end{bmatrix} V^{-1} A_L V \zeta = \begin{bmatrix} e_1^{\mathsf{T}} \\ e_2^{\mathsf{T}} \end{bmatrix} V^{-1} A_L \begin{bmatrix} v^{(1)} \; v^{(2)} \; 0 \cdots 0 \end{bmatrix} A_L^{-1} b$$
$$= \begin{bmatrix} \hat{A}_L \; 0 \cdots 0 \end{bmatrix} A_L^{-1} b \, .$$

Equation (A.58) is the left half-flow of (1.10) restricted to W^c. It indeed is of the form (1.11) as stated in the theorem. Finally, due to condition (ii), it is easily verified that at $\mu = \eta = 0$

$$\hat{\varrho}^{\mathsf{T}} \hat{b} = \frac{\det(\hat{A}_L)}{\det(A_L)} \varrho^{\mathsf{T}} b \neq 0 \, , \tag{A.60}$$

where $\hat{\varrho}^{\mathsf{T}} = e_1^{\mathsf{T}} \mathrm{adj}(\hat{A}_L)$. By (A.59), $\hat{A}_L(0,0)$ has eigenvalues $\nu(0,0) \pm \mathrm{i}\omega(0,0)$, thus (A.58) is observable, by Lemma 1.2. Consequently, since we also have (A.60), (A.58) may be transformed to the observer canonical form and by condition (iii), (3.9) and condition (i) of Theorem 3.2 may be satisfied. □

Proof. [Theorem 3.4] Suppose $K_L < K_R$ (the case $K_L > K_R$ follows by a reversal of the direction of time). Let

$$\omega^{(i)} = \sqrt{\delta^{(i)} - \frac{1}{4} \tau^{(i)2}} \, .$$

Let $\mathcal{P}_L$ and $\mathcal{P}_R$ be the same maps that were defined in Sec. 2.2 and $\mathcal{P} = \mathcal{P}_L \circ \mathcal{P}_R$, (2.9)-(2.11), see Fig. 2.1. $\mathcal{P}$ is a Poincaré map from the positive y-axis to itself. As detailed in the proof of Theorem 2.1, $\mathcal{P}_L$ and $\mathcal{P}_R$ are well-defined for small $\mu \leq 0$, and by considering C^k extensions, we may assume $\mathcal{P}_L$ and $\mathcal{P}_R$ are C^k functions in a neighborhood of $(y; \mu, \eta) = (0; 0, 0)$. Since $\mathcal{P}_R(0; \mu, \eta) \equiv 0$, we may write

$$\mathcal{P}_R(y_0; \mu, \eta) = \alpha_2 y_0 + \alpha_4 \mu y_0 + \alpha_5 y_0^2 + O(|y_0, \mu|^3) \, ,$$

where the coefficients, α_i, vary with η. Similarly, since $\mathcal{P}_L(0; 0, \eta) \equiv 0$ and $\frac{\partial \mathcal{P}_L}{\partial \mu}(0; 0, \eta) \equiv -\gamma_L$ we may write

$$\mathcal{P}_L(y_1; \mu, \eta) = -\gamma_L \mu + \beta_2 y_1 + \beta_3 \mu^2 + \beta_4 \mu y_1 + \beta_5 y_1^2 + O(|y_1, \mu|^3) \, ,$$

where each β_i varies with η. Then

$$\mathcal{P}(y; \mu, \eta) = -\gamma_L \mu + \alpha_2 \beta_2 y + \beta_3 \mu^2 + (\alpha_4 \beta_2 + \alpha_2 \beta_4) \mu y + (\alpha_5 \beta_2 + \alpha_2^2 \beta_5) y^2 + O(|y, \mu|^3) \, . \tag{A.61}$$

To compute fixed points of $\mathcal{P}$ (these correspond to periodic orbits of (1.10)), we first need to obtain expressions for some of the coefficients α_i and β_i.

Below we determine α_2 and α_5 by deriving explicit equations for the right half-flow when $\mu = 0$. By symmetry, expressions for β_2 and β_5 will be identical except with "R"'s replaced with "L"'s. Expressions for the remaining coefficients $(\alpha_4, \beta_3, \beta_4)$ are not required for this proof.

For ease of notation we expand the vector field in the right half-plane to second order in x and y

$$\begin{aligned} f^{(R)}(x, y; \mu, \eta) &= \tau^{(R)} x + y + a_1 x^2 + a_2 xy + a_3 y^2 + O(|x, y|^3) , \\ g^{(R)}(x, y; \mu, \eta) &= -\delta^{(R)} x + b_1 x^2 + b_2 xy + b_3 y^2 + O(|x, y|^3) . \end{aligned} \tag{A.62}$$

Let

$$\begin{bmatrix} X(y_0, t; \eta) \\ Y(y_0, t; \eta) \end{bmatrix} = \varphi_t^{(R)}(0, y_0; 0, \eta) = \begin{bmatrix} A_1 y_0 + A_2 y_0^2 \\ B_1 y_0 + B_2 y_0^2 \end{bmatrix} + O(y_0^3) ,$$

denote the C^k right half-flow for an initial condition $(0, y_0)$ with $y_0 > 0$ when $\mu = 0$. The coefficients A_i and B_i vary with t and η. Using the relations $\frac{\partial X}{\partial t} = f^{(R)}(X, Y; 0, \eta)$ and $\frac{\partial Y}{\partial t} = g^{(R)}(X, Y; 0, \eta)$ we obtain

$$\begin{aligned} &(\dot{A}_1 - \tau A_1 - B_1) y_0 \\ &+ (\dot{A}_2 - \tau A_2 - B_2 - a_1 A_1^2 - a_2 A_1 B_1 - a_3 B_1^2) y_0^2 = O(y_0^3) , \\ &(\dot{B}_1 + \delta A_1) y_0 + (\dot{B}_2 + \delta A_2 - b_1 A_1^2 - b_2 A_1 B_1 - b_3 B_1^2) y_0^2 = O(y_0^3) , \end{aligned}$$

where $\tau = \tau^{(R)}(0, \eta)$ and $\delta = \delta^{(R)}(0, \eta)$. A_1 and B_1 may be determined by solving the two-dimensional, homogeneous, initial value problem

$$\begin{aligned} \dot{A}_1 &= \tau A_1 + B_1 , \qquad & A_1(0) &= 0 , \\ \dot{B}_1 &= -\delta A_1 , \qquad & B_1(0) &= 1 . \end{aligned}$$

We obtain

$$\begin{aligned} A_1(t) &= \frac{1}{\omega} \mathrm{e}^{\frac{t\tau}{2}} \sin(\omega t) , \\ B_1(t) &= \mathrm{e}^{\frac{t\tau}{2}} (\cos(\omega t) - \frac{\tau}{2\omega} \sin(\omega t)) , \end{aligned} \tag{A.63}$$

where $\omega = \omega^{(R)}(0, \eta)$. To determine A_2 and B_2 we are required to solve the two-dimensional, non-homogeneous, initial value problem

$$\begin{aligned} \dot{A}_2 &= \tau A_2 + B_2 + a_1 A_1^2 + a_2 A_1 B_1 + a_3 B_1^2 , \qquad & A_2(0) &= 0 , \\ \dot{B}_2 &= -\delta A_2 + b_1 A_1^2 + b_2 A_1 B_1 + b_3 B_1^2 , \qquad & B_2(0) &= 0 . \end{aligned}$$

Write

$$\begin{aligned} a_1 A_1^2 + a_2 A_1 B_1 + a_3 B_1^2 &= \mathrm{e}^{t\tau} \left(p_1 + p_2 \cos(2\omega t) + p_3 \sin(2\omega t) \right) , \\ b_1 A_1^2 + b_2 A_1 B_1 + b_3 B_1^2 &= \mathrm{e}^{t\tau} \left(q_1 + q_2 \cos(2\omega t) + q_3 \sin(2\omega t) \right) , \end{aligned}$$

where

$$\begin{bmatrix} p_1 \\ p_2 \\ p_3 \\ q_1 \\ q_2 \\ q_3 \end{bmatrix} = \begin{bmatrix} \frac{1}{2\omega^2} & -\frac{\tau}{4\omega^2} & \frac{\delta}{2\omega^2} & 0 & 0 & 0 \\ -\frac{1}{2\omega^2} & \frac{\tau}{4\omega^2} & 1-\frac{\delta}{2\omega^2} & 0 & 0 & 0 \\ 0 & -\frac{1}{2\omega} & -\frac{\tau}{2\omega} & 0 & 0 & 0 \\ 0 & 0 & 0 & \frac{1}{2\omega^2} & -\frac{\tau}{4\omega^2} & \frac{\delta}{2\omega^2} \\ 0 & 0 & 0 & -\frac{1}{2\omega^2} & \frac{\tau}{4\omega^2} & 1-\frac{\delta}{2\omega^2} \\ 0 & 0 & 0 & 0 & -\frac{1}{2\omega} & -\frac{\tau}{2\omega} \end{bmatrix} \begin{bmatrix} a_1 \\ a_2 \\ a_3 \\ b_1 \\ b_2 \\ b_3 \end{bmatrix} . \tag{A.64}$$

Then

$$\begin{aligned} A_2(t) &= \mathrm{e}^{\frac{t\tau}{2}} \left(-(c_1+c_2)(\cos(\omega t) + \frac{\tau}{2\omega}\sin(\omega t)) - \frac{1}{\omega}(d_1+d_2)\sin(\omega t) \right) \\ &\quad + \mathrm{e}^{t\tau}\left(c_1 + c_2\cos(2\omega t) + c_3\sin(2\omega t)\right) , \\ B_2(t) &= \mathrm{e}^{\frac{t\tau}{2}} \left(\frac{\delta}{\omega}(c_1+c_2)\sin(\omega t) - (d_1+d_2)(\cos(\omega t) - \frac{\tau}{2\omega}\sin(\omega t)) \right) \\ &\quad + \mathrm{e}^{t\tau}\left(d_1 + d_2\cos(2\omega t) + d_3\sin(2\omega t)\right) , \end{aligned}$$

where the coefficients c_1, c_2, c_3, d_1, d_2, d_3, may be determined by solving the nonsingular, linear system

$$\begin{bmatrix} p_1 \\ p_2 \\ p_3 \\ q_1 \\ q_2 \\ q_3 \end{bmatrix} = \begin{bmatrix} 0 & 0 & 0 & -1 & 0 & 0 \\ 0 & 0 & 2\omega & 0 & -1 & 0 \\ 0 & -2\omega & 0 & 0 & 0 & -1 \\ \delta & 0 & 0 & \tau & 0 & 0 \\ 0 & \delta & 0 & 0 & \tau & 2\omega \\ 0 & 0 & \delta & 0 & -2\omega & \tau \end{bmatrix} \begin{bmatrix} c_1 \\ c_2 \\ c_3 \\ d_1 \\ d_2 \\ d_3 \end{bmatrix} . \tag{A.65}$$

In particular we obtain (using (A.64))

$$\begin{aligned} d_1 + d_2 = \frac{1}{9\delta^2 - 2\delta\tau^2} \Big(&- 6\delta a_1 + 4\delta\tau a_2 - \delta(3\delta + 2\tau^2)a_3 \\ &- 2\tau b_1 + (-3\delta + 2\tau^2)b_2 + \tau(5\delta - 2\tau^2)b_3 \Big) . \end{aligned} \tag{A.66}$$

By solving $X(y_0, \mathcal{T}_R; \eta) = 0$ it is easily determined that the transition time for the Poincaré map $\mathcal{P}_R$, when $\mu = 0$, is a C^k function

$$\mathcal{T}_R(y_0; \eta) = \frac{\pi}{\omega} + O(y_0) .$$

Then

$$\begin{aligned} \mathcal{P}_R(y_0; \eta) &= Y(y_0, \mathcal{T}_R(y_0; \eta); \eta) \\ &= B_1(\mathcal{T}_R)y_0 + B_2(\mathcal{T}_R)y_0^2 + O(y_0^3) \\ &= -\mathrm{e}^{\frac{\pi\tau}{2\omega}} y_0 + (d_1+d_2)\mathrm{e}^{\frac{\pi\tau}{2\omega}}(1 + \mathrm{e}^{\frac{\pi\tau}{2\omega}})y_0^2 + O(y_0^3) , \end{aligned}$$

since, in particular, $B_1(\mathcal{T}_R) = -\mathrm{e}^{\frac{\pi\tau}{2\omega}} + O(y_0^2)$. Therefore we have obtained

$$\alpha_2 = -\mathrm{e}^{\frac{\pi\tau}{2\omega}} \,,$$
$$\alpha_5 = (d_1 + d_2)\mathrm{e}^{\frac{\pi\tau}{2\omega}} \left(1 + \mathrm{e}^{\frac{\pi\tau}{2\omega}}\right) \,.$$

Thus

$$\alpha_2(0) = -\mathrm{e}^{\frac{\pi\tau_R}{2\omega_R}} \,,$$
$$\alpha_5(0) = K_R \mathrm{e}^{\frac{\pi\tau_R}{2\omega_R}} \left(1 + \mathrm{e}^{\frac{\pi\tau_R}{2\omega_R}}\right) \,,$$

and similarly

$$\beta_2(0) = -\mathrm{e}^{\frac{\pi\tau_L}{2\omega_L}} \,,$$
$$\beta_5(0) = K_L \mathrm{e}^{\frac{\pi\tau_L}{2\omega_L}} \left(1 + \mathrm{e}^{\frac{\pi\tau_L}{2\omega_L}}\right) \,.$$

Thus the full Poincaré map, $\mathcal{P}$, is

$$\mathcal{P}(y_0;\mu,\eta) = \zeta_1\mu + \zeta_2 y_0 + \zeta_3\mu^2 + \zeta_4\mu y_0 + \zeta_5 y_0^2 + \zeta_6\mu\eta + \zeta_7\eta y_0 + O(3) \,, \quad \text{(A.67)}$$

where in particular

$$\begin{aligned}
\zeta_1 &= -\gamma_L(0) \,, \\
\zeta_2 &= \alpha_2(0)\beta_2(0) = \mathrm{e}^{\pi\Lambda(0,0)} = 1 \,, \\
\zeta_5 &= \alpha_5(0)\beta_2(0) + \alpha_2^2(0)\beta_5(0) = (K_L - K_R)\left(1 + \mathrm{e}^{\frac{\pi\tau_R}{2\omega_R}}\right) \\
&= (K_L - K_R)\left(1 + \mathrm{e}^{-\frac{\pi\tau_L}{2\omega_L}}\right) \,, \\
\zeta_7 &= \tfrac{\mathrm{d}}{\mathrm{d}\eta}\alpha_2\beta_2\big|_{\eta=0} = \pi\Lambda_\eta(0,0) = \pi \,.
\end{aligned}$$

Fixed points of (A.67) may be found by solving $\mathcal{P}(y_0;\mu,\eta) = y_0$. Since $\zeta_1 \neq 0$ we may use the implicit function theorem to solve for μ. We obtain

$$\mu_{\mathrm{FP}}(y_0,\eta) = -\frac{\zeta_5}{\zeta_1}y_0^2 - \frac{\zeta_7}{\zeta_1}y_0\eta + O(3) \,.$$

Saddle-node bifurcations occur when

$$\begin{aligned}
1 &= \frac{\partial\mathcal{P}(y_0;\mu_{\mathrm{FP}}(y_0,\eta),\eta)}{\partial y_0} \\
&= 1 + 2\zeta_5 y_0 + \zeta_7\eta + O(2) \,,
\end{aligned}$$

hence

$$y_0^{\mathrm{SN}}(\eta) = -\frac{\zeta_7}{2\zeta_5}\eta + O(\eta^2) \,, \quad \text{(A.68)}$$

since $\zeta_5 \neq 0$. Thus saddle-node bifurcations occur along the C^{k-1} curve

$$\begin{aligned}\mu &= h(\eta) \\ &= \mu_{\rm FP}(y_0^{\rm SN}(\eta),\eta) \\ &= \frac{\zeta_7^2}{4\zeta_1\zeta_5}\eta^2 + O(\eta^3)\,,\end{aligned}$$

which verifies (3.30). Here $\zeta_5 < 0$, thus by (A.68), $y_0^{\rm SN}(\eta)$ is positive (corresponding to an admissible saddle-node bifurcation) exactly when η is positive, in agreement with the final statement of the theorem. □

Proof. [Theorem 5.2] The theorem is proved in five steps. First, the function h_1 is calculated by finding where $x^{*(L)}$ has an associated multiplier of -1. Second, h_2 is calculated from an expression for ${f^{(L)}}^2$. Third, an explicit computation of the piecewise-smooth, second iterate map allows the border-collision bifurcation of the two-cycle that occurs along $\eta = h_2(\mu)$ to be classified. Finally, in steps four and five we prove parts (4) and (5) respectively.

Since $f^{(L)}$ (5.9) is C^k and $k \geq 3$, we can write

$$\begin{aligned}a_L(\mu,\eta) &= -1 + \alpha_1\mu + \eta + \alpha_3\mu^2 + \alpha_4\mu\eta + \alpha_5\eta^2 + o(2)\,, \\ b(\mu,\eta) &= 1 + \beta_1\mu + \beta_2\eta + O(2)\,, \\ p(\mu,\eta) &= \gamma_0 + \gamma_1\mu + \gamma_2\eta + o(1)\,, \\ q(\mu,\eta) &= \delta_0 + o(0)\,,\end{aligned} \tag{A.69}$$

where we have used hypotheses of the theorem to simplify the coefficients.

Step 1: *Compute the function h_1 which corresponds to the existence of a fixed point of the left half-map with multiplier -1.*
By the implicit function theorem the left half-map has the fixed point

$$x^{*(L)}(\mu,\eta) = \frac{1}{2}\mu + k_1\mu^2 + k_2\mu\eta + k_3\eta^2 + O(3)\,, \tag{A.70}$$

which is a C^k function of μ and η and locally satisfies $f^{(L)}(x^{*(L)}(\mu,\eta);\mu,\eta) - x^{*(L)}(\mu,\eta) \equiv 0$. By a second order expansion, it is determined that the scalar coefficients, k_i, have the values

$$\begin{aligned}k_1 &= \frac{\beta_1}{2} + \frac{\alpha_1}{4} + \frac{\gamma_0}{8}\,, \\ k_2 &= \frac{\beta_2}{2} + \frac{1}{4}\,, \\ k_3 &= 0\,.\end{aligned} \tag{A.71}$$

The multiplier associated with $x^{*(L)}(\mu,\eta)$ is found by substituting (A.70) into $D_x f^{(L)}(x;\mu,\eta)$ (see (5.9)). We obtain

$$\begin{aligned} D_x f^{(L)}(x^{*(L)}(\mu,\eta);\mu,\eta) &= -1 + (\alpha_1+\gamma_0)\mu + \eta + (\alpha_4 + 2k_2\gamma_0 + \gamma_2)\mu\eta \\ &+ \left(\alpha_3 + 2k_1\gamma_0 + \gamma_1 + \frac{3\delta_0}{4}\right)\mu^2 \\ &+ \alpha_5\eta^2 + o(2)\,. \end{aligned} \tag{A.72}$$

The implicit function theorem implies that

$$h_1(\mu) = -(\alpha_1+\gamma_0)\mu + l_1\mu^2 + o(\mu^2)\,, \tag{A.73}$$

is a C^{k-1} function that locally satisfies $D_x f^{(L)}(x^{*(L)}(\mu,h_1(\mu));\mu,h_1(\mu)) + 1 \equiv 0$, where l_1 is found by substituting (A.73) into (A.72),

$$l_1 = -\left(\alpha_3 + 2k_1\gamma_0 + \gamma_1 + \frac{3\delta_0}{4}\right) + (\alpha_4+2k_2\gamma_0+\gamma_2)(\alpha_1+\gamma_0) - \alpha_5(\alpha_1+\gamma_0)^2\,. \tag{A.74}$$

When $c_0 \neq 0$, the existence of period-doubling bifurcations along $\eta = h_1(\mu)$ for small μ is verified by checking the three conditions of the standard theorem, see for instance [Guckenheimer and Holmes (1986)]:

i) (singularity) by construction, $D_x f^{(L)}(x^{*(L)}(\mu,h_1(\mu));\mu,h_1(\mu)) \equiv -1$,

ii) (transversality) $\left(\frac{\partial f^{(L)}}{\partial\eta}\frac{\partial^2 f^{(L)}}{\partial x^2} + 2\frac{\partial^2 f^{(L)}}{\partial x\partial\eta}\right)\Big|_{\eta=h_1(\mu)} = 2 + O(\mu) \neq 0$,

iii) (non-degeneracy) $\left(\frac{1}{2}\left(\frac{\partial^2 f^{(L)}}{\partial x^2}\right)^2 + \frac{1}{3}\frac{\partial^3 f^{(L)}}{\partial x^3}\right)\Big|_{\eta=h_1(\mu)} = 2c_0 + O(\mu) \neq 0$.

Consequently we have proved (1) of the theorem.

Step 2: *Compute the function h_2 which corresponds to the existence of a period-two orbit of the left half-map at $x=0$.*

To determine h_2, we compute the C^k function $f^{(L)^2}(0;\mu,\eta) = f^{(L)}(\mu b(\mu,\eta);\mu,\eta)$ which may be written as $f^{(L)^2}(0;\mu,\eta) = \mu g(\mu,\eta)$ where

$$g(\mu,\eta) = b(\mu,\eta)\Big(1 + a_L(\mu,\eta) + \mu b(\mu,\eta)p(\mu,\eta) + \mu^2 b^2(\mu,\eta)q(\mu,\eta)\Big) + o(2)\,, \tag{A.75}$$

is C^{k-1}. Substitution of (A.69) into (A.75) produces

$$\begin{aligned} g(\mu,\eta) &= (\alpha_1+\gamma_0)\mu + \eta + (\alpha_1\beta_1 + \alpha_3 + 2\beta_1\gamma_0 + \gamma_1 + \delta_0)\mu^2 \\ &+ (\beta_1 + \alpha_1\beta_2 + \alpha_4 + 2\beta_2\gamma_0 + \gamma_2)\mu\eta \\ &+ (\beta_2 + \alpha_5)\eta^2 + o(2)\,. \end{aligned} \tag{A.76}$$

The implicit function theorem implies that

$$h_2(\mu) = -(\alpha_1 + \gamma_0)\mu + l_2\mu^2 + o(\mu^2) , \tag{A.77}$$

is a C^{k-1} function that locally satisfies $g(\mu, h_2(\mu)) \equiv 0$. By substituting (A.77) into (A.76) we deduce

$$\begin{aligned} l_2 = &-(\alpha_1\beta_1 + \alpha_3 + 2\beta_1\gamma_0 + \gamma_1 + \delta_0) - (\beta_2 + \alpha_5)(\alpha_1 + \gamma_0)^2 \\ &+ (\beta_1 + \alpha_1\beta_2 + \alpha_4 + 2\beta_2\gamma_0 + \gamma_2)(\alpha_1 + \gamma_0) , \end{aligned} \tag{A.78}$$

Subtracting (A.74) from (A.78) produces (after algebraic simplification)

$$l_2 - l_1 = -\frac{\gamma_0^2 + \delta_0}{4} ,$$

which proves (5.14) in the statement of the theorem.

Step 3: *Determine all two-cycles to verify the phase portraits in Fig. 5.2.*

For small, fixed $\mu < 0$, the two-cycle generated in a period-doubling bifurcation at $\eta = h_1(\mu)$ undergoes a border-collision bifurcation when it collides with the switching manifold at $\eta = h_2(\mu)$. The stability and relative admissibility of two-cycles emanating from this border-collision bifurcation is found by determining a map of the form (1.39) that describes the bifurcation. Such a map may be obtained by computing the second iterate of (5.8) and replacing η with the new parameter

$$\hat{\eta} = \eta - h_2(\mu) ,$$

which controls the border-collision. Near $\hat{\eta} = 0$, when $\mu < 0$ period-two orbits are guaranteed to have one negative point, so we compute $f^{(L)} \circ f^{(L)}$ and $f^{(L)} \circ f^{(R)}$ which leads to

$$f^2(x; \mu, \hat{\eta}) = \begin{cases} (\mu + O(\mu^2))\hat{\eta} + (1 - c_0\mu^2 + o(\mu^2))x + O(|x, \hat{\eta}|^2), & x \le 0 \\ (\mu + O(\mu^2))\hat{\eta} + (-a_0^{(R)} + O(\mu))x + O(|x, \hat{\eta}|^2), & x \ge 0 \end{cases} \tag{A.79}$$

A fixed point, $x^{*(LL)}$, of the left half-map of (A.79) corresponds to a two-cycle of (5.8) with both points negative-valued, when admissible. By (A.79), if $c_0 < 0$ [$c_0 > 0$], then $x^{*(LL)}$ is unstable and admissible when $\hat{\eta} < 0$ [stable and admissible when $\hat{\eta} > 0$]. Similarly, a fixed point, $x^{*(RL)}$ of the right half-map of (A.79) corresponds to a two-cycle of (5.8) comprised of two points with different signs, when admissible. If $|a_0^{(R)}| < 1$, then $x^{*(RL)}$ is stable, otherwise it is unstable. Furthermore if $a_0^{(R)} < -1$ [$a_0^{(R)} > -1$], then $x^{*(RL)}$ is admissible when $\hat{\eta} > 0$ [$\hat{\eta} < 0$]. Lastly, since $a_0^{(R)} \neq -1$, the fixed point, $x^{*(RL)}$, is unique for small μ and η. Therefore the two-cycle created at $\mu = 0$ collides with the switching manifold at $\eta = h_2(\mu)$. These statements verify all two-cycles shown in Fig. 5.2.

Step 4: *Show that f, (5.8), has no n-cycles with $n \geq 3$, verifying part (4).*

Except when $a_0^{(R)} = 0$ our proof is founded on the knowledge that a one-dimensional map cannot have an n-cycle with $n \geq 3$ contained in an interval on which the map is monotone. For the special case $a_0^{(R)} = 0$ we used several additional logical arguments including the contraction mapping theorem.

Case I: $a_0^{(R)} < 0$.

In this case there exists $\delta > 0$ such that $\forall \mu, \eta \in [-\delta, \delta]$, f is decreasing on $[-\delta, \delta]$. Consequently f has no n-cycles with $n \geq 3$ on $[-\delta, \delta]$.

Case II: $a_0^{(R)} = 0$.

First we construct an interval containing the origin on which f is forward invariant. Since the two components of f are differentiable, there exists $\hat{\delta} > 0$ such that $\forall \mu, \eta, x \in [-\hat{\delta}, \hat{\delta}]$,

$$\left| \frac{\partial f^{(R)}}{\partial x} \right|, \left| \frac{\partial f^{(L)}}{\partial x} + 1 \right|, |b(\mu, \eta) - 1| \leq \frac{1}{3} \, .$$

Let $\delta = \frac{1}{8}\hat{\delta}$, let $I = [-4\delta, 8\delta]$ and assume $\mu, \eta \in [-\delta, \delta]$. Then $f(I) \subset I$ because for any $x \in I$, we may assume $x > 0$ when calculating a lower bound for $f(x)$: $f(x) \geq \mu b - \frac{1}{3}x \geq -\frac{4}{3}\delta - \frac{8}{3}\delta = -4\delta$, and we may assume $x < 0$ when calculating an upper bound: $f(x) \leq \mu b - \frac{4}{3}x \leq \frac{4}{3}\delta + \frac{4}{3}4\delta < 8\delta$.

If $\mu \geq 0$ then the image of any $x < 0$ under f is positive and since the right half-map is contracting and the slope of the left half-map is near -1, f^2 is a contraction on I, indeed $|D_x f^2(x)| \leq \frac{4}{3}\frac{1}{3} < 1$. By the contraction mapping theorem, iterates of f^2 approach a unique fixed point, which must be $x^{*(R)}$, therefore f cannot have an n-cycle with $n \geq 3$ on $\mathcal{N}$ (5.12).

So we now assume $\mu < 0$. Recall that when $\eta = h_2(\mu)$, ${f^{(L)}}^2(0) = 0$, i.e. $f(0) = {f^{(L)}}^{-1}(0)$, and when $\eta > h_2(\mu)$, $f(0) > {f^{(L)}}^{-1}(0)$. Therefore if $\eta \geq h_2(\mu)$, see Fig. A.2(a), the interval $[{f^{(L)}}^{-1}(0), 0]$ is forward invariant. Since f is monotone on the interval

$$J = [f(0), 0] \, , \tag{A.80}$$

it contains no n-cycles with $n \geq 3$ here. On the other hand if $\eta < h_2(\mu)$ as in Fig. A.2(b), the forward orbit of any point in $[{f^{(L)}}^{-1}(0), 0]$ is either contained in this interval (where f is monotone) or enters $[f(0), {f^{(L)}}^{-1}(0)]$ and approaches the stable two-cycle (this may be shown formally by the contraction mapping theorem). In any case, an n-cycle of f with $n \geq 3$ cannot include a point in $J = [f(0), 0]$.

Suppose for a contradiction f has an n-cycle with $n \geq 3$, call it $\{x_0 \ldots x_{n-1}\} \subset [-\delta, \delta] \setminus J$. We now show that the restriction of f^2 to this n-cycle is a contraction giving a contradiction according to the contraction mapping theorem. Specifically we will argue that for any i and j,

$$|f^2(x_j) - f^2(x_i)| < \frac{4}{9}|x_j - x_i| . \tag{A.81}$$

There are several cases to consider. For example, suppose that $x_j > 0$, $x_i < 0$ and $f^2(x_j) < f^2(x_i)$. Then $f(x_j) > f(0) - \frac{1}{3}x_j$ thus $f^2(x_j) > f(0) + \frac{1}{3}f(0) - \frac{1}{9}x_j$. Also $f(x_i) < f(0) - \frac{4}{3}x_i$ thus $f^2(x_i) < f(0) + \frac{1}{3}f(0) - \frac{4}{9}x_i$. Subtracting these inequalities leads to (A.81). The remaining cases can be shown similarly.

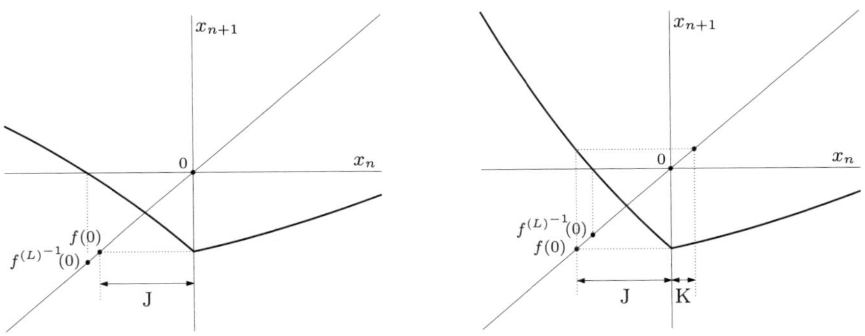

(a) $\eta > h_2(\mu)$ (b) $\eta < h_2(\mu)$

Fig. A.2: Schematic diagrams illustrating the map f, (5.8), when $\mu < 0$ and $\eta > h_2(\mu)$ in panel (a) and $\eta < h_2(\mu)$ in panel (b). When $\eta = h_2(\mu)$, $f^{(L)^{-1}}(0) = f(0)$.

Case III: $a_0^{(R)} > 0$.

Case IIIa: First suppose that $\mu \geq 0$. There exists $\delta_1 > 0$ such that whenever $\mu \in [0, \delta_1]$ and $\eta \in [-\delta_1, \delta_1]$, f is decreasing on $[-\delta_1, 0]$ and increasing on $[0, \delta_1]$. Since $\mu \geq 0$, $\forall x \in [-\delta_1, \delta_1]$, $f(x) \geq 0$. Thus any periodic solution of f in $[-\delta_1, \delta_1]$ must lie entirely in $[0, \delta_1]$. But f is increasing on this interval thus has no n-cycles with $n \geq 3$ here.

Case IIIb: Now suppose $\mu < 0$ and $\eta \geq h_2(\mu)$. There exists $\delta_2 > 0$ such that whenever $\mu \in [-\delta_2, 0)$ and $\eta \in [h_2(\mu), \delta_2]$, f is decreasing on $[-\delta_2, 0]$ and increasing on $[0, \delta_2]$, see Fig. A.2(a). Suppose for a contradiction, f has an n-cycle with $n \geq 3$ on $[-\delta_2, \delta_2]$. Such an n-cycle must enter both $[-\delta_2, 0)$ and $(0, \delta_2]$, and so it includes a point $x \in (0, \delta_2]$ with $f(x) \in [-\delta_2, 0)$. But

$f(x)$ lies in J, (A.80), and since $\eta \geq h_2(\mu)$, $f^2(0) \leq 0$, hence $f(J) \subset J$. Thus the forward orbit of x cannot return to $(0, \delta_2]$ which contradicts the assumption that x belongs to an n-cycle.

Case IIIc: Finally suppose $\mu < 0$, $\eta < h_2(\mu)$ and $0 < a_0^{(R)} < 1$. Then there exists $\delta_3 > 0$ such that whenever $\mu \in [-\delta_3, 0)$, $\eta \in [-\delta_3, h_2(\mu))$ and $x \in [-\delta_3, \delta_3]$,

$$-\left(1 + \frac{1 - a_0^{(R)}}{2(1 + a_0^{(R)})}\right) \leq f^{(L)\prime}(x) \leq -\frac{1}{2}\,,$$

$$\frac{a_0^{(R)}}{2} \leq f^{(R)\prime}(x) \leq \frac{1 + a_0^{(R)}}{2}\,,$$

where the particular non-symmetric $a_0^{(R)}$-dependent bounds on the slopes have been chosen to simplify the subsequent analysis. Suppose for a contradiction f has an n-cycle with $n \geq 3$ in $[-\delta_3, \delta_3]$. As before, such an n-cycle must include a point $x \in (0, \delta_3]$ with $f(x) \in [-\delta_3, 0)$. Again $f(x) \in J$, but here $\eta < h_2(\mu)$, thus the forward orbit of x may return to the right of the origin but must enter the interval $K = [0, f^2(0)]$, see Fig. A.2(b). Notice $f^3(0) \leq f(0) + \frac{1+a_0^{(R)}}{2} f^2(0)$. Then by evaluating the first-order expansion of $f^{(L)}$ about $f(0)$ at $f^3(0)$, we arrive at $f^4(0) \geq f^2(0) - \left(1 + \frac{1}{2}\frac{1-a_0^{(R)}}{1+a_0^{(R)}}\right)\frac{1+a_0^{(R)}}{2} f^2(0) > 0$, therefore $f^2(K) \subset K$. Then $\forall y \in K$, $|{f^2}'(y)| \leq \left(1 + \frac{1}{2}\frac{1-a_0^{(R)}}{1+a_0^{(R)}}\right)\frac{1+a_0^{(R)}}{2} < 1$. Thus $f^2 : K \to K$ is a contraction mapping and therefore iterates of f that enter K cannot belong to an n-cycle with $n \geq 3$.

Finally let $\delta = \min(\delta_1, \delta_2, \delta_3)$, then the result, (5.8), is proved.

Step 5: *Prove the existence of chaos when $\mu < 0$, $\eta < h_2(\mu)$ and $a_0^{(R)} > 1$.*

We first construct a trapping set, T, for f^2. Let $\rho(\mu, \eta) = f^2(0; \mu, \eta)$. Since $\mu < 0$, by (A.79), $\rho(\mu, \eta) = O(2)$. Since $\eta < h_2(\mu)$, $\rho > 0$. Therefore

$$f(2\rho) = f(0) + (2a_0^{(R)} + O(1))\rho + O(\rho^2) < 0\,.$$

Consequently,

$$f^2(2\rho) = (1 - 2a_0^{(R)} + O(1))\rho + O(\rho^2)\,, \tag{A.82}$$

$${f^{(L)}}^{-1}(f(2\rho)) = (-2a_0^{(R)} + O(1))\rho + O(\rho^2)\,. \tag{A.83}$$

Let

$$T = \left[{f^{(L)}}^{-1}(f(2\rho)), 2\rho\right]\,,$$

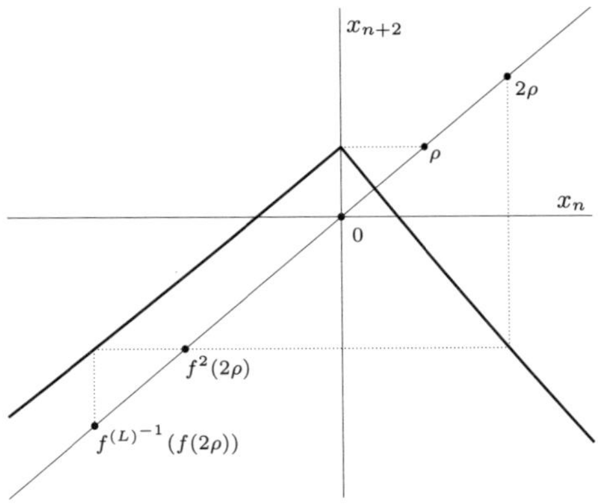

Fig. A.3: The second iterate of (5.8), f^2, near $x = 0$ when $\mu < 0$, $\eta < h_2(\mu)$ and $a_0^{(R)} > 1$. The slope of this map is approximately unity left of the origin and approximately $-a_0^{(R)}$ right of the origin. As a consequence, see the text, the map exhibits chaos in $T = \left[f^{(L)^{-1}}(f(2\rho)), 2\rho\right]$.

It follows, see Fig. A.3, that $\forall x \in T$, $f^2(2\rho) \leq f^2(x) \leq \rho$. But $\rho, f^2(2\rho) \in \text{int}(T)$, thus $f^2(T) \subset \text{int}(T)$, i.e. T is a trapping set for f^2. Furthermore, $\bigcap_{i=0}^{\infty} f^{2i}(T)$, is an attracting set for f^2.

Let $M \in \mathbb{Z}$ with $M \geq 2a_0^{(R)} + 2$. We now show that $\forall x \in T$, $\exists j \in \mathbb{Z}$ with $0 \leq j < M$ such that $f^{2j}(x) > 0$. Suppose otherwise. Then whenever $0 \leq j < M$, $f^{2j}(x) \leq 0$ and thus

$$f^{2j}(x) = (j + O(1))\rho + (1 + O(1))x + O(x^2) . \tag{A.84}$$

Since f^2 is increasing on T_-, it suffices to consider $x = f^{(L)^{-1}}(f(2\rho))$. By combining (A.83) and (A.84) we find

$$\begin{aligned} f^{2(M-1)}(x) &\geq (2a_0^{(R)} + 1 + O(1))\rho + (1 + O(1))(-2a_0^{(R)} + O(1))\rho + O(\rho^2) \\ &= (1 + O(1))\rho + O(\rho^2) > 0 , \end{aligned}$$

which is a contradiction.

Now consider the map $f^{2M} : T \to T$. Note that f^2 has one critical point on T, namely 0. Thus $y \in T$ is a critical point of f^{2M} if $f^{2j}(y) = 0$ for some $0 \leq j < M$. Consequently f^{2M} has at most $2^M - 1$ critical points, y_i, and f^{2M} is differentiable on T except at each y_i. For all $x \in T$, $x \neq y_i$,

$$f^{2M'}(x) = \prod_{j=0}^{M-1} f^{2'}(f^{2j}(x)) .$$

Let $m \geq 1$ be the number of iterates, $f^{2j}(x)$, that are positive. Since left of zero the slope of f^2 is near 1 and right of zero the slope is near $-a_0^{(R)}$, we have

$$\left| f^{2M'}(x) \right| = a_0^{(R)^m} + O(1) > \frac{1 + a_0^{(R)}}{2} > 1 \, ,$$

for any $x \neq y_i$, for sufficiently small μ, η. Consequently $f^{2M} : T \to T$ is *piecewise expanding* [Robinson (2004)]. By the theorem of Li and Yorke [Li and Yorke (1978)], f^{2M} is chaotic on T. Thus f is chaotic on $T \cup f(T)$. □

Proof. [Theorem 5.3] The theorem is proved in three steps. In the first two steps a center manifold, W^c, for the left half-map, $f^{(L)}$, is constructed and the lowest order terms of the restriction of $f^{(L)}$ to W^c are calculated. In the third step the 2-cycle that has points in each half-plane is computed explicitly.

Step 1: *Compute W^c.*

Let

$$F = \begin{bmatrix} x' \\ \mu' \\ \eta' \end{bmatrix} = \begin{bmatrix} f^{(L)}(x;\mu,\eta) \\ \mu \\ \eta \end{bmatrix} , \tag{A.85}$$

denote the $(N+2)$-dimensional, C^k, extended map. Then

$$DF(0;0,0) = \left[\begin{array}{c|c|c} A_L(0,0) & b(0,0) & 0 \\ \hline 0 & 1 & 0 \\ 0 & 0 & 1 \end{array} \right] ,$$

has a three-dimensional center space

$$E^c = \text{span} \left\{ \begin{bmatrix} v(0,0) \\ 0 \\ 0 \end{bmatrix} , \begin{bmatrix} \varphi \\ 1 \\ 0 \end{bmatrix} , e_{N+2} \right\} ,$$

where

$$\varphi = (I - A_L(0,0))^{-1} b(0,0) \, . \tag{A.86}$$

Notice

$$e_1^{\mathsf{T}} \varphi = \frac{\varrho^{\mathsf{T}}(0,0) b(0,0)}{\det(I - A_L(0,0))} \neq 0 \, . \tag{A.87}$$

Since $e_1^{\mathsf{T}} v(0,0) = 1 \neq 0$, the center manifold, W^c, may be expressed locally in terms of $s = e_1^{\mathsf{T}} x$, μ and η. On E^c

$$\begin{bmatrix} x \\ \mu \\ \eta \end{bmatrix} = (s - e_1^{\mathsf{T}}\varphi\mu) \begin{bmatrix} v(0,0) \\ 0 \\ 0 \end{bmatrix} + \mu \begin{bmatrix} \varphi \\ 1 \\ 0 \end{bmatrix} + \eta e_{N+2} ,$$

thus by the center manifold theorem, there exists a C^{k-1} function, H, such that on W^c

$$x = H(s;\mu,\eta) = v(0,0)s + \zeta\mu + O(2) \tag{A.88}$$

where

$$\zeta = \varphi - v(0,0)e_1^{\mathsf{T}}\varphi . \tag{A.89}$$

Step 2: *Determine an expression for $f^{(L)}$ on W^c and verify conditions (i)-(iii) of Theorem 5.2.*
On W^c

$$\begin{aligned} s' &= e_1^{\mathsf{T}} f^{(L)}(H(s;\mu,\eta);\mu,\eta) \\ &= e_1^{\mathsf{T}} b(\mu,\eta)\mu + e_1^{\mathsf{T}} A_L(\mu,\eta) H(s;\mu,\eta) + e_1^{\mathsf{T}} g^{(L)}(H(s;\mu,\eta);\mu,\eta) , \end{aligned} \tag{A.90}$$

where $g^{(L)}$ denotes all terms of $f^{(L)}$ that are nonlinear in x. With the given "hatted" variables (A.90) satisfies condition (ii) of Theorem 5.2 because

$$\begin{aligned} \left.\frac{\partial s'}{\partial \mu}\right|_{(0;0,0)} &= e_1^{\mathsf{T}} b(0,0) + e_1^{\mathsf{T}} A_L(0,0)\zeta \\ &= e_1^{\mathsf{T}} b(0,0) + e_1^{\mathsf{T}} A_L(0,0)\varphi - e_1^{\mathsf{T}} A_L(0,0)v(0,0)e_1^{\mathsf{T}}\varphi \quad \text{(by (A.89))} \\ &= e_1^{\mathsf{T}}\varphi + e_1^{\mathsf{T}} v(0,0)e_1^{\mathsf{T}}\varphi \quad \text{(by (A.86))} \\ &= 2e_1^{\mathsf{T}}\varphi , \end{aligned}$$

which with $\hat{\mu} = 2e_1^{\mathsf{T}}\varphi\mu$, in view of (A.87), produces

$$\left.\frac{\partial s'}{\partial \hat{\mu}}\right|_{(0;0,0)} = 1 .$$

A similar verification of (i) and (iii) of Theorem 5.2 is now given. By scaling, we may assume that for small η, $e_1^{\mathsf{T}} v(0,\eta) \equiv 1$. Let $l(\eta) = \frac{\partial H}{\partial s}(0;0,\eta)$. By matching two different expressions for $f^{(L)}(s;0,\eta)$ to first order in s, one may obtain

$$A(0,\eta)l(\eta) = l(\eta)e_1^{\mathsf{T}} A(0,\eta)l(\eta) .$$

Since $l(0) = v(0,0)$, it follows $l(\eta) \equiv v(0,\eta)$. Therefore

$$H(s;0,\eta) = v(0,\eta)s + O(s^2) . \tag{A.91}$$

Consequently

$$\begin{aligned} s'|_{\mu=0} &= e_1^{\mathsf{T}} A(0,\eta) v(0,\eta) s + O(s^2) \\ &= \lambda(0,\eta) s + O(s^2) \,. \end{aligned}$$

Thus

$$\begin{aligned} \left.\frac{\partial s'}{\partial s}\right|_{(0;0,0)} &= -1 \,, \\ \left.\frac{\partial^2 s'}{\partial \eta \partial s}\right|_{(0;0,0)} &= 1 \,, \end{aligned}$$

as required.

Step 3: *Compute the 2-cycle that has points in each half-plane to verify the final statement of the theorem.*

We have

$$\begin{aligned} f^{(L)}(f^{(R)}(x;\mu,\eta);\mu,\eta) &= (I + A_L(0,\eta)) b(0,\eta)\mu \\ &\quad + A_L(0,\eta) A_R(0,\eta) x + O(|\mu, x|^2) \,. \end{aligned}$$

By the implicit function theorem and condition (v) of the theorem, the 2-cycle with points in each half-plane exists and is unique for small μ and η. When admissible, the point of this cycle in the right half-plane is given by a C^k function

$$x^{*(RL)}(\mu,\eta) = (I - A_L A_R)^{-1}(I + A_L) b\Big|_{\mu=0} \mu + O(\mu^2) \,.$$

Then

$$\left.\frac{\partial s^{*(RL)}}{\partial \mu}\right|_{\mu=0} = e_1^{\mathsf{T}}(I - A_L A_R)^{-1}(I + A_L) b\Big|_{\mu=0} \,.$$

Following the same steps as in Sec. 6.2 to obtain (6.17) (here $\mathcal{S} = RL$) we arrive at

$$\begin{aligned} \left.\frac{\partial s^{*(RL)}}{\partial \mu}\right|_{\mu=0} &= \left.\frac{\det(I + A_L)}{\det(I - A_L A_R)} \varrho^{\mathsf{T}} b\right|_{\mu=0} \\ &= \left.\frac{\frac{\partial}{\partial \eta}\det(I + A_L)}{\det(I - A_L A_R)} \varrho^{\mathsf{T}} b\right|_{(0,0)} \eta + O(\eta^2) \,. \end{aligned}$$

By Theorem 5.2, there exists a C^{k-2} curve, $\eta = h_2(\hat{\mu})$, along which, $s^{*(RL)} = 0$. Furthermore, when $\hat{\mu} \le 0$, the 2-cycle is admissible along this curve. If $\tilde{\eta} = \eta - h_2(\hat{\mu})$, then

$$s^{*(RL)}(\hat{\mu},\tilde{\eta}) = \left.\frac{\det(I - A_L)\frac{\partial}{\partial \eta}\det(I + A_L)}{2\det(I - A_L A_R)}\right|_{(0,0)} \hat{\mu}\tilde{\eta} + O(3) \,, \qquad \text{(A.92)}$$

which confirms the final statement of the theorem. □

Proof. [Lemma 6.9] Since the definitions of non-terminating and terminating shrinking points vary drastically, it is necessary to treat the two classes of shrinking points separately.

For non-terminating shrinking points.

(a) First, $P_{\check{\mathcal{S}}} = P_{\mathcal{S}}$ is singular by assumption, thus by Corollary 6.1, $t = 0$. Second, by (6.29), $P_{\check{\mathcal{S}}^{(ld)}}$ is singular thus by Corollary 6.1, $t_{ld} = 0$.

(b) Suppose for a contradiction, $t_d = 0$. Since also $t_{ld} = 0$ from (a), by a double application of Lemma 6.5, p solves the n-cycle solution system of $\check{\mathcal{S}}^{\overline{d}\,\overline{ld}}$. From the definition, (6.18), it is seen that $\check{\mathcal{S}}^{\overline{d}\,\overline{ld}} = \check{\mathcal{S}}^{(-d)}$. Therefore, p_d solves the n-cycle solution system of $\check{\mathcal{S}}$, hence $p = p_d$. The point p is admissible by assumption, therefore $p = p_{kd}$ for any integer k. Putting $k = m$ we obtain $p = p_1$, hence p is a fixed point of (1.19) and lies on the switching manifold. However (1.19) cannot have a fixed point on the switching manifold because by assumption $\mu \neq 0$ and $\varrho^{\mathsf{T}} b \neq 0$, so we have a contradiction. Thus $t_d \neq 0$. The remaining three points may be proved nonzero in a similar fashion. The given signs follow immediately from admissibility.

(c) Suppose for a contradiction, $\{p_i\}$ is of period $\tilde{n} < n$. Clearly $\tilde{n}$ divides n. Then $p_{i\tilde{n}+k} = p_k$ for any integers i and k. But from part (b), $p_d = p_{i\tilde{n}+d}$ lies left of the switching manifold. Thus $\check{\mathcal{S}}_{i\tilde{n}+d} = L$ for $i = 0, \ldots, \frac{n}{\tilde{n}} - 1$, since the orbit is admissible. But $\gcd(d, \tilde{n}) = 1$, hence in view of (6.18), we must have $l \geq 2 + (\frac{n}{\tilde{n}} - 1)\tilde{n} = n - \tilde{n} + 2 \geq \frac{n}{2} + 2$. Similarly, also from part (b), $p_{-d} = p_{i\tilde{n}-d}$ lies right of the switching manifold and hence $\check{\mathcal{S}}_{i\tilde{n}-d} = R$ for $i = 0, \ldots, \frac{n}{\tilde{n}} - 1$. Here it follows $l < \tilde{n} \leq \frac{n}{2}$ which provides a contradiction.

(d) From part (a), $t = 0$, thus by Lemma 6.5, p solves the n-cycle solution system of $\check{\mathcal{S}}^{\overline{0}} = \mathcal{S}$. Similarly, since $t_{ld} = 0$, p solves the n-cycle solution system of $\check{\mathcal{S}}^{\overline{ld}} = \mathcal{S}^{(-d)}$ by (6.18) and therefore p_d solves the n-cycle solution system of $\mathcal{S}$. Let $w(\tau) = \tau p + (1 - \tau) p_d$. Since $\{p_i\}$ is assumed to be admissible, $\{w_i(\tau)\}$ is an admissible $\mathcal{S}$-cycle for all $\tau \in [0, 1)$.

From part (c), the n points $\{p_i\}$ are distinct. We now show that the union of all the $\{w_i(\tau)\}$, call it $\mathcal{P}$, has no self-intersections. Suppose for a contradiction that $w_{i_1}(\tau_1) = w_{i_2}(\tau_2)$ for some $i_1 \neq i_2$ and $0 < \tau_1, \tau_2 < 1$. Let $x = w_0(\tau_1)$, then $\{x_i\}$ is an admissible $\mathcal{S}$-cycle and an admissible $\mathcal{S}^{(i_2 - i_1)}$-cycle. Since $\tau_1 \neq 0, 1$, by part (b), $s_0 < 0$ and $s_{-d} > 0$. Therefore $\mathcal{S}_{i_2-i_1} = L$ and $\mathcal{S}_{i_2-i_1-d} = R$. By Lemma 6.8b, $i_2 - i_1 = 0$, which is a contradiction. Hence $\mathcal{P}$ has no self-intersections and is therefore an invariant, nonplanar n-gon. To relate (1.19) to a map on the unit circle we describe a bijection between $\mathbb{S}^1$ and $\mathcal{P}$. The angular coordinate, $\theta \in [0, 2\pi)$, uniquely describes a point on $\mathbb{S}^1$. Let $j(\theta) = \lfloor \frac{n\theta}{2\pi} \rfloor$ and $\tau(\theta) = \frac{n\theta}{2\pi} - j(\theta)$. Let $z : \mathbb{S}^1 \to \mathcal{P}$ be defined by $z(\theta) = w_{j(\theta)d}(\tau(\theta))$. It is easily verified the function z is a bijection. The induced map on $\mathbb{S}^1$ is $g(\theta) = (z^{-1} \circ f \circ z)(\theta) = \theta + \frac{2\pi m}{n}$, i.e. rigid rotation with rotation number m/n.

For terminating shrinking points.
Let $v = y + \mathrm{i}z$ be an eigenvector corresponding to the multiplier $\lambda = \mathrm{e}^{\pm\frac{2\pi i m}{n}}$ for the matrix A_L. Let $D = \begin{bmatrix} \cos(\frac{2\pi m}{n}) & \sin(\frac{2\pi m}{n}) \\ -\sin(\frac{2\pi m}{n}) & \cos(\frac{2\pi m}{n}) \end{bmatrix}$. Then $D^j = \begin{bmatrix} \cos(\frac{2\pi jm}{n}) & \sin(\frac{2\pi jm}{n}) \\ -\sin(\frac{2\pi jm}{n}) & \cos(\frac{2\pi jm}{n}) \end{bmatrix}$ and $A_L^j[y\ z] = [y\ z]D^j$. Notice also $D^n = I$. The first component of v must be nonzero because otherwise $A_R[y\ z] = [y\ z]D$ and hence $M_{\check{\mathcal{S}}}[y\ z] = [y\ z]D^n = [y\ z]$, violating the assumption that $(I - M_{\check{\mathcal{S}}})$ is nonsingular. Thus E_c intersects the switching manifold. Furthermore, E_c is two-dimensional, because otherwise there exists another eigenvector for $\lambda = \mathrm{e}^{\pm\frac{2\pi i m}{n}}$, call it $\hat{v}$, linearly independent to v, and the linear combination, $ve_1^{\mathsf{T}}\hat{v} - \hat{v}e_1^{\mathsf{T}}v$, is an eigenvector with zero first component contradicting the previous argument.

Let $x = x_0 = \alpha y + \beta z + x^{*(L)}$ for some as yet undetermined scalars α and β. Let x_i denote the ith iterate of x via the symbol sequence $\hat{\mathcal{S}} = L^n$. Then $x_i = [y\ z]D^i[\alpha\ \beta]^{\mathsf{T}} + x^{*(L)}$ and hence $\{x_i\}$ is an n-cycle. We wish to choose α and β such that $s_0 = e_1^{\mathsf{T}}[y\ z][\alpha\ \beta]^{\mathsf{T}} + s^{*(L)} = 0$ and $s_{-d} = e_1^{\mathsf{T}}[y\ z]D^{-d}[\alpha\ \beta]^{\mathsf{T}} + s^{*(L)} = 0$. Combining these two equations yields the linear system, $X[\alpha\ \beta]^{\mathsf{T}} = -s^{*(L)}[1\ 1]^{\mathsf{T}}$ where $X = \begin{bmatrix} e_1^{\mathsf{T}}y & e_1^{\mathsf{T}}z \\ e_1^{\mathsf{T}}y\cos(\frac{2\pi}{n}) + e_1^{\mathsf{T}}z\sin(\frac{2\pi}{n}) & -e_1^{\mathsf{T}}y\sin(\frac{2\pi}{n}) + e_1^{\mathsf{T}}z\cos(\frac{2\pi}{n}) \end{bmatrix}$ has determinant, $\det(X) = -\left((e_1^{\mathsf{T}}y)^2 + (e_1^{\mathsf{T}}z)^2\right)\sin(\frac{2\pi}{n}) \neq 0$. Thus we let $[\alpha\ \beta]^{\mathsf{T}} = -s^{*(L)}X^{-1}[1\ 1]^{\mathsf{T}}$. It follows that

$$s_{id} = s^{*(L)}\left(1 - \frac{\cos(\frac{2\pi(i+\frac{1}{2})}{n})}{\cos(\frac{\pi}{n})}\right). \tag{A.93}$$

Thus the n-cycle, $\{x_i\}$, is admissible and solves the n-cycle solution system of $\check{\mathcal{S}}$. Therefore $x = p$. Thus $\{p_i\}$ has period n which verifies part

(c) of the lemma. Also $t_i = s_i$, for each i, thus by (A.93) we have parts (a) and (b). To verify part (d), note that as in the non-terminating case, p and p_d are admissible solutions to the n-cycle solution system of $\mathcal{S}$. Thus $w(\tau) = \tau p + (1-\tau)p_d$ is an admissible solution to the n-cycle solution system of $\mathcal{S}$ for all $\tau \in [0,1)$. The union of all such cycles is an invariant, $\mathcal{P}$, which lies on E_c, so is planar. The points, p_i, lie on an ellipse and are ordered so that $\mathcal{P}$ has no self-intersections and is an n-gon. As before, a bijection between $\mathbb{S}^1$ and $\mathcal{P}$ may be constructed to show that the restriction of (6.2) to $\mathcal{P}$ is homeomorphic to rigid rotation with rotation number m/n.

Proof. [Theorem 6.1] Write □

$$\det(I - M_{\check{\mathcal{S}}}(\eta,\nu)) = k_3 + O(1) \,, \tag{A.94}$$

$$\det(I - M_{\hat{\mathcal{S}}}(\eta,\nu)) = k_4 + O(1) \,, \tag{A.95}$$

where $k_3, k_4 \neq 0$ by assumption. We have

$$x(\eta,\nu) = \mu(I - M_{\mathcal{S}}(\eta,\nu))^{-1} P_{\mathcal{S}}(\eta,\nu) b \,, \tag{A.96}$$

whenever $\nu \neq h(\eta)$, (6.33). For small $\nu \neq 0$, if $\eta = 0$ then x is defined and coincides with $\check{x}$ because $\det(P_{\mathcal{S}}) = 0$. Therefore $x_{id}(0,\nu) = \check{x}_{id}(0,\nu)$, for all i. In particular,

$$s_{id}(0,\nu) = t_{id} + O(\nu) \,, \tag{A.97}$$

so that by (6.30) and (6.31)

$$\begin{aligned} s(0,\nu) &= 0 \,, \\ s_{ld}(0,\nu) &= \nu + O(\nu^2) \,. \end{aligned} \tag{A.98}$$

Similarly for small $\eta \neq 0$, if $\nu = 0$ then x is defined and $\det(P_{\mathcal{S}^{((l-1)d)}}) = 0$, thus x solves the n-cycle solution system of $\mathcal{S}^{\overline{(l-1)d}} = \check{\mathcal{S}}^{(d)}$. Therefore $x_{id}(\eta,0) = \check{x}_{(i+1)d}(\eta,0)$, for all i. In particular,

$$s_{id}(\eta,0) = t_{(i+1)d} + O(\eta) \,, \tag{A.99}$$

so that

$$\begin{aligned} s_{-d}(\eta,0) &= \eta + O(\eta^2) \,, \\ s_{(l-1)d}(\eta,0) &= 0 \,. \end{aligned}$$

The lowest order terms in the Taylor series of $\hat{s}_{ld}(\eta,\nu)$ are now computed. An application of Lemma 6.6 to the symbol sequence $\mathcal{S}^{(ld)}$, using also Lemma 6.7, produces

$$\det(I - M_{\mathcal{S}}(\eta,\nu)) s_{ld}(\eta,\nu) = \det(I - M_{\hat{\mathcal{S}}}(\eta,\nu)) \hat{s}_{ld}(\eta,\nu) \,.$$

Using (6.32), (A.95), (A.98) and (A.99) we obtain

$$\hat{s}_{ld}(0,\nu) = \frac{k_2}{k_4}\nu^2 + O(\nu^3) ,$$

$$\hat{s}_{ld}(\eta,0) = \frac{k_1 t_{(l+1)d}}{k_4}\eta + O(\eta^2) . \tag{A.100}$$

By the implicit function theorem, there exists a unique C^k function $g_1 : \mathbb{R} \to \mathbb{R}$ such that for small ν, $\hat{s}_{ld}(g_1(\nu),\nu) = 0$ and

$$g_1(\nu) = -\frac{k_2}{k_1 t_{(l+1)d}}\nu^2 + O(\nu^3) . \tag{A.101}$$

In a similar fashion, by applying Lemma 6.6 to the symbol sequence $\mathcal{S}^{(-d)}$ and noting $\mathcal{S}^{(-d)\overline{0}} = \hat{\mathcal{S}}$, we obtain

$$\hat{s}(0,\nu) = \frac{k_2 t_{-d}}{k_4}\nu + O(\nu^2) , \tag{A.102}$$

$$\hat{s}(\eta,0) = \frac{k_1}{k_4}\eta^2 + O(\eta^3) ,$$

so that the implicit function theorem gives the function

$$g_2(\eta) = -\frac{k_1}{k_2 t_{-d}}\eta^2 + O(\eta^3) . \tag{A.103}$$

The inequality, $k_1 k_2 > 0$, is now demonstrated. Let $\Sigma_i \subset \mathbb{R}^2$ denote the intersection of the interior of the $i^{\rm th}$ quadrant with a sufficiently small neighborhood of the origin ($i = 1,\ldots,4$). Suppose for a contradiction, $k_1 k_2 < 0$. Then, by (6.32), $(I - M_{\mathcal{S}})$ is nonsingular throughout Σ_2 and hence s is continuous throughout Σ_2. Notice $s = 0$ only when $\eta = 0$, thus the sign of s is constant in Σ_2. When $\nu = 0$, by admissibility and using (A.99) we have $s(\eta,0) = t_d + O(1) < 0$, thus s is negative throughout Σ_2. In particular, s is negative when $\nu = \frac{k_1\eta}{k_2}$ for small $\eta < 0$. By applying Lemma 6.6 to $\mathcal{S}$, we obtain

$$\det(I - M_{\mathcal{S}}(\eta,\nu))s(\eta,\nu) = \det(I - M_{\check{\mathcal{S}}}(\eta,\nu))\check{s}(\eta,\nu) . \tag{A.104}$$

Substituting $\nu = \frac{k_1\eta}{k_2}$ and using (6.30), (6.32) and (A.94) gives

$$(2k_1\eta + O(\eta^2))s(\eta,\frac{k_1\eta}{k_2}) = (k_3 + O(\eta))(\eta + O(\eta^2)) ,$$

$$\Rightarrow \quad 2k_1 s(\eta,\frac{k_1\eta}{k_2}) = k_3 + O(\eta) ,$$

$$\Rightarrow \quad k_1 k_3 < 0 . \tag{A.105}$$

Via a similar argument with $s_{(l-1)d}$, we find $k_2 k_3 < 0$. This inequality provides a contradiction with (A.105), thus $k_1 k_2 > 0$. Hence s and $s_{(l-1)d}$

are in fact continuous and negative throughout Σ_1 and Σ_3, thus $k_1 k_3 < 0$. Via similar arguments we find $k_1 k_4 > 0$. Consequently we have

$$\text{sgn}(k_1) = \text{sgn}(k_2) = -\text{sgn}(k_3) = \text{sgn}(k_4) \,. \tag{A.106}$$

Furthermore, from (A.101) and (A.103) we have

$$g_1''(0), g_2''(0) < 0 \,, \tag{A.107}$$

since $t_{(l+1)d}, t_{-d} > 0$.

Near $(\eta, \nu) = (0, 0)$, since $\{p_i\}$ is admissible, $\check{\mathcal{S}}$-cycles are admissible if and only if $\check{s}, \check{s}_{ld} \geq 0$, thus only in Ψ_1 by (6.30) and (6.31). Similarly $\hat{\mathcal{S}}$-cycles are admissible only if $\hat{s}, \hat{s}_{ld} \leq 0$. By (A.102), $\frac{\partial \hat{s}}{\partial \nu}(0,0) = \frac{k_2 t_{-d}}{k_4}$ is positive by (A.106) and admissibility. Thus $\hat{s}(\eta, \nu) \leq 0$ when $\nu \leq h_2(\eta)$. Similarly, by (A.100), $\hat{s}_{ld}(\eta, \nu) \leq 0$ when $\eta \leq h_1(\nu)$. Therefore $\hat{\mathcal{S}}$-cycles are admissible in Ψ_2. The result for $\mathcal{S}$-cycles follows by looking at the signs of each s_i on the boundaries of Ψ_1 and Ψ_2. □

Proof. [Lemma 7.1] The lemma is proved in five steps. In Step 1 it is shown that $0 < \Delta\phi(z;\mu) < \pi$. Consequently $0 \leq \rho(z;\mu) \leq \frac{1}{2}$ and it remains to show that these inequalities may be replaced by strict inequalities. Let

$$a_0(\theta) = \Delta\phi\left(\begin{bmatrix} \cos(\theta) \\ \sin(\theta) \end{bmatrix}; 0\right) \,. \tag{A.108}$$

In Step 2 it is shown that there exists $a_0^{\min} > 0$ and $a_0^{\max} < \pi$ such that for any θ, $a_0^{\min} \leq a_0(\theta) \leq a_0^{\max}$. Step 3 looks at the case that z is very near z^* and shows that here $a_0^{\min} \leq \Delta\phi(z;\mu) \leq a_0^{\max}$. Step 4 looks at the case that $|z|$ is very large and shows that for fixed μ, $\Delta\phi\left(\begin{bmatrix} R\cos(\theta) \\ R\sin(\theta) \end{bmatrix}; \mu\right) \to a_0(\theta)$ as $R \to \infty$. Step 5 shows that the remaining intermediate values of z lie in a compact set within which $\Delta\phi$ is continuous. Consequently $\Delta\phi$ has an absolute minimum, $b_0^{\min} > 0$, and an absolute maximum, $b_0^{\max} < \pi$, and the result follows.

Step 1: *Show* $0 < \Delta\phi(z;\mu) < \pi$.

Let

$$(u, v)^{\mathsf{T}} = z - z^* \,, \tag{A.109}$$

$$(u', v')^{\mathsf{T}} = f(z) - z^* \,. \tag{A.110}$$

Via simple geometry it may be shown that

$$\sin(\Delta\phi(z;\mu)) = \frac{u'v - uv'}{\sqrt{u^2 + v^2}\sqrt{u'^2 + v'^2}} \,.$$

In this step it remains to show that $u'v - uv'$ is positive, for then $\sin(\Delta\phi(z;\mu))$ will be positive and therefore $0 < \Delta\phi(z;\mu) < \pi$.

For ease of notation let $\alpha_i = 2\cos(2\pi\omega_i)$, for $i = L, R$, let $\zeta_L = 1 - \alpha_L r_L + r_L^2$ and $\zeta_R = 1 - \alpha_R s_R + s_R^2$. By assumption, $-2 < \alpha_i < 2$ and $\zeta_i > 0$ for each $i = L, R$. Suppose $x \leq 0$ and $\mu \leq 0$. Then

$$u'v - uv' = (\mu + \alpha_L r_L x + y - \tfrac{\mu}{\zeta_L})(y + \tfrac{\mu r_L^2}{\zeta_L}) - (x - \tfrac{\mu}{\zeta_L})(-r_L^2 x + \tfrac{\mu r_L^2}{\zeta_L}) .$$

Upon expansion,

$$\begin{aligned}\zeta_L(u'v - uv') &= r_L^2 x^2 - \alpha_L r_L^3 x^2 + r_L^4 x^2 + y^2 - \alpha_L r_L y^2 + r_L^2 y^2 + \alpha_L r_L xy \\ &\quad - \alpha_L^2 r_L^2 xy + \alpha_L r_L^3 xy + \mu\alpha_L r_L^3 x - 2\mu r_L^2 x + 2\mu r_L^2 y \\ &\quad - \mu\alpha_L r_L y + \mu^2 r_L^2 \\ &= \left(\tfrac{\alpha_L}{2}(r_L^2 x - y) + r_L(-x + y + \mu)\right)^2 \\ &\quad + (1 - \tfrac{1}{4}\alpha_L^2)(r_L^2 x + y)^2 . \qquad\qquad (A.111)\end{aligned}$$

The right-hand side of (A.111) is positive, hence in this case $u'v - uv' > 0$. Now suppose $x \leq 0$ and $\mu > 0$. Here

$$u'v - uv' = (\mu + \alpha_L r_L x + y - \tfrac{\mu s_R^2}{\zeta_R})(y + \tfrac{\mu}{\zeta_R}) - (x - \tfrac{\mu s_R^2}{\zeta_R})(-r_L^2 x + \tfrac{\mu}{\zeta_R}) . \quad (A.112)$$

Upon expansion,

$$\begin{aligned}\zeta_R(u'v - uv') &= r_L^2 x^2 - \alpha_R r_L^2 s_R x^2 + r_L^2 s_R^2 x^2 + y^2 - \alpha_R s_R y^2 + s_R^2 y^2 \\ &\quad + \alpha_L r_L xy - \alpha_L \alpha_R r_L s_R xy + \alpha_L r_L s_R^2 xy + \mu\alpha_L r_L x - \mu x \\ &\quad - \mu r_L^2 s_R^2 x + 2\mu y - \mu\alpha_R s_R y + \mu^2 \\ &= \left((\tfrac{1}{2}\alpha_L - s_R(1 + \tfrac{1}{2}\alpha_L - \tfrac{1}{2}\alpha_R))r_L x + (1 - \tfrac{1}{2}\alpha_R s_R)y + \mu\right)^2 \\ &\quad + (1 - \tfrac{1}{4}\alpha_L^2)(1 - s_R)^2 r_L^2 x^2 + (1 - \tfrac{1}{4}\alpha_R^2)s_R^2(r_L x - y)^2 \\ &\quad + 2(1 + \tfrac{1}{2}\alpha_L)(1 - \tfrac{1}{2}\alpha_R)s_R(1 - s_R)r_L^2 x^2 \\ &\quad + 2(1 + \tfrac{1}{2}\alpha_L)(1 - \tfrac{1}{2}\alpha_R)s_R(1 + s_R)r_L xy \\ &\quad - (1 - r_L s_R)^2 \mu x - (\alpha_R - \alpha_L) r_L s_R \mu x . \qquad\qquad (A.113)\end{aligned}$$

If $\alpha_L \leq \alpha_R$ and $y \leq 0$, then the right-hand side of (A.113) is positive (it is nonzero because x, y and μ cannot all be zero), thus $u'v - uv' > 0$. For other signs of $(\alpha_L - \alpha_R)$ and y, $u'v - uv'$ may be shown positive via alternate expressions like (A.113). Lastly if $x > 0$, the result follows by use of (7.4).

Step 2: *Show* $a_0^{\min} \leq a_0(\theta) \leq a_0^{\max}$.

By the previous step, $0 < a_0(\theta) < \pi$, for any θ. Since a_0 is continuous and defined on a compact set, it has a minimum value $a_0^{\min} > 0$ and a maximum value $a_0^{\max} < \pi$.

Step 3: *Consider z near z^*.*
Let

$$r_\mu = \begin{cases} 1, & \mu = 0 \\ \frac{|x^*|}{2}, & \mu \neq 0 \end{cases}, \tag{A.114}$$

Suppose $|z-z^*| < r_\mu$ and $z \neq z^*$. Write $z = (r\cos(\theta), r\sin(\theta))^\mathsf{T}$. It is easily shown that $\Delta\phi(z;0) = a_0(\theta)$. If $\mu \neq 0$, then z lies in the same half-plane as z^*. It follows that $\Delta\phi(z;\mu)$ is equal to either $a_0(\theta)$ or $a_0((\theta+\pi) \bmod 2\pi)$. Thus for any value of μ, $a_0^{\min} \leq \Delta\phi(z;\mu) \leq a_0^{\max}$.

Step 4: *Consider large z.*
Via simple geometry, with the coordinates (A.109) and (A.110),

$$\Delta\phi(z;\mu) = \cos^{-1}\left(\frac{uu' + vv'}{\sqrt{u^2+v^2}\sqrt{u'^2+v'^2}}\right) \equiv a^*(u,v,u',v')$$

If $\cos(\theta) \leq 0$, then

$$\begin{aligned}
\Delta\phi\left(\begin{bmatrix} R\cos(\theta) \\ R\sin(\theta)\end{bmatrix};\mu\right) &= a^*(R\cos(\theta) - x^*, R\sin(\theta) - y^*, \mu + \alpha_L r_L R\cos(\theta) \\
&\quad + R\sin(\theta) - x^*, -r_L^2 R\cos(\theta) - y^*) \\
&= a^*(\cos(\theta) - \tfrac{x^*}{R}, \sin(\theta) - \tfrac{y^*}{R}, \tfrac{\mu}{R} + \alpha_L r_L \cos(\theta) \\
&\quad + \sin(\theta) - \tfrac{x^*}{R}, -r_L^2\cos(\theta) - \tfrac{y^*}{R}) \\
&\to a^*(\cos(\theta), \sin(\theta), \alpha_L r_L \cos(\theta) + \sin(\theta), \\
&\qquad -r_L^2\cos(\theta)) \\
&\quad \text{as } R \to \infty \\
&= a_0(\theta)
\end{aligned}$$

If $\cos(\theta) > 0$, the result follows similarly.

Step 5: *Show $b_0^{\min} \leq \Delta\phi(z;\mu) \leq b_0^{\max}$ and complete the result.*
From the previous step, $\Delta\phi((R\cos(\theta), R\sin(\theta))^\mathsf{T};\mu) \to a_0(\theta)$ pointwise in θ. Since the values, θ, are taken from a compact set, $\mathbb{S}^1$, and a_0 is continuous, the convergence is also uniform (see for instance [White (1968)]). Thus there exists $M_0 \in \mathbb{R}$ such that for all $z = (R\cos(\theta), R\sin(\theta))^\mathsf{T}$ with $R \geq M_0$, $\frac{a_0^{\min}}{2} \leq \Delta\phi(z;\mu) \leq \frac{a_0^{\max}+\pi}{2}$. From Step 3, for any $z \neq z^*$ with $|z - z^*| \leq r_\mu$ (A.114), we have $a_0^{\min} \leq \Delta\phi(z;\mu) \leq a_0^{\max}$. The remaining points, z, may be bounded by a compact set, Ω, that does not include z^*. Hence $\Delta\phi(z;\mu)$ is continuous in Ω, thus $\Delta\phi$ has minimum and maximum values on Ω. Thus $0 < b_0^{\min} \leq \Delta\phi(z;\mu) \leq b_0^{\max} < \pi$ for some $b_0^{\min}$ and $b_0^{\max}$ and therefore $0 < \rho(z;\mu) < \pi$. □

Appendix B

Additional Figures

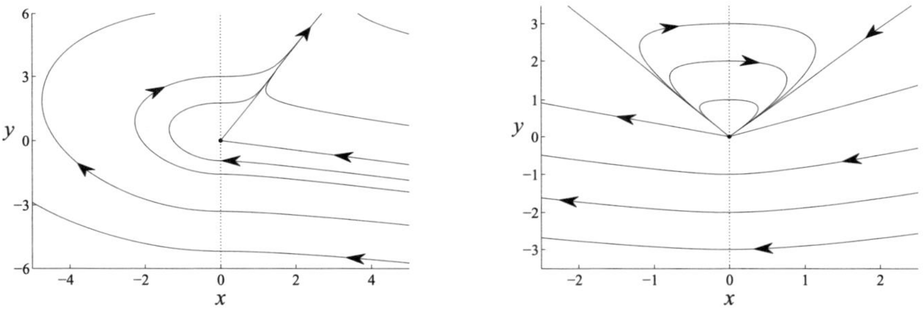

(a)

(b)

Fig. B.1: Phase portraits of (2.1) when $\mu = 0$ with $(\tau_L, \tau_R, \delta_L, \delta_R) = (0.4, -2, 1, -0.5)$ in panel (a) and $(\tau_L, \tau_R, \delta_L, \delta_R) = (2, -2, 0.6, 0.8)$ in panel (b). Phase portraits with the same parameter values except $\mu \neq 0$ are shown in Figs. 2.9 and 2.10.

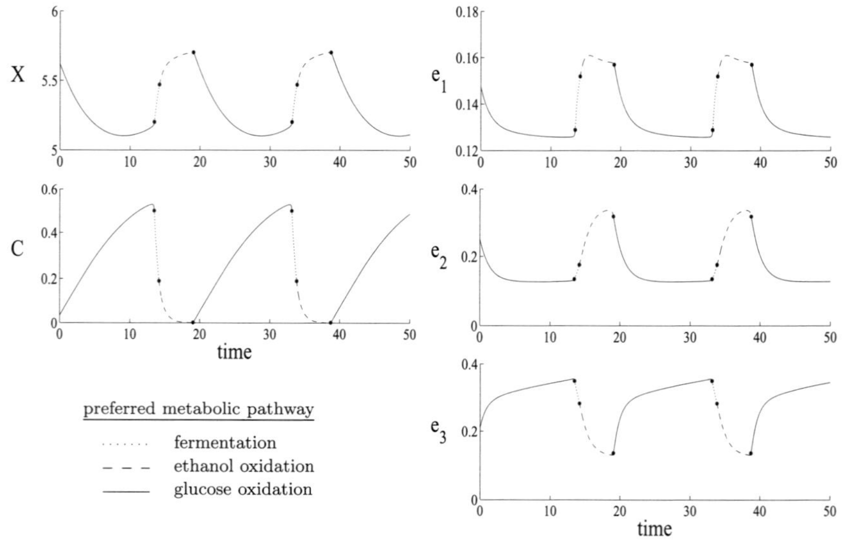

Fig. B.2: Time series of (4.1)-(4.8) showing the variables that were not included in Fig. 4.9. Here $D = 0.1$ and $k_L a = 150$.

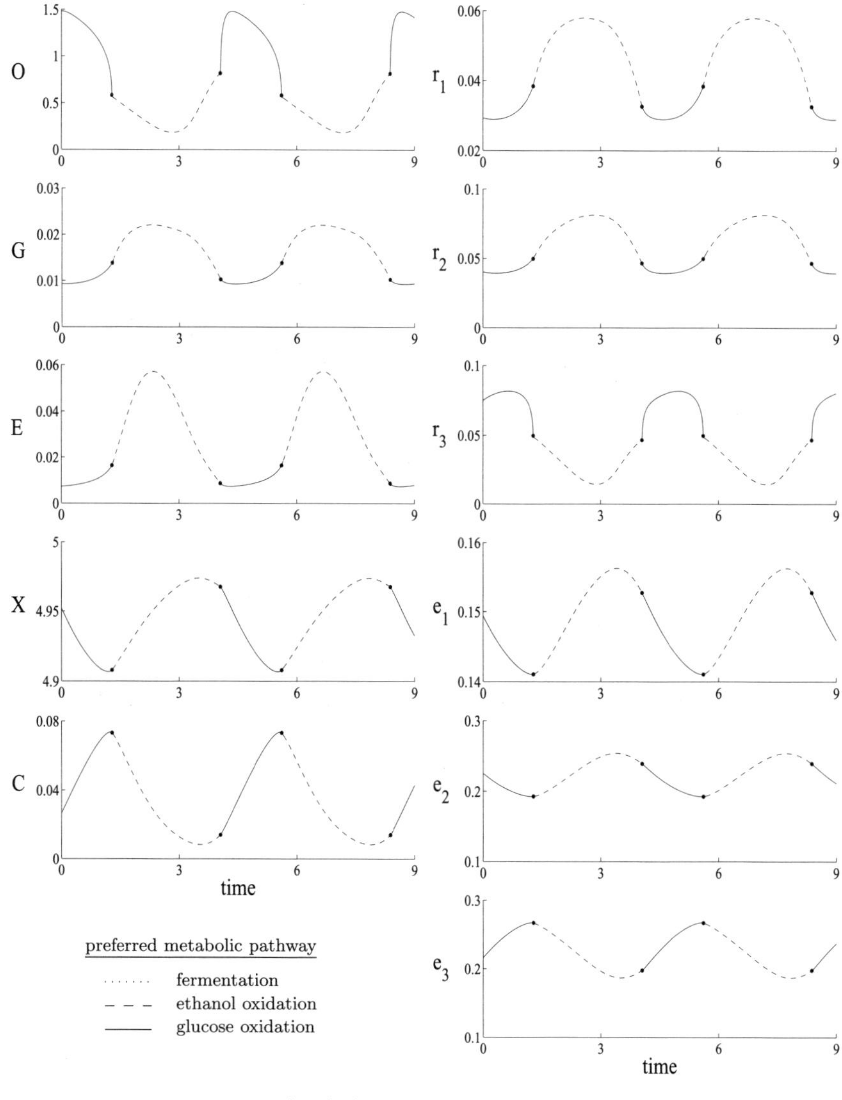

Fig. B.3: Time series of (4.1)-(4.8) when $D = 0.12$ and $k_L a = 150$. Note that fermentation is not a preferred pathway at any time during the oscillations.

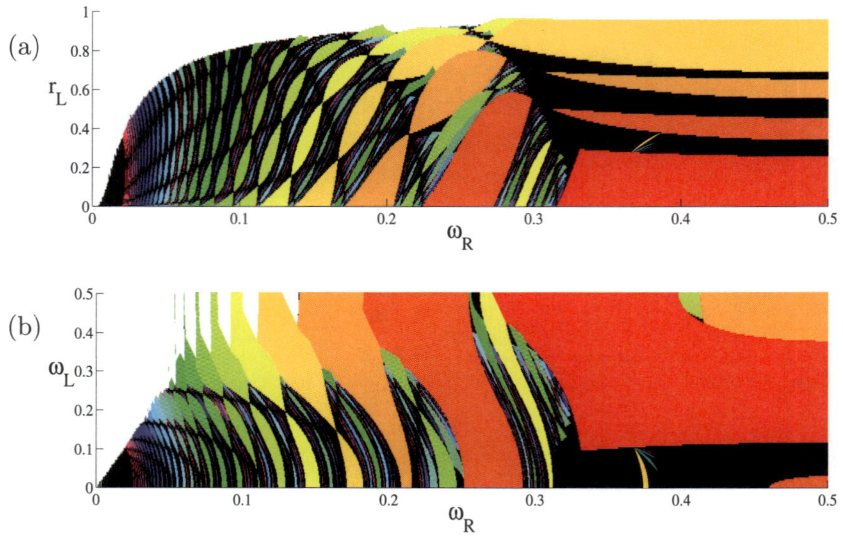

Fig. B.4: (continued on the next page) Resonance tongues of (7.5) when $\mu = 1$. In each panel two parameters are varied and two parameters are fixed at the following values: $(r_L, s_R, \omega_L, \omega_R) = (0.3, 0.8, 0.09, 0.12)$. There are six possible pairs of parameters to choose on the axes; this figure shows five, the sixth is shown in Fig. 7.6(a).

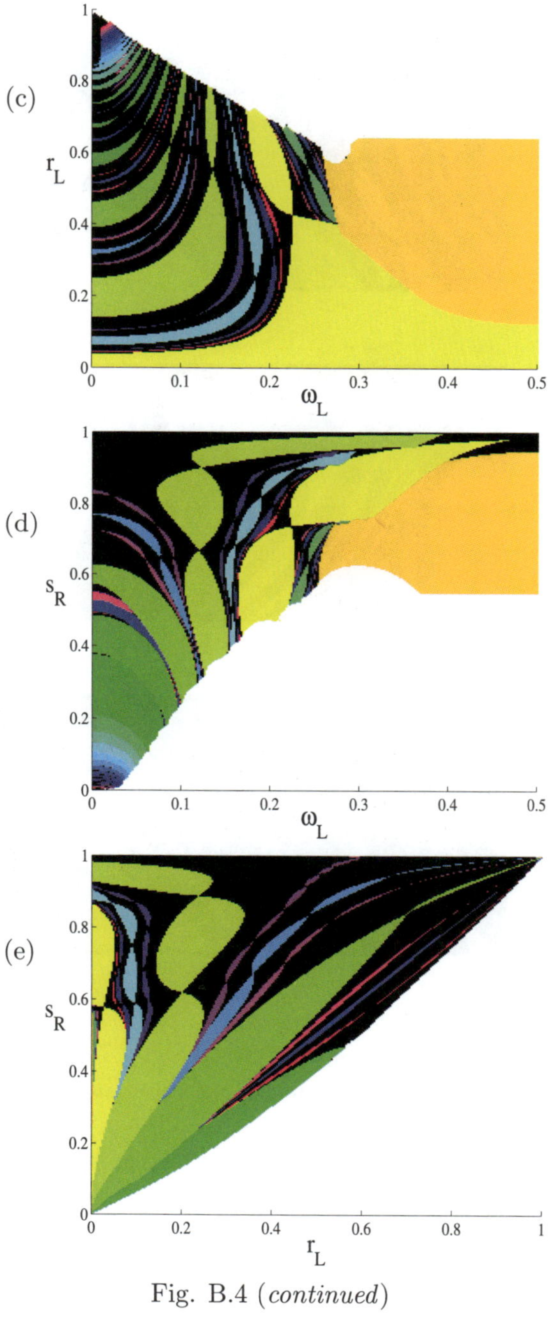

Fig. B.4 (*continued*)

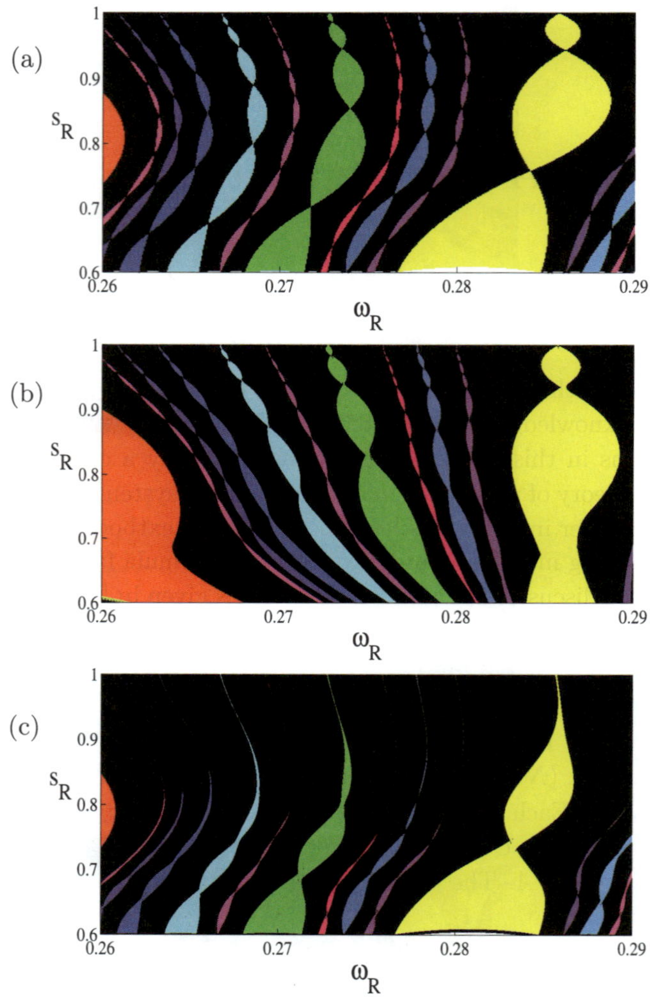

Fig. B.5: Panel (a) is a magnified view of Fig. 7.7(a). Panel (b) shows resonance tongues computed over the same parameter range when the nonlinear term $[0, -\frac{xy}{10}]^{\mathsf{T}}$ is added to the left half-map of (7.5). By adding nonlinearity shrinking points appear to break apart on one side. Panel (c) shows resonance tongues for a smooth approximation to (7.5) constructed by replacing $|x_1|$ in the alternate form (1.15) with $x_1 \tanh(Kx_1)$, where $K = 2$. Here resonance tongues are extremely skinny near $s_R = 1$.

Appendix C

Adjugate Matrices

The purpose of this appendix is to provide the reader with the necessary background knowledge of adjugate matrices to follow the linear algebra manipulations in this book. Adjugate matrices play a crucial role in the bifurcation theory of piecewise-smooth, continuous systems, see for instance Sec. 1.3. However in most standard linear algebra textbooks they are given only as a passing mention in order to provide a formula for the inverse of a matrix. Some discussion of adjugate matrices is given in [Berberian (1992); Kolman (1996)]. Note that the adjugate is sometimes instead called the "adjunct" or the "classical adjoint".

Definition C.1. Let A be an $N \times N$ matrix. Let m_{ij} be the determinant of the $(N-1) \times (N-1)$ matrix formed by removing the $i^{\rm th}$ row and $j^{\rm th}$ column from A. Each m_{ij} is called a *minor* of A. The *cofactor matrix* of A is the $N \times N$ matrix C, whose elements are $c_{ij} = (-1)^{i+j} m_{ij}$. Each c_{ij} is called a *cofactor* of A. The *adjugate* of A, denoted $\mathrm{adj}(A)$, is the transpose of C.

From the definition it is easy to see that $\mathrm{adj}(I) = I$ and $\mathrm{adj}(A^{\mathsf{T}}) = \mathrm{adj}(A)^{\mathsf{T}}$. Also, if the rank of A is less than $N-1$, then every minor, m_{ij}, is zero and consequently the adjugate of A is the zero matrix.

As every undergraduate mathematics student should know, if a_{ij} denotes the (i,j)-element of A, then the sum, $\sum_{i=1}^{N} a_{ij} c_{ij}$, is independent of j and equal to $\sum_{j=1}^{N} a_{ij} c_{ij}$ for any i, and that this constant is the determinant of A. Now consider the matrix $\hat{A}$ formed by replacing the $j^{\rm th}$ column of a matrix A with the $k^{\rm th}$ column of A, where $j \neq k$. Then $\hat{A}$ has two identical columns, thus $\det(\hat{A}) = 0$. But $\det(\hat{A}) = \sum_{i=1}^{N} a_{ik} c_{ij}$. Therefore $\sum_{i=1}^{N} a_{ik} c_{ij} = 0$ for any $j \neq k$. Similarly $\sum_{j=1}^{N} a_{kj} c_{ij} = 0$ for any $i \neq k$.

The above arguments amount to the following fundamental identity:

$$\mathrm{adj}(A)A = A\,\mathrm{adj}(A) = \det(A)I\,. \tag{C.1}$$

Consequently if A is nonsingular then $A^{-1} = \frac{\mathrm{adj}(A)}{\det(A)}$. Also (C.1) implies $\det(\mathrm{adj}(A)) = \det(A)^{N-1}$, hence $\mathrm{adj}(A)$ is nonsingular if and only if A is nonsingular.

The following identity is used in Chapter 6:

$$\mathrm{adj}(AB) = \mathrm{adj}(B)\mathrm{adj}(A)\,. \tag{C.2}$$

If A and B are both nonsingular, then (C.2) follows from $(AB)^{-1} = B^{-1}A^{-1}$. Since nonsingular matrices are dense in the set of all matrices and the adjugate of a matrix is a continuous function of the elements of that matrix, (C.2) holds in general.

The last result discussed here is arguably the most useful in regards to piecewise-smooth bifurcation theory. For example it is essential in the derivation of Feigin's results [di Bernardo *et al.* (1999)]. It is also elementary. For any $\xi \in \mathbb{R}^N$, the matrix $B = A + \xi e_1^{\mathsf{T}}$ differs from A in only the first column. Therefore the first columns of the cofactor matrices of A and B are identical. Equivalently the first rows of $\mathrm{adj}(A)$ and $\mathrm{adj}(B)$ are the same, i.e.

$$e_1^{\mathsf{T}}\mathrm{adj}(A + \xi e_1^{\mathsf{T}}) = e_1^{\mathsf{T}}\mathrm{adj}(A)\,. \tag{C.3}$$

Appendix D

Parameter Values for the *Saccharomyces cerevisiae* model

Table D.1: Parameter values of the model of Jones & Kompala [Jones and Kompala (1999)] used for all numerical investigations in this book.

G_0	10 $\mathrm{gL^{-1}}$	K_1	0.05 $\mathrm{gL^{-1}}$
Y_1	0.16 $\mathrm{gg^{-1}}$	K_2	0.01 $\mathrm{gL^{-1}}$
Y_2	0.75 $\mathrm{gg^{-1}}$	K_3	0.001 $\mathrm{gL^{-1}}$
Y_3	0.60 $\mathrm{gg^{-1}}$	K_{O_2}	0.01 $\mathrm{mgL^{-1}}$
ϕ_1	0.403	K_{O_3}	2.2 $\mathrm{mgL^{-1}}$
ϕ_2	2000 $\mathrm{mgg^{-1}}$	γ_1	10
ϕ_3	1000 $\mathrm{mgg^{-1}}$	γ_2	10
ϕ_4	0.95	γ_3	0.8 $\mathrm{gg^{-1}}$
O^*	7.5 $\mathrm{mgL^{-1}}$	$\mu_{1,\max}$	0.44 $\mathrm{h^{-1}}$
α	0.3 $\mathrm{gg^{-1}h^{-1}}$	$\mu_{2,\max}$	0.19 $\mathrm{h^{-1}}$
α^*	0.1 $\mathrm{gg^{-1}h^{-1}}$	$\mu_{3,\max}$	0.36 $\mathrm{h^{-1}}$
β	0.7 $\mathrm{h^{-1}}$		

Bibliography

Abadie, M. (2000). Dynamical simulation of rigid bodies: Modelling of frictional contact. in B. Brogliato (ed.), *Impacts in Mechanical Systems: Analysis and Modelling.* (Springer-Verlag, New York). [7]

Agliari, A., Bischi, G., Dieci, R. and Gardini, L. (2005). Global bifurcations of closed invariant curves in two-dimensional maps: A computer assisted study. *Int. J. Bifurcation Chaos* **15**, 4, pp. 1285–1328. [151]

Aizerman, M. and Gantmakher, F. (1958). On the stability of periodic motions. *J. Appl. Math. Mech.* **22**, 6, pp. 750–758. [25]

Alur, R., Belta, C., Ivančić, F., Kumar, V., Mintz, M., Pappas, G., Rubin, H. and Schug, J. (2001). Hybrid modeling and simulation of biomolecular networks. in M. Di Benedetto and A. Sangiovanni-Vincentelli (eds.), *Hybrid Systems: Computation and Control., Lecture Notes in Computer Science.*, Vol. 2034 (Springer-Verlag, New York), pp. 19–32. [7]

Andrecut, M. and Ali, M. (2001). Robust chaos in smooth unimodal maps. *Phys. Rev. E* **64**, 2, p. 025203. [23]

Angulo, F., di Bernardo, M., Hogan, S., Kowalczyk, P. and Olivar, G. (2005). Two-parameter non-smooth bifurcations in power converters. *IEEE Int. Symp. Circ. Syst.* , pp. 1485–1488 Kobe, Japan, 23-26 May. [99]

Angulo, F., Olivar, G. and di Bernardo, M. (2008). Two-parameter discontinuity-induced bifurcation curves in a ZAD-strategy-controlled DC-DC buck converter. *IEEE Trans. Circuits Systems I Fund. Theory Appl.* **55**, 8, pp. 2392–2401. [99]

Aoi, S. and Tsuchiya, K. (2006). Bifurcation and chaos of a simple walking model driven by a rhythmic signal. *Int. J. Non-Linear Mech.* **41**, pp. 438–446. [5]

Arnold, V. (1988). *Geometrical Methods in the Theory of Ordinary Differ-*

ential Equations. (Springer-Verlag, New York). [14, 139]

Aronson, D., Chory, M., Hall, G. and McGehee, R. (1982). Bifurcations from an invariant circle for two-parameter families of maps of the plane: A computer-assisted study. *Comm. Math. Phys.* **83**, pp. 303–354. [151, 157]

Avrutin, V., Eckstein, B. and Schanz, M. (2008). The bandcount increment scenario. I. Basic structures. *Proc. R. Soc. A* **464**, pp. 1867–1883. [152]

Avrutin, V. and Schanz, M. (2005a). On special types of two- and three-parametric bifurcations in piecewise-smooth dynamical systems. *WSEAS Trans. Math.* **4**, 3, pp. 224–230. [95]

Avrutin, V. and Schanz, M. (2005b). Period-doubling scenario without flip bifurcations in a one-dimensional map. *Int. J. Bifurcation Chaos* **15**, 4, pp. 1267–1284. [122]

Avrutin, V. and Schanz, M. (2008). On the fully developed bandcount adding scenario. *Nonlinearity* **21**, pp. 1077–1103. [111, 152]

Avrutin, V., Schanz, M. and Banerjee, S. (2006). Multi-parametric bifurcations in a piecewise-linear discontinuous map. *Nonlinearity* **19**, pp. 1875–1906. [95]

Avrutin, V., Schanz, M. and Banerjee, S. (2007). Codimension-three bifurcations: Explanation of the complex one-, two-, and three-dimensional bifurcation structures in nonsmooth maps. *Phys. Rev. E.* **75**, p. 066205. [vi, 27, 31, 95]

Awrejcewicz, J., Kudra, G. and Lamarque, C. (2004). Investigation of triple pendulum with impacts using fundamental solution matrices. *Int. J. Bifurcation Chaos* **14**, 12, pp. 4191–4213. [5]

Awrejcewicz, J. and Lamarque, C. (2003). *Bifurcation and Chaos in Nonsmooth Mechanical Systems.* (World Scientific, Singapore). [4, 7]

Aziz-Alaoui, M., Robert, C. and Grebogi, C. (2001). Dynamics of a Hénon-Lozi-type map. *Chaos Solitons Fractals* **12**, pp. 2323–2341. [31]

Bailey, J. and Ollis, D. (1986). *Biochemical Engineering Fundamentals.* (McGraw-Hill, New York). [75]

Banerjee, S. and Grebogi, C. (1999). Border collision bifurcations in two-dimensional piecewise smooth maps. *Phys. Rev. E* **59**, 4, pp. 4052–4061. [23]

Banerjee, S., Karthik, M., Yuan, G. and Yorke, J. (2000a). Bifurcations in one-dimensional piecewise smooth maps - Theory and applications in switching circuits. *IEEE Trans. Circuits Systems I Fund. Theory Appl.* **47**, 3, pp. 389–394. [22]

Banerjee, S., Ranjan, P. and Grebogi, C. (2000b). Bifurcations in two-dimensional piecewise smooth maps - Theory and applications in switching circuits. *IEEE Trans. Circuits Systems I Fund. Theory Appl.* **47**, 5, pp. 633–643. [23]

Banerjee, S. and Verghese, G. (eds.) (2001). *Nonlinear Phenomena in Power Electronics.* (IEEE Press, New York). [1, 4, 222]

Banerjee, S., Yorke, J. and Grebogi, C. (1998). Robust chaos. *Phys. Rev. Lett.* **80**, 14, pp. 3049–3052. [23]

Barnett, S. (1975). *Introduction to Mathematical Control Theory.* (Clarendon Press, Oxford). [16]

Barton, D., Krauskopf, B. and Wilson, R. (2006). Periodic solutions and their bifurcations in a non-smooth second-order delay differential equation. *Dyn. Syst.* **21**, 3, pp. 289–311. [2]

Batako, A., Lalor, M. and Piiroinen, P. (2007). Numerical bifurcation analysis of a friction-driven vibro-impact system. *J. Sound Vib.* **308**, pp. 392–404. [7, 86]

Belghith, S. (2000). Symbolic dynamics in nondifferentiable system originating in R-L-diode driven circuit. *Disc. Cont. Dyn. Sys.* **6**, 2, pp. 275–292. [23, 111]

Bemporad, A., Ferrari-Trecate, G. and Morari, M. (2000). Observability and controllability of piecewise affine and hybrid systems. *IEEE Trans. Automat. Contr.* **45**, 10, pp. 1864–1876. [14]

Bemporad, A. and Morari, M. (1999). Control of systems integrating logic, dynamics, and constraints. *Automatica* **35**, pp. 407–427. [2]

Benadero, L., El Aroundi, A., Olivar, G., Toribio, E. and Gómez, E. (2003). Two-dimensional bifurcation diagrams. Background pattern of fundamental DC-DC converters with PWM control. *Int. J. Bifurcation Chaos* **13**, 2, pp. 427–451. [4]

Berberian, S. (1992). *Linear Algebra.* (Oxford University Press, New York). [211]

Bergami, M., Bizzarri, F., Carlevaro, A. and Storace, M. (2006). Structurally stable PWL approximation of nonlinear dynamical systems admitting limit cycles: An example. *IEICE Trans. Fund.* **E89-A**, 10, pp. 2759–2766. [vi]

Blazejczyk-Okolewska, B., Czolczynski, K., Kapitaniak, T. and Wojewoda, J. (1999). *Chaotic Mechanics in Systems with Impacts and Friction* (World Scientic, Singapore). [1, 4]

Bonatto, C. and Gallas, J. (2008). Periodicity hub and nested spirals in the phase diagram of a simple resistive circuit. *Phys. Rev. Lett.* **101**, 5, p. 054101. [23]

Boyland, P. (1986). Bifurcations of circle maps: Arnol'd tongues, bistability and rotation intervals. *Comm. Math. Phys.* **106**, 3, pp. 353–381. [118]

Branicky, M. (1998). Multiple Lyapunov functions and other analysis tools for switched and hybrid systems. *IEEE Trans. Automat. Contr.* **43**, 4, pp. 475–482. [11]

Breban, R., Vardavas, R. and Blower, S. (2007). Mean-field analysis of an inductive reasoning game: Application to influenza. *Phys. Rev. E* **76**, p. 031127. [8]

Brogliato, B. (1999). *Nonsmooth Mechanics: Models, Dynamics and Control.* (Springer-Verlag, New York). [1, 4]

Budd, C. and Piiroinen, P. (2006). Corner bifurcations in non-smoothly forced impact oscillators. *Phys. D* **220**, pp. 127–145. [29]

Buzzi, C., da Silva, P. and Teixeira, M. (2006). A singular approach to discontinuous vector fields on the plane. *J. Differential Equations* **231**, pp. 633–655. [3]

Caballé, J., Jarque, X. and Michetti, E. (2006). Chaotic dynamics in credit constrained emerging economies. *J. Econ. Dyn. Control* **30**, pp. 1261–1275. [7]

Carmona, V., Freire, E., Ponce, E., Ros, J. and Torres, F. (2005a). Limit cycle bifurcation in 3d continuous piecewise linear systems with two zones. Application to Chua's circuit. *Int. J. Bifurcation Chaos* **15**, 10, pp. 3153–3164. [4]

Carmona, V., Freire, E., Ponce, E. and Torres, F. (2002). On simplifying and classifying piecewise-linear systems. *IEEE Trans. Circuits Systems I Fund. Theory Appl.* **49**, 5, pp. 609–620. [4, 10, 14, 15]

Carmona, V., Freire, E., Ponce, E. and Torres, F. (2005b). Bifurcation of invariant cones in piecewise linear homogeneous systems. *Int. J. Bifurcation Chaos* **15**, 8, pp. 2469–2484. [2, 11, 46]

Carmona, V., Freire, E., Ponce, E. and Torres, F. (2006). The continuous matching of two stable linear systems can be unstable, *Disc. Cont. Dyn. Sys.* **16**, 3, pp. 689–703. [11]

Casey, R., de Jong, H. and Gouzé, J. (2006). Piecewise-linear models of genetic regulatory networks: Equilibria and their stability. *J. Math. Biol.* **52**, pp. 27–56. [7, 11]

Casini, P., Giannini, O. and Vestroni, F. (2006). Experimental evidence of non-standard bifurcations in non-smooth oscillator dynamics. *Nonlinear Dyn.* **46**, 3, pp. 259–272. [27]

Chang, H. and Juang, J. (2004). Piecewise two-dimensional maps and applications to cellular neural networks. *Int. J. Bifurcation Chaos* **14**, 7, pp. 2223–2228. [111]

Charalambides, C. (2002). *Enumerative Combinatorics.* (Chapman & Hall/CRC, Boca Raton, FL). [125]

Chin, W., Ott, E., Nusse, H. and Grebogi, C. (1994). Grazing bifurcations in impact oscillators, *Phys. Rev. E* **50**, 6, pp. 4427–4450. [29]

Chua, L. (1994). Chua's circuit 10 years later. *Internat. J. Circuit Theory Appl.* **22**, 4, pp. 279–305. [4]

Chua, L. and Deng, A. (1988). Canonical piecewise-linear representation. *IEEE Trans. Circuits Systems* **35**, 1, pp. 101–111. [10, 31]

Chua, L. and Lin, G. (1990). Canonical realization of Chua's circuit family. *IEEE Trans. Circuits Systems* **37**, 7, pp. 885–902. [4]

Chua, L. and Lin, G. (1991). Intermittency in a piecewise-linear circuit. *IEEE Trans. Circuits Systems* **38**, 5, pp. 510–520. [4]

Colombo, A. and Dercole, F. (unpublished). Discontinuity induced bifurcations of non-hyperbolic cycles in nonsmooth systems. Submitted to: *SIAM J. Appl. Dyn. Sys.* [96]

Cone, K. and Zadoks, R. (1995). A numerical study of an impact oscillator with the addition of dry friction. *J. Sound Vib.* **188**, 5, pp. 659–683. [7]

Coombes, S. and Osbaldestin, A. (2000). Period-adding bifurcations and chaos in a periodically stimulated excitable neural relaxation oscillator. *Phys. Rev. E* **62**, 3, pp. 4057–4066. [31]

Cornfeld, I., Fomin, S. and Sinai, Y. (1982). *Ergodic Theory.* (Springer-Verlag, New York). [138, 139, 157]

Csernák, G. and Stépán, G. (2006). On the periodic response of a harmonically excited dry friction oscillator. *J. Sound Vibrat.* **295**, pp. 649–658. [7]

Cunha, F., Pagano, D. and Moreno, U. (2003). Sliding bifurcations of equilibria in planar variable structure systems. *IEEE Trans. Circuits Systems I Fund. Theory Appl.* **50**, 8, pp. 1129–1134. [27]

Dai, D., Li, S., Ma, X. and Tse, C. (2007). Slow-scale instability of single-stage power-factor-correction power supplies. *IEEE Trans. Circuits Systems I Fund. Theory Appl.* **54**, 8, pp. 1724–1735. [135]

Dankowicz, H. and Nordmark, A. (2000). On the origin and bifurcations of stick-slip oscillations. *Phys. D* **136**, pp. 280–302. [7, 26, 66]

Dankowicz, H. and Piiroinen, P. (2002). Exploiting discontinuities for stabilization of recurrent motions. *Dynamical Systems* **17**, 4, pp. 317–342. [5]

Danø, S., Hynne, F., De Monte, S., d'Ovidio, F., Sørensen, P. and Westerhoff, H. (2001). Synchronization of glycolytic oscillations in a yeast cell population. *Faraday Discuss.* **120**, pp. 261–276. [76]

Deimling, K. (1992). *Multivalued Differential Equations.* (W. de Gruyter, New York). [2]

Dercole, F. (2005). Border collision bifurcations in the evolution of mutualistic interactions. *Int. J. Bifurcation Chaos* **15**, 7, pp. 2179–2190. [8]

Dercole, F. and Maggi, S. (2005). Detection and continuation of a border collision bifurcation in a forest fire model. *Appl. Math. Comput.* **186**, pp. 623–635. [8]

di Bernardo, M., Budd, C. and Champneys, A. (1998a). Grazing, skipping and sliding: Analysis of the non-smooth dynamics of the DC/DC buck converter. *Nonlinearity* **11**, pp. 859–890. [29]

di Bernardo, M., Budd, C. and Champneys, A. (2001a). Corner collision implies border-collision bifurcation. *Phys. D* **154**, pp. 171–194. [29, 99]

di Bernardo, M., Budd, C. and Champneys, A. (2001b). Normal form maps for grazing bifurcations in n-dimensional piecewise-smooth dynamical systems. *Phys. D* **160**, pp. 222–254. [vii, 8, 26, 29, 66]

di Bernardo, M., Budd, C., Champneys, A. and Kowalczyk, P. (2008a). *Piecewise-smooth Dynamical Systems. Theory and Applications.* (Springer-Verlag, New York). [v, 1, 3, 12, 14, 21, 22, 25, 27, 29]

di Bernardo, M., Feigin, M., Hogan, S. and Homer, M. (1999). Local analysis of C-bifurcations in n-dimensional piecewise-smooth dynamical systems. *Chaos Solitons Fractals* **10**, 11, pp. 1881–1908. [21, 22, 23, 105, 108, 117, 133, 137, 212]

di Bernardo, M., Fossas, E., Olivar, G. and Vasca, F. (1997). Secondary bifurcations and high periodic orbits in voltage controlled buck converter. *Int. J. Bifurcation Chaos* **7**, pp. 2755–2771. [29]

di Bernardo, M., Garofalo, F., Glielmo, L. and Vasca, F. (1998b). Switchings, bifurcations and chaos in DC/DC converters. *IEEE Trans. Circuits Systems I Fund. Theory Appl.* **45**, 2, pp. 133–141. [v]

di Bernardo, M., Garofalo, F., Iannelli, L. and Vasca, F. (2002a). Bifurcations in piecewise-smooth feedback systems, *Internat. J. Control* **75**, 16-17, pp. 1243–1259. [4]

di Bernardo, M., Johansson, K. and Vasca, F. (2001c). Self-oscillations and sliding in relay feedback systems: Symmetry and bifurcations. *Int J. Bifurcation Chaos* **11**, 4, pp. 1121–1140. [27]

di Bernardo, M., Kowalczyk, P. and Nordmark, A. (2002b). Bifurcations of dynamical systems with sliding: Derivation of normal-form mappings. *Phys. D* **170**, pp. 175–205. [28, 99]

di Bernardo, M., Kowalczyk, P. and Nordmark, A. (2003). Sliding bifurcations: A novel mechanism for the sudden onset of chaos in dry friction oscillators. *Int. J. Bifurcation Chaos* **13**, 10, pp. 2935–2948. [7, 27, 28]

di Bernardo, M., Nordmark, A. and Olivar, G. (2008b). Discontinuity-induced bifurcations of equilibria in piecewise-smooth and impacting dynamical systems. *Phys. D* **237**, pp. 119–136. [3]

Do, Y. (2007). A mechanism for dangerous border collision bifurcations. *Chaos Solitons Fractals* **32**, pp. 352–362. [11]

Do, Y. and Baek, H. (2006). Dangerous border-collision bifurcations of a piecewise-smooth map. *Comm. Pure Appl. Anal.* **5**, 3, pp. 493–503. [11]

Do, Y., Kim, S. and Kim, P. (2008). Stability of fixed points placed on the border in the piecewise linear systems. *Chaos Solitons Fractals* **38**, 2, pp. 391–399. [11]

Doedel, E., Champneys, A., Fairgrieve, T., Kuznetsov, Y., Oldeman, B., Paffenroth, R., Sandstede, B., Wang, X. and Zhang, C. (2007). *AUTO-07P: Continuation and Bifurcation Software for Ordinary Differential Equations.*,
http://indy.cs.concordia.ca/auto. [83]

Dullin, H., Meiss, J. and Sterling, D. (2005). Symbolic codes for rotational orbits. *SIAM J. Appl. Dyn. Syst.* **4**, 3, pp. 515–562. [111]

Dullin, H., Sterling, D. and Meiss, J. (2000). Self-rotation number using the turning angle. *Phys. D* **145**, pp. 25–46. [138]

Dummit, D. and Foote, R. (2004). *Abstract Algebra.* (Wiley, Hoboken, NJ). [14, 15]

Dutta, M., Nusse, H., Ott, E., Yorke, J. and Yuan, G. (1999). Multiple attractor bifurcations: A source of unpredictability in piecewise smooth systems. *Phys. Rev. Lett.* **83**, 21, pp. 4281–4284. [23, 162]

Dutta, P., Routroy, B., Banerjee, S. and Alam, S. (2008). On the existence of low-period orbits in n-dimensional piecewise linear discontinuous maps. *Nonlinear Dynamics* **53**, pp. 369–380. [21]

Edoh, K. and Lorenz, J. (2003). Numerical approximation of rough invariant curves of planar maps, *SIAM J. Sci. Comput.* **25**, 1, pp. 213–223. [154]

Edwards, R. (2000). Analysis of continuous-time switching networks. *Phys. D* **146**, pp. 165–199. [8]

Elnashaie, S., Harraz, H. and Abashar, M. (2001). Homoclinical chaos and the period-adding route to complex non-chaotic attractors in fluidized bed catalytic reactors. *Chaos Solitons Fractals* **12**, pp. 1761–1792. [30]

Fairman, F. (1998). *Linear Control Theory. The State Space Approach.* (Wiley, New York). [15, 17]

Fang, C. and Abed, E. (2001). Saddle-node and Neimark bifurcations in PWM converters. In [Banerjee and Verghese (2001)]. [4]

Feigin, M. (1970). Doubling of the oscillation period with C-bifurcations in piecewise continuous systems. *J. Appl. Math. Mech.* **34**, 5, pp. 822–830. [20]

Feigin, M. (1974). On the generation of sets of subharmonic modes in a piecewise continuous system. *J. Appl. Math. Mech.* **38**, 5, pp. 759–767. [13]

Feigin, M. (1978). On the structure of C-bifurcation boundaries of piecewise-continuous systems. *J. Appl. Math. Mech.* **42**, 5, pp. 885–895. [13]

Feigin, M. (1995). The increasingly complex structure of the bifurcation tree of a piecewise-smooth system. *J. Appl. Math. Mech.* **59**, 6, pp. 853–863. [13]

Fenichel, N. (1971). Persistence and smoothness of invariant manifolds for flows. *Indiana Univ. Math. J.* **21**, 3. [151]

Ferrari-Trecate, G., Cuzzola, F. and Morari, M. (2003). Lagrange stability and performance analysis of discrete-time piecewise affine systems with logic states. *Int. J. Control* **76**, 16, pp. 1585–1598. [2]

Fiechter, H., A. von Meyenburg (1968). Regulatory properties of growing cell populations of *Saccharomyces cerevisiae* in a continuous culture system. in A. Kocková-Kratochrílová (ed.), *Yeasts. Proceedings of the 2nd Symposium on Yeasts held in Bratislava, 16-21 July 1966.* (Slovenskej Akadémie Vied, Bratislava), pp. 387–398. [75]

Filippov, A. (1964). Differential equations with discontinuous right-hand side. in *American Mathematical Society Translations.*, Vol. 42 (AMS, Ann Arbor), pp. 199–231. [1, 27]

Filippov, A. (1988). *Differential Equations with Discontinuous Righthand Sides.* (Kluwer Academic Publishers., Norwell). [1]

Fogg, N. (2002). *Substitutions in Dynamics, Arithmetics and Combinatorics.* (Springer-Verlag, New York). [122]

Fossas, E. and Olivar, G. (1996). Study of chaos in the buck converter. *IEEE Trans. Circuits Systems I Fund. Theory Appl.* **43**, 1, pp. 13–25. [v]

Franks, J. (1990). Periodic points and rotation numbers for area preserving diffeomorphisms of the plane. *Inst. Hautes Études Sci. Publ. Math.* **71**, pp. 105–120. [138]

Fredriksson, M. and Nordmark, A. (1997). Bifurcations caused by grazing incidence in many degrees of freedom impact oscillators. *Proc. R. Soc. A* **453**, pp. 1261–1276. [29]

Freire, E., Ponce, E., Rodrigo, F. and Torres, F. (1998). Bifurcation sets of continuous piecewise linear systems with two zones. *Int. J. Bifurcation Chaos* **8**, 11, pp. 2073–2097. [20, 33, 35, 36, 39]

Freire, E., Ponce, E., Rodrigo, F. and Torres, F. (2004). A piecewise linear electronic circuit with a multiplicity of bifurcations. *Int. J. Bifurcation Chaos* **14**, 11, pp. 3871–3881. [46]

Freire, E., Ponce, E. and Ros, J. (2007). A biparametric bifurcation in 3D continuous piecewise linear systems with two zones. Application to Chua's circuit. *Int. J. Bifurcation Chaos.* **17**, 2, pp. 445–457. [4]

Freire, E., Ponce, E. and Torres, F. (1997). Hopf-like bifurcations in planar piecewise linear systems, *Publicacions Matemátiques* **41**, pp. 131–148. [41]

Fu, X. and Ashwin, P. (2003). Symbolic analysis for some planar piecewise linear maps. *Discrete Contin. Dyn. Syst.* **9**, 6, pp. 1533–1548. [111]

Gallegati, M., Gardini, L., Puu, T. and Sushko, I. (2003). Hicks' trade cycle revisited: Cycles and bifurcations. *Math. Comput. Simulation* **63**, pp. 505–527. [145]

Gallian, J. (1998). *Contemporary Abstract Algebra.* (Houghton Mifflin, Boston). [122]

Ganguli, A. and Banerjee, S. (2005). Dangerous bifurcation at border collision: When does it occur? *Phys. Rev. E.* **71**, 5, p. 057202. [11]

Garcia, M., Chatterjee, A., Ruina, A. and Coleman, M. (1998). The simplest walking model: stability, complexity, and scaling. *ASME J. Biomechanical Eng.* **120**, 2, pp. 281–288. [5]

Gardini, L. (2008). *private communication.* [162]

Gardini, L., Puu, T. and Sushko, I. (2006a). A Goodwin-type model with a piecewise linear investment function. In [Puu and Sushko (2006)], pages 317-333. [12]

Gardini, L., Puu, T. and Sushko, I. (2006b). The Hicksian model with investment floor and income ceiling. In [Puu and Sushko (2006)], pages 179-191. [24]

Gear, C. and Østerby, O. (1984). Solving ordinary differential equations with discontinuities. *ACM. Trans. Math. Soft.* **10**, 1, pp. 23–44. [83]

Giannakopoulos, F. and Pliete, K. (2001). Planar systems of piecewise linear differential equations with a line of discontinuity. *Nonlinearity* **14**, pp. 1611–1632. [3]

Glendinning, P. (1999). *Stability, Instability and Chaos: An Introduction to the Theory of Nonlinear Differential Equations.* (Cambridge University Press, New York). [24, 34, 72, 175, 176]

Gonçalves, J. (2005). Regions of stability for limit cycle oscillations in piecewise linear systems. *IEEE Trans. Automat. Contr.* **50**, 11, pp. 1877–1882. [33]

Gonçalves, J., Megretski, A. and Dahleh, M. (2003). Global analysis of piecewise linear systems using impact maps and surface Lyapunov functions. *IEEE Trans. Automat. Contr.* **48**, 12, pp. 2089–2106. [11]

Gouzé, J. and Sari, T. (2002). A class of piecewise linear differential equations arising in biological models. *Dyn. Syst.* **17**, 4, pp. 299–316. [8]

Grebogi, C., Ott, E. and Yorke, J. (1983). Crises, sudden changes in chaotic attractors, and transient chaos. *Phys. D* **7**, pp. 181–200. [90]

Griffiths, R. and Chou, W. (1986). Effective potentials: A new approach and new results for one-dimensional systems with competing length scales. *Phys. Rev. Lett.* **56**, 18, pp. 1929–1931. [108]

Guckenheimer, J. and Holmes, P. (1986). *Nonlinear Oscillations, Dynamical Systems, and Bifurcations of Vector Fields.* (Springer-Verlag, New York). [34, 57, 70, 72, 173, 175, 176, 188]

Hadeler, K. (1992). Periodic solutions of homogeneous equations. *J. Diff. Eq.* **95**, 1, pp. 183–202. [11]

Halse, C., Homer, M. and di Bernardo, M. (2003). C-bifurcations and period-adding in one-dimensional piecewise-smooth maps. *Chaos Solitons Fractals* **18**, pp. 953–976. [vi, 31]

Halse, C., Wilson, R., di Bernardo, M. and Homer, M. (2007). Coexisting solutions and bifurcations in mechanical oscillations with backlash. *J. Sound. Vib.* **305**, pp. 854–885. [5]

Hao, B. and Zheng, W. (1998). *Applied Symbolic Dynamics and Chaos.* (World Scientific, Singapore). [111, 126]

Hartley, B. and Hawkes, T. (1970). *Rings, Modules and Linear Algebra.* (Chapman and Hall., London). [14, 15]

Hartman, P. (1964). *Ordinary Differential Equations.* (Wiley, New York). [167]

Hassouneh, M., Abed, E. and Nusse, H. (2004). Robust dangerous border-collision bifurcations in piecewise smooth systems, *Phys. Rev. Lett.* **92**, p. 070201. [11]

Heemels, W. and Brogliato, B. (2003). The complementarity class of hybrid dynamical systems. *Euro. J. Control* **9**, pp. 322–360. [2]

Heemels, W., Schumacher, J. and Weiland, S. (2000). Linear complementarity systems. *SIAM. J. Appl. Math.* **60**, 4, pp. 1234–1269. [2]

Henson, M. (2003). Dynamic modeling and control of yeast cell populations in continuous biochemical reactors. *Comput. Chem. Eng.* **27**, pp. 1185–

1199. [76]

Hirsch, M., Pugh, C. and Shub, M. (1977). *Invariant Manifolds.*, *Lecture Notes in Mathematics.*, Vol. 583 (Springer-Verlag, New York). [151]

Hogan, S. (1994). Slender rigid block motion. *J. Eng. Mech.* **120**, 1, pp. 11–24. [5]

Hogan, S., Higham, L. and Griffin, T. (2007). Dynamics of a piecewise linear map with a gap. *Proc. R. Soc. A* **463**, pp. 49–65. [27]

Holden, A. and Fan, Y. (1992). From simple to complex oscillatory behaviour via intermittent chaos in the Rose-Hindmarsh model for neuronal activity. *Chaos Solitons Fractals* **2**, 4, pp. 349–369. [30]

Hommes, C. and Nusse, H. (1991). "Period three to period two" bifurcation for piecewise linear models. *J. Economics* **54**, 2, pp. 157–169. [20]

Hommes, C., Nusse, H. and Simonovits, A. (1995). Cycles and chaos in a socialist economy. *J. Econ. Dyn. Control* **19**, pp. 155–179. [7]

Ivanov, A. (2000). Stability of periodic motions with impacts. in B. Brogliato (ed.), *Impacts in Mechanical Systems: Analysis and Modelling.* (Springer-Verlag, New York). [5]

Iwatani, Y. and Shinji, H. (2006). Stability tests and stabilization for piecewise linear systems based on poles and zeros of subsystems. *Automatica* **42**, pp. 1685–1695. [11]

Jain, P. and Banerjee, S. (2003). Border-collision bifurcations in one-dimensional discontinuous maps. *Int. J. Bifurcation Chaos* **13**, 11, pp. 3341–3351. [27]

Johansson, M. (2003). *Piecewise Linear Control Systems.*, *Lecture Notes in Control and Information Sciences.*, Vol. 284 (Springer-Verlag, New York). [1]

Jones, K. (1995). *Cybernetic Modeling of Spontaneous Oscillations in Continuous Cultures of* Saccharomyces cerevisiae., Master's thesis, University of Colorado, Boulder, CO. [78, 81]

Jones, K. and Kompala, D. (1999). Cybernetic model of the growth dynamics of *Saccharomyces cerevisiae* in batch and continuous cultures, *J. Biotech.* **71**, pp. 105–131. [vii, 76, 77, 78, 80, 81, 213]

Julián, P., Desages, A. and Agamennoni, O. (1999). High-level canonical piecewise linear representation using a simplicial partition. *IEEE Trans. Circuits Systems I Fund. Theory Appl.* **46**, 4, pp. 463–480. [10]

Kang, S. and Chua, L. (1978). A global representation of multidimensional piecewise-linear functions with linear partitions. *IEEE Trans. Circuits Systems* **CAS-25**, 11, pp. 938–940. [10, 31]

Kapitaniak, T. and Maistrenko, Y. (1998). Multiple choice bifurcations as a source of unpredictability in dynamical systems. *Phys. Rev. E* **58**, 4, pp. 5161–5163. [162]

Kapitaniak, T. and Maistrenko, Y. (1999). Riddling bifurcations in coupled piecewise linear maps. *Phys. D* **126**, pp. 18–26. [12]

Kawczyński, A. and Strizhak, P. (2000). Period adding and broken Farey tree sequence of bifurcations for mixed-mode oscillations and chaos in the simplest three-variable nonlinear system. *J. Chem. Phys.* **112**, 14, pp. 6122–6130. [vi, 31]

Keener, J. and Sneyd, J. (1998). *Mathematical Physiology* (Springer-Verlag, New York). [8]

Killough, D. and Edwards, R. (2005). Bifurcations in glass networks. *Int. J. Bifurcation Chaos* **15**, 2, pp. 395–423. [8]

Kolman, B. (1996). *Elementary Linear Algebra.* (Prentice Hall, Upper Saddle River, NJ). [211]

Kompala, D., Ramkrishna, D., Jansen, N. and Tsao, G. (1986). Investigation of bacterial growth on mixed substrates: Experimental evaluation of cybernetic models. *Biotech. Bioeng.* **28**, pp. 1044–1055. [78, 80, 81]

Kompala, D., Ramkrishna, D. and Tsao, G. (1984). Cybernetic modelling of microbial growth on multiple substrates. *Biotech. Bioeng.* **26**, pp. 1272–1281. [78, 80]

Kowalczyk, P. (2005). Robust chaos and border-collision bifurcations in non-invertible piecewise-linear maps. *Nonlinearity* **18**, pp. 485–504. [27]

Kowalczyk, P. and di Bernardo, M. (2005). Two-parameter degenerate sliding bifurcations in Filippov systems. *Phys. D* **204**, pp. 204–229. [95]

Kowalczyk, P., di Bernardo, M., Champneys, A., Hogan, S., Homer, M., Piiroinen, P., Kuznetsov, Y. and Nordmark, A. (2006). Two-parameter discontinuity-induced bifurcations of limit cycles: Classification and open problems. *Int. J. Bifurcation Chaos* **16**, 3, pp. 601–629. [95, 99]

Kowalczyk, P. and Piiroinen, P. (2008). Two-parameter sliding bifurcations of periodic solutions in a dry-friction oscillator. *Phys. D* **237**, pp. 1053–1073. [95]

Krupa, M. and Szmolyan, P. (2001). Relaxation oscillation and canard explosion. *J. Diff. Eq.* **174**, pp. 312–368. [93]

Kumar, M., Kienle, A., Zeyer, K. and Pushpavanam, S. (2008). Nonlinear analysis of the effect of maintenance in continuous cell cultures. *Math. Comput. Simulation* **79**, 3, pp. 728–748. [82]

Küpper, T. (2008). Invariant cones for non-smooth dynamical systems. *Math. Comput. Simulation* **79**, 4, pp. 1396–1408. [11]

Küpper, T. and Moritz, S. (2001). Generalized Hopf bifurcation for nonsmooth planar systems. *Phil. Trans. R. Soc. Lond. A* **359**, pp. 2483–2496. [46]

Kuznetsov, Y. (2004). *Elements of Bifurcation Theory.*, *Appl. Math. Sci.*, Vol. 112 (Springer-Verlag, New York). [vi, 14, 43, 70, 72, 87, 96, 138, 139, 175, 176]

Kuznetsov, Y., Rinaldi, S. and Gragnani, A. (2003). One-parameter bifurcations in planar Filippov systems. *Int. J. Bifurcation Chaos* **13**, 8, pp. 2157–2188. [3, 8]

Lagarias, J. and Rains, E. (2005a). Dynamics of a family of piecewise-linear area-preserving plane maps. I. rational rotation numbers. *J. Difference Eqn. Appl.* **11**, 12, pp. 1089–1108. [139]

Lagarias, J. and Rains, E. (2005b). Dynamics of a family of piecewise-linear area-preserving plane maps. II. invariant circles. *J. Difference Eqn. Appl.* **11**, 13, pp. 1137–1163. [141]

Lagarias, J. and Tresser, C. (1995). A walk along the branches of the extended Farey tree. *IBM J. Res. Dev.* **39**, 3, pp. 283–294. [122]

Laugesen, J. and Mosekilde, E. (2006). Border-collision bifurcations in a dynamic management game. *Comput. Oper. Res.* **33**, pp. 464–478. [7]

Lázaro, M., Santamaría, I., Pantaleón, C., Sánchez, A., Puente, A. and Fernández, T. (2001). Smoothing the canonical piecewise-linear model: An efficient and derivable large-signal model for MESFET/HEMT transistors. *IEEE Trans. Circuits Systems I Fund. Theory Appl.* **48**, 2, pp. 184–192. [31]

Le Calvez, P. (2001). Rotation numbers in the infinite annulus. *Proc. Amer. Math. Soc.* **129**, 11, pp. 3221–3230. [138]

Lee, C., Collins, P., Krauskopf, B. and Osinga, H. (2008). Tangency bifurcations of global Poincaré maps. *SIAM J. Appl. Dyn. Sys.* **7**, 3, pp. 712–754. [24]

Lee, J. (1992). *Biochemical Engineering.* (Prentice Hall, Englewood Cliffs, NJ). [79]

Leine, R. (2006). Bifurcations of equilibria in non-smooth continuous systems. *Phys. D* **223**, pp. 121–137. [8, 12, 18, 19, 31]

Leine, R. and Nijmeijer, H. (2004). *Dynamics and Bifurcations of Non-smooth Mechanical systems*, *Lecture Notes in Applied and Computational Mathematics*, Vol. 18 (Springer-Verlag, Berlin). [1, 4, 8, 12, 18, 19, 31, 47]

Leine, R. and Van Campen, D. (2002). Discontinuous bifurcations of periodic solutions. *Math. Comput. Model.* **36**, pp. 259–273. [7, 29]

Leine, R., Van Campen, D. and Van de Vrande, B. (2000). Bifurcations in nonlinear discontinuous systems. *Nonlinear Dynamics* **23**, pp. 105–164. [7, 27, 29, 31]

Lenci, S. and Rega, G. (2005). Heteroclinic bifurcations and optimal control in the nonlinear rocking dynamics of generic and slender rigid blocks. *Int. J. Bifurcation Chaos* **15**, 6, pp. 1901–1918. [5]

Li, T. and Yorke, J. (1975). Period three implies chaos. *Amer. Math. Monthly.* **82**, 10, pp. 985–992. [23, 32]

Li, T. and Yorke, J. (1978). Ergodic transformations from an interval into itself. *Trans. Amer. Math. Soc.* **235**, 1, pp. 183–192. [194]

Li, X., Wang, S. and Yin, W. (2001). A canonical representation of high-dimensional continuous piecewise-linear functions. *IEEE Trans. Circuits Systems I Fund. Theory Appl.* **48**, 11, pp. 1347–1351. [10]

Lipscombe, P. and Pellegrino, S. (1993). Free rocking of prismatic blocks. *J. Eng. Mech.* **119**, 7, pp. 1387–1410. [5]

Llibre, J., Ponce, E. and Teruel, A. (2007). Horseshoes near homoclinic orbits for piecewise linear differential systems in $\mathbb{R}^3$. *Int. J. Bifurcation Chaos* **17**, 4, pp. 1171–1184. [33, 46]

Llibre, J. and Sotomayor, J. (1996). Phase portraits of planar control systems. *Nonlin. Anal. TMA* **27**, 10, pp. 1177–1197. [35]

Llibre, J. and Teruel, A. (2004). Existence of Poincaré maps in piecewise linear differential systems in $\mathbb{R}^n$. *Int. J. Bifurcation Chaos* **14**, 8, pp. 2843–2851. [25]

López, R., Rodríguez, M., Granero, M., Ojeda, J. and Bruzón, E. (2005). A symmetric piecewise-linear chaotic system with a single equilibrium point. *Int. J. Bifurcation Chaos* **15**, 4, pp. 1411–1415. [4]

Lothaire, M. (1983). *Combinatorics on Words.*, *Encyclopedia of Mathematics and Its Applications*, Vol. 17 (Addison-Wesley, Reading, MA). [125, 126]

Lum, R. and Chua, L. (1991). Global properties of continuous piecewise linear vector-fields. 1. Simplest case in $\mathbb{R}^2$. *Int. J. Circ. Theor. Appl.* **19**, 3, pp. 251–307. [33, 36]

Luo, A. and Han, R. (1996). The dynamics of a bouncing ball with a sinusoidally vibrating table revisited. *Nonlinear Dynamics* **10**, 1, pp. 1–18. [5]

Ma, Y., Agarwal, M. and Banerjee, S. (2006). Border collision bifurcations in a soft impact system. *Phys. Lett. A* **354**, pp. 281–287. [7]

Ma, Y., Ing, J., Banerjee, S., Wiercigroch, M. and Pavlovskaia, E. (2008). The nature of the normal form map for soft impacting systems. *Int. J. Nonlinear Mech.* **43**, pp. 504–513. [7]

Maggio, G., di Bernardo, M. and Kennedy, M. (2000). Nonsmooth bifurcations in a piecewise-linear model of the Colpitts oscillator. *IEEE Trans. Circuits Systems I Fund. Theory Appl.* **47**, 8. [4]

Maistrenko, Y., Maistrenko, V., Vikul, S. and Chua, L. (1995). Bifurcations of attracting cycles from time-delayed Chua's circiut. *Int. J. Bifurcation Chaos.* **5**, 3, pp. 653–671. [108]

Maistrenko, Y., Sushko, I. and Gardini, L. (1998). About two mechanisms of reunion of chaotic attractors. *Chaos Solitons Fractals* **9**, 8, pp. 1373–1390. [151, 152, 162]

Marsden, J. and McCracken, M. (1976). *The Hopf Bifurcation and its Applications* (Springer-Verlag, New York). [vi]

Metropolis, N., Stein, M. and Stein, P. (1973). On finite limit sets for transformations on the unit interval. *J. Comb. Theor.* **A15**, p. 25. [110]

Milnor, J. and Thurston, W. (1988). On iterated maps of the interval. in A. Dold and B. Eckmann (eds.), *Lect. Notes. in Math.*, Vol. 1342 (Springer-Verlag, New York), pp. 465–563. [111]

Mitrovski, C. and Kocarev, L. (2001). Periodic trajectories in piecewise-linear maps. *IEEE Trans. Circuits Systems I Fund. Theory Appl.* **48**, 10, pp. 1244–1246. [33]

Morse, M. and Hedlund, G. (1940). Symbolic dynamics II. Sturmian trajectories. *Am. J. Math.* **62**, pp. 1–42. [122]

Mosekilde, E. and Laugesen, J. (2007). Nonlinear dynamic phenomena in the beer model. *Syst. Dyn. Rev.* **23**, pp. 229–252. [7]

Najafpour, G. (2007). *Biochemical Engineering and Biotechnology.* (Elsevier, Amsterdam). [75]

Namjoshi, A. and Ramkrishna, D. (2001). Multiplicity and stability of steady states in continuous bioreactors: Dissection of cybernetic models. *Chem. Eng. Sci.* **56**, 19, pp. 5593–5607. [81]

Nayfeh, A. (1973). *Perturbation Methods.* (Wiley, New York). [46]

Nordmark, A. (1997). Universal limit mapping in grazing bifurcations. *Phys. Rev. E* **55**, 1, pp. 266–270. [29]

Nordmark, A. (2001). Existence of periodic orbits in grazing bifurcations of impacting mechanical oscillators. *Nonlinearity* **14**, pp. 1517–1542. [29]

Nordmark, A. and Kowalczyk, P. (2006). A codimension-two scenario of sliding solution in grazing-sliding bifurcations. *Nonlinearity* **19**, 1, pp. 1–26. [95]

Nusse, H., Ott, E. and Yorke, J. (1994). Border-collision bifurcations: An explanation for observed bifurcation phenomena. *Phys. Rev. E* **49**, 2, pp. 1073–1076. [23, 162]

Nusse, H. and Yorke, J. (1992). Border-collision bifurcations including "period two to period three" for piecewise smooth systems. *Phys. D* **57**, pp. 39–57. [20, 142]

Nusse, H. and Yorke, J. (1995). Border-collision bifurcations for piecewise-smooth one-dimensional maps. *Int. J. Bifurcation Chaos.* **5**, 1, pp. 189–207. [22]

Osorio, G., di Bernardo, M. and Santini, S. (2008). Corner-impact bifurcations: A novel class of discontinuity-induced bifurcations in cam-follower systems. *SIAM J. Appl. Dyn. Sys.* **7**, 1, pp. 18–38. [29]

Ott, E. (1993). *Chaos in Dynamical Systems.* (Cambridge University Press, New York). [90]

Peckham, B. (1990). The necessity of the Hopf bifurcation for periodically forced oscillators. *Nonlinearity* **3**, 2, pp. 261–280. [138]

Perko, L. (2001). *Differential Equations and Dynamical Systems.* (Springer-Verlag, New York). [143]

Peterka, F. and Vacik, J. (1992). Transition to chaotic motion in mechanical systems with impacts. *J. Sound Vib.* **154**, 1, pp. 95–115. [5]

Piassi, V., Tufaile, A. and Sartorelli, J. (2004). Period-adding bifurcations and chaos in a bubble column. *Chaos* **14**, 2, pp. 477–486. [31]

Piiroinen, P., Virgin, L. and Champneys, A. (2004). Chaos and period-adding: Experimental and numerical verification of the grazing bifurcation. *J. Nonlinear Sci.* **14**, 4, pp. 383–404. [5, 31, 86]

Plahte, E. and Kjøglum., S. (2005). Analysis and generic properties of gene regulatory networks with graded response functions. *Phys. D* **201**, pp. 150–176. [8]

Popp, K. (2000). Non-smooth mechanical systems. *J. Appl. Math. Mech.* **64**, 5, pp. 765–772. [1, 4, 7]

Porro, D., Martegani, E., Ranzi, B. and Alberghina, L. (1988). Oscillations in continuous cultures of budding yeast: A segregated parameter anaylsis. *Biotech. Bioeng.* **32**, pp. 411–417. [75]

Puu, T., Gardini, L. and Sushko, I. (2006). On the change of periodicities in the Hicksian multiplier-accelerator model with a consumption floor. *Chaos Solitons Fractals* **29**, pp. 681–696. [145]

Puu, T. and Sushko, I. (eds.) (2006). *Business Cycle Dynamics: Models and Tools.* (Springer-Verlag, New York). [7, 145, 223, 232]

Rakshit, B., Apratim, M., Jain, P. and Banerjee, S. (2008). Bifurcation phenomena in two-dimensional piecewise smooth discontinuous maps. *http://arxiv.org/abs/0809.3121.* [27]

Ramkrishna, D., Kompala, D. and Tsao, G. (1987). Are microbes optimal strategists? *Biotech. Prog.* **3**, pp. 121–126. [80]

Richard, P. (2003). The rhythm of yeast. *FEMS Microbiol. Rev.* **27**, p. 547:557. [76]

Robinson, C. (1999). *Dynamical Systems. Stability, Symbolic Dynamics, and Chaos.* (CRC Press, Boca Raton, FL). [143, 151]

Robinson, R. (2004). *An Introduction to Dynamical Systems. Continuous and Discrete.* (Prentice Hall, Upper Saddle River, NJ). [24, 96, 194]

Rosen, R. (1970). *Dynamical System Theory in Biology.* (Wiley-Interscience, New York). [8]

Roy, I. and Roy, A. (2008). Border collision bifurcations in three-dimensional piecewise smooth systems. *Int. J. Bifurcation Chaos* **18**, 2, pp. 577–586. [23]

Satroutdinov, A., Kuriyama, H. and Kobayashi, H. (1992). Oscillatory metabolism of *Saccharomyces cerevisiae* in continuous culture. *FEMS Microbiol. Lett.* **98**, pp. 261–268. [75]

Schwartzman, S. (1957). Asymptotic cycles. *Annals of Math.* **66**, 2, pp. 270–284. [138]

Senthilkumar, D. and Lakshmanan, M. (2005). Bifurcations and chaos in time delayed piecewise linear dynamical systems. *Int. J. Bifurcation Chaos* **15**, 9, pp. 2895–2912. [2]

Seydel, R. (1994). *Practical bifurcation and stability analysis: From equilibrium to chaos.* (Springer-Verlag, New York). [47]

Sharkovskii, A. (1995). Co-existence of cycles of a continuous map of the line to itself. *Int. J. Bifurcation Chaos* **5**, 5, pp. 1263–1273, translation of *Ukrain. Mat. Z.*, 16:61-71, 1964. [23, 32]

Sieber, J. (2006). Dynamics of delayed relay systems. *Nonlinearity* **19**, 11, pp. 2489–2527. [2]

Sieber, J., Kowalczyk, P., Hogan, S. and di Bernardo, M. (unpublished). Dynamics of symmetric dynamical systems with delayed switching. To appear: *J. Vib. Control.* [2, 24, 29]

Siegel, R., Tresser, C. and Zettler, G. (1992). A decoding problem in dynamics and in number theory. *Chaos* **2**, 4, pp. 473–493. [122]

Simpson, D. (2008). *Bifurcations in Piecewise-Smooth, Continuous Systems.*, Ph.D. thesis, The University of Colorado. [vi]

Simpson, D. and Kompala, D. (2008). Spontaneous oscillations in yeast chemostat cultures. *http://demonstrations.wolfram.com/ SpontaneousOscillationsInYeastChemostatCultures.* [77]

Simpson, D., Kompala, D. and Meiss, J. (2009). Discontinuity induced bifurcations in a model of *Saccharomyces cerevisiae*. *Math. Biosci.* **218**,

1, pp. 40–49. [ix]
Simpson, D. and Meiss, J. (2007). Andronov-Hopf bifurcations in planar, piecewise-smooth, continuous flows. *Phys. Lett. A* **371**, 3, pp. 213–220. [ix]
Simpson, D. and Meiss, J. (2008a). Neimark-Sacker bifurcations in planar, piecewise-smooth, continuous maps. *SIAM J. Appl. Dyn. Sys.* **7**, 3, pp. 795–824. [ix, 135]
Simpson, D. and Meiss, J. (2008b). Unfolding a codimension-two discontinuous Andronov-Hopf bifurcation. *Chaos* **18**, 3, p. 033125. [ix]
Simpson, D. and Meiss, J. (2009a). Shrinking point bifurcations of resonance tongues for piecewise-smooth, continuous maps. *Nonlinearity* **22**, 5, pp. 1123–1144. [ix, 131]
Simpson, D. and Meiss, J. (2009b). Simultaneous border-collision and period-doubling bifurcations. *Chaos* **19**, 3, p. 033146. [ix]
Sontag, E. (1998). *Mathematical Control Theory.* (Springer-Verlag, New York). [15]
Sparrow, C. (1981). Chaos in a three-dimensional single loop feedback system with a piecewise linear feedback function. *J. Math. Anal. Appl.* **83**, pp. 275–291. [11, 33, 46]
Sushko, I., Agliari, A. and Gardini, L. (2005). Bistability and border-collision bifurcations for a family of unimodal piecewise smooth maps. *Discrete Contin. Dyn. Syst. Ser. B* **5**, 3, pp. 881–897. [111]
Sushko, I. and Gardini, L. (2006). Center bifurcation for a two-dimensional piecewise linear map. In [Puu and Sushko (2006)], pages 49-78. [vi, 12, 23, 24, 129, 133, 145, 151]
Sushko, I. and Gardini, L. (2008). Center bifurcation for two-dimensional border-collision normal form. *Int. J. Bifurcation Chaos* **18**, 4, pp. 1029–1050. [129, 133, 145, 148, 151, 156, 162]
Sushko, I., Gardini, L. and Puu, T. (2004). Tongues of periodicity in a family of two-dimensional discontinuous maps of real Möbius type. *Chaos Solitons Fractals* **21**, pp. 403–412. [24]
Szalai, R. and Osinga, H. (2008). Invariant polygons in systems with grazing-sliding. *Chaos* **18**, 2, p. 023121. [7, 27]
Tanaka, G., Murashige, S. and Aihara, K. (2003). Bifurcation structures of period-adding phenomena in an ocean internal wave model. *Int. J. Bifurcation Chaos* **13**, 11, pp. 3409–3424. [30]
Teixeira, M. (1993). Generic bifurcation of sliding vector fields. *J. Math. Anal. Appl.* **176**, pp. 436–457. [27]
Thamilmaran, K., Lakshmanan, M. and Venkatesan, A. (2004). Hyperchaos

in a modified canonical Chua's circuit. *Int. J. Bifurcation Chaos* **14**, 1, pp. 221–243. [23]

Tomita, K. and Tsuda, I. (1980). Towards the interpretation of the global bifurcation structure of the Lorenz system. A simple one-dimensional model. *Suppl. Prog. Theor. Phys.* **69**, pp. 185–199. [30]

Trentelman, H., Stoorvogel, A. and Hautus, M. (2001). *Control Theory for Linear Systems.* (Springer-Verlag, New York). [15]

Tse, C. (2003). *Complex Behavior of Switching Power Converters.* (CRC Press, Boca Raton, FL). [1, 4]

Tse, C., Dai, D. and Ma, X. (2005). Symbolic analysis of bifurcation in switching power converters: Practical alternative viewpoint of border collision. *Int. J. Bifurcation Chaos* **15**, 7, pp. 2263–2270. [111]

Tung, P. and Shaw, S. (1988). The dynamics of an impact print hammer. *J. Vibr. Acoust. Stress, Reliability Design* **110**, 2, pp. 193–200. [5]

Turnbull, H. and Aitken, A. (1932). *An Introduction to the Theory of Canonical Matrices.* (Blackie, London). [15]

Van der Schaft, A. and Schumacher, J. (2000). *An Introduction to Hybrid Dynamical Systems.* (Springer-Verlag, New York). [1]

van Lint, J. and Wilson, R. (2001). *A Course in Combinatorics.* (Cambridge University Press, New York). [125]

Vincent, T. and Grantham, W. (1997). *Nonlinear and Optimal Control Systems.* (Wiley, New York). [15]

von Meyenburg, H. (1973). Stable synchrony oscillations in continuous cultures of *Saccharomyces cerevisiae* under glucose limitation. in B. Chance, E. Pye, T. Ghosh and B. Hess (eds.), *Biological and Biochemical Oscillators.* (Academic Press, New York), pp. 411–417. [75]

Wechselberger, M. (2005). Existence and bifurcation of canards in $\mathbb{R}^3$ in the case of a folded node. *SIAM J. Appl. Dyn. Sys.* **4**, 1, pp. 101–139. [93]

White, A. (1968). *Real Analysis: An Introduction.* (Addison Wesley, Boston). [203]

Wiercigroch, M. and De Kraker, B. (eds.) (2000). *Applied Nonlinear Dynamics and Chaos of Mechanical Systems with Discontinuities.* (World Scientific, Singapore). [1, 4, 5, 7]

Wiggins, S. (2003). *Introduction to Applied Nonlinear Dynamical Systems and Chaos.*, *Texts in Appl. Math.*, Vol. 2 (Springer-Verlag, New York). [24, 43]

Yang, T. and Chua, L. (2000). Piecewise-linear chaotic systems with a single equilibrium point. *Int. J. Bifurcation Chaos* **10**, 9, pp. 2015–2060. [4]

Yang, W.-M. and Hao, B.-L. (1987). How the Arnol'd tongues become sausages in a piecewise linear circle map, *Comm. Theoret. Phys.* **8**, pp. 1–15. [24, 108, 126, 134, 154]

Yang, Z., Lu, Q. and Li, L. (2006). The genesis of period-adding bursting without bursting-chaos in the Chay model. *Chaos Solitons Fractals* **27**, pp. 689–697. [30, 94]

Yuan, G., Banerjee, S., Ott, E. and Yorke, J. (1998). Border-collision bifurcations in the buck converter. *IEEE Trans. Circuits Systems I Fund. Theory Appl.* **45**, 7, pp. 707–716. [v]

Zhang, Y., Zamamiri, A., Henson, M. and Hjortsø, M. (2002). Cell population models for bifurcation analysis and nonlinear control of continuous yeast bioreactors. *J. Process Control* **12**, pp. 721–734. [76]

Zhao, X. and Dankowicz, H. (2006). Unfolding degenerate grazing dynamics in impact actuators. *Nonlinearity* **19**, 2, pp. 399–418. [95]

Zhao, X. and Schaeffer, D. (2007). Alternate pacing of border-collision period-doubling bifurcations. *Nonlinear Dyn.* **50**, 3, pp. 733–742. [21]

Zhao, X., Schaeffer, D., Berger, C., Krassowska, W. and Gauthier, D. (2008). Cardiac alternans arising from an unfolded border-collision bifurcation. *J. Comput. Nonlinear Dynam.* **3**, 4, p. 041004. [7, 21, 31]

Zhusubaliyev, Z. and Mosekilde, E. (2003). *Bifurcations and Chaos in Piecewise-Smooth Dynamical Systems.* (World Scientific, Singapore). [vi, 1, 4, 145]

Zhusubaliyev, Z. and Mosekilde, E. (2006a). Birth of bilayered torus and torus breakdown in a piecewise-smooth dynamical system. *Phys. Lett. A* **351**, 3, pp. 167–174. [4]

Zhusubaliyev, Z. and Mosekilde, E. (2006b). Torus birth bifurcations in a DC/DC convertor. *IEEE Trans. Circuits Systems I Fund. Theory Appl.* **53**, 8. [4, 24, 111]

Zhusubaliyev, Z. and Mosekilde, E. (2008a). Direct transition from a stable equilibrium to quasiperiodicity in non-smooth systems. *Phys. Lett. A* **372**, pp. 2237–2246. [4, 5]

Zhusubaliyev, Z. and Mosekilde, E. (2008b). Equilibrium-torus bifurcation in nonsmooth systems. *Phys. D* **237**, pp. 930–936. [24, 133]

Zhusubaliyev, Z., Mosekilde, E., De, S. and Banerjee, S. (2008). Transitions from phase-locked dynamics to chaos in a piecewise-linear map. *Phys. Rev. E* **77**, p. 026206. [23, 135, 151]

Zhusubaliyev, Z., Mosekilde, E., Maity, S., Mohanan, S. and Banerjee, S. (2006). Border collision route to quasiperiodicity: Numerical investigation and experimental confirmation. *Chaos* **16**, 2, p. 023122. [4, 23, 24,

135, 151, 162]

Zhusubaliyev, Z., Soukhoterin, E. and Mosekilde, E. (2002). Border-collision bifurcations on a two-dimensional torus. *Chaos Solitons Fractals* **13**, 9, pp. 1889–1915. [24]

Zhusubaliyev, Z., Soukhoterin, E. and Mosekilde, E. (2003). Quasi-periodicity and border-collision bifurcations in a DC-DC converter with pulsewidth modulation. *IEEE Trans. Circuits Systems I Fund. Theory Appl.* **50**, 8, pp. 1047–1057. [24]

Zhusubaliyev, Z., Soukhoterin, E. and Mosekilde, E. (2007). Quasiperiodicity and torus breakdown in a power electronic DC/DC converter. *Math. Comput. Simulation* **73**, pp. 364–377. [4, 151]

Zou, Y. and Küpper, T. (2005). Generalized Hopf bifurcation emanated from a corner for piecewise smooth planar systems, *Non. Anal.* **62**, 1, pp. 1–17. [46]

Zou, Y., Küpper, T. and Beyn, W.-J. (2006). Generalized Hopf bifurcation for planar Filippov systems continuous at the origin, *J. Nonlinear Sci.* **16**, pp. 159–177. [46]

Index